Algorithmic Learning

# Graduate Texts in Computer Science

*Editors*

D. M. Gabbay, C. L. Hankin, and T. S. E. Maibaum

# Algorithmic Learning

ALAN HUTCHINSON

*Department of Computer Science*
*King's College London*

CLARENDON PRESS · OXFORD

*Oxford University Press, Walton Street, Oxford OX2 6DP*
*Oxford New York*
*Athens Auckland Bangkok Bombay*
*Calcutta Cape Town Dar es Salaam Delhi*
*Florence Hong Kong Istanbul Karachi*
*Kuala Lumpur Madras Madrid Melbourne*
*Mexico City Nairobi Paris Singapore*
*Tapei Tokyo Toronto*
*and associated companies in*
*Berlin Ibadan*

*Oxford is a trade mark of Oxford University Press*

*Published in the United States*
*by Oxford University Press Inc., New York*



*A catalogue record for this book is available from the British Library*

*Library of Congress Cataloging in Publication Data*
*Hutchinson, Alan, 1948–*
*Algorithmic learning/Alan Hutchinson.*
*(Graduate texts in computer science; 2)*
*1. Machine learning.   2. Algorithms.   I. Title.   II. Series*
*Q325.5.H88   1994      006.3'1'015118–dc20      93–29815*
*ISBN 0 19 853848 0 (Pbk)*
*ISBN 0 19 853766 2 (Hbk)*

*Printed and bound in Great Britain by*
*Biddles Ltd, Guildford and King's Lynn*

*To Mother and Father*

習

# Preface

Do you believe in fairies? Do you believe in machine learning? Until not very long ago, the questions were not so different. For the past thirty years, people have been inventing new learning algorithms and testing them with computer programs, but only in the past decade has the subject blossomed into anything like a coherent discipline.

When I first had a chance to investigate machine learning, a wise man warned me that it is a very difficult field. He is right, but I am glad that he did not put me off. Thanks to the contributions of many researchers, it is perhaps easier now than it was then, because more examples have been investigated and we have a better notion of what we are trying to do, how far the subject has gone, and how far it has yet to go.

There are four groups of people who might wish to understand learning algorithms. The first includes *everybody*. The subject is profound, fascinating, and vital. It impinges on the development of our culture and of us ourselves. Beyond that group, *programmers* encode algorithms and use them for practical purposes. My first objective is to present enough algorithms in enough detail so that this part of a programmer's job will be somewhat easier. *Students* and their teachers are not necessarily expected to encode algorithms, though I hope that some academic readers will. Their first objective is to categorize algorithms and recognize their similarities and differences, so that if there is need for a new algorithm, they will have some idea how to design it.

The last group are the learning *theorists*. They are likely to be least satisfied with this presentation. Such people tend to have different priorities from experimenters like me. At a recent meeting on an obscure branch of theoretical computer science, a naïve doctoral student was trying to compare mathematics, statistics, and computing. I remarked that computing is about helping people. He disagreed! This book is definitely not for the likes of him.

Chapter 8 is an introduction to some theoretical issues. If the subject were mature then perhaps this chapter should be put first. In the present situation, though, that would be a mistake. Theory is still not yet addressing all the issues which matter to people who use learning algorithms, and if theory were put before practice then the whole subject would be distorted. In current theories of learning, by and large:

(1) all terms are defined rigorously in terms of the soundest possible mathematics;

(2) all data types are explicit;

(3) the objective of a theory is often an estimate of the complexity of an algorithm or the probability of its success in 'random' situations;

(4)    theory is about algorithms which produce results which can be classified as either 'right' or 'wrong';

(5)    there is usually no consideration of meta-level issues. The theory is designed to cover learning at just one level;

(6)    theories of learning tend to begin from assumptions which make theory simple, rather from assumptions which make learning simple. Gold's theorems are about infallible learning algorithms. PAC theory is about learning from randomly chosen training data.

In practice, the situation is quite different.

1.    Most terms used in experimental algorithms can be defined adequately. They are not always defined as well as they could be and I shall be as pleased as anyone when the situation improves, but the current imprecision is not a fundamental limitation on the subject.

2.    In the real world, the data types one meets are not simple. It is not always obvious what they are. How can one describe all the data types of all phenomena which might bear on predicting the weather? Perhaps the question can be answered, but the answer would obscure the really interesting aspects of the learning situation. Theoretical rigour sometimes hides the wood with lots of trees.

3.    Such theoretical objectives are laudable, but they are not the only ones. Often, learning is only practical when training sets are graded in increasing complexity or when the training sets are simple. Teachers deliberately give their pupils atypical graded training sets, with small simple examples first, to avoid problems of complexity.

4.    There are some seemingly worthwhile learning algorithms whose outputs are themselves components of heuristic algorithms. It is very hard to decide whether such an output is 'right' or 'wrong', so current theories do not obviously apply to them.

5.    As yet, theory cannot cope with all the subtleties of the subject. There are real situations when an algorithm can learn at more than one level. The 'SOAR' algorithm is one such. It would be ridiculous to refrain from discussing such algorithms simply because we do not yet have adequate theories of them.

6.    Experimenters do not expect their learning algorithms to be infallible, so Gold's hypotheses are inappropriate. Similarly, PAC theory, although profound and beautiful, is sometimes irrelevant. Teachers do not present random examples and exercises, be the learner human or artificial. Our universe has low entropy. Experimenters take advantage of it, whereas much current theory assumes the opposite.

Theory is essential, but it should not be allowed to stifle discussion of experiment. Of course, I hope that even the most erudite theoretician will enjoy this text, and perhaps find it helpful too.

The assumed prerequisites include a nodding acquaintance with first-order logic and some skill in programming. A little knowledge of probability and the notions of continuity and smoothness will also be helpful. Much of the text can be read without even that. Brief notes on graphs, search, metrics, experimental backing, and complexity are provided in Chapter 2. Other topics are mentioned as needed. Chapter 5 explains entropy. Unification and functional programming in the FP language are introduced in Chapter 6, and Chapter 7 summarizes proof by resolution.

In a book of this size, it would be quite impossible to include more than a selection of all published algorithms. The particular ones chosen by any particular author are bound to be a matter of taste. Some are so well known and standard that there would be raised eyebrows if they did not appear. Among these are back-propagation, some form of learner for decision trees, an example of similarity-based learning, and explanation-based generalization. As well as these, I have tried to include a selection of others which show the diversity of learning in each class of methods. All these algorithms are effective, but they should be regarded as paradigmic prototypes and not necessarily as the ultimate exemplars of efficient learning.

Comparisons between algorithms are usually only possible when the algorithms involve data with similar structure. Individual real numbers have negligible structure, so the algorithms of Chapter 3 cannot be compared. Classifiers, discussed in Chapter 5, lend themselves to comparison much more naturally, and so also do rule-based learners which are the subject of Chapter 7. I have not tried to give more than a simple introduction to neural nets, although Chapter 4 still includes a broad range of several varieties. It would be interesting if there were more known kinds of syntactic learners. The four presented in Chapter 6 are both simple and powerful, but they serve quite different functions.

The classification used here is a rather *ad hoc* association between paradigms, algorithms, and forms of data. I have adopted the obvious algorithmic paradigms: learning as optimization, as correlation, as inductive inference, and as accumulation of heuristic information. The more the subject develops, the more this seems appropriate. One can envisage a composite learner which takes very low-grade input on a kind of retina and classifies it with a neural net of some sort, such as the one devised by Kohonen. This yields crude clusters, which might then be named and subjected to statistical analysis for correlations. The correlator could invent relations by minimizing entropy of experiments and build hierarchies of relations by the Scott–Markovitch process. These could form suitable data for similarity-based learning of IF ... THEN rules, after which the learner could apply all the paraphernalia of EBG, rule ordering, and other high-level methods. As yet, this kind of combined learning is only a dream, but there is nothing implausible about it.

Writing this text has been much harder than one might expect. An obvious reason is the need for and lack of a generally accepted classification scheme for algorithms. The one outlined above is not perfect. The other big problem has been the nature of the source material. Artificial intelligence is a young subject. Machine learning is an even younger exotic branch of it. People have spent centuries developing styles of presentation in the hard sciences and mathematics, and no doubt one day there will be an accepted good style in machine learning too. In the meantime, the going is rough. Many authors quoted below write beautifully and they have obviously put much profitable thought into their work. It is a pleasure to read and understand much of it, and the effort and originality which have gone into it are admirable. On the other hand, there seems to be a tradition among artificial intelligence researchers that they refuse to adopt the most elementary techniques of good practice in program documentation. In many conference proceedings, for example, it is quite exceptional to find a description of an algorithm which includes a straightforward list of its inputs and outputs and their types. When algorithms are stated explicitly, they often appear as incomplete snippets of code in Lisp. Both the syntax and semantics of Lisp are notoriously ill-defined. The subject would benefit immensely if authors could bring themselves to write straightforward pseudocode in some notation resembling Pascal, without procedure or function calls; but then, if people did that, perhaps the subject would lose its mystique.

Some researchers may find that the descriptions below of 'their' algorithms are not what they would expect. If so, will they please understand that I have done my level best to read their papers. When I could not make complete sense of them, it seemed best to describe a plausible algorithm of some sort and not to confuse readers with a preamble about obscure sources.

Under these circumstances, readers will understand that this is not a complete survey. There are many aspects of learning not mentioned here and, even for the topics covered, the references cited are only a selection. (The ones marked [†] are accepted references which I cannot remember ever having seen.) There is now a range of textbooks given over, at least in part, to machine learning, such as those by Kodratoff (1988), Schalkoff (1992), Thornton (1992), and Winston (1992). Anyone who wants to read further could begin with the journals *Machine Learning*, *Artificial Intelligence*, and *Neural Computation*. The next obvious sources are conference reports, particularly the series IJCAI, ECAI, the European Working Sessions on Learning, the International Workshops on Machine Learning, the Workshops on Computational Learning Theory, and meetings and newsletters of societies such as AAAI, the ACM SIGART, and AISB. (Note that series of conferences have the habit of changing their names.) Some researchers make their computational experiments available over the academic network through the Unix 'ftp' program. After that, the search lies among random books, theses, and ephemera. Happy hunting!

I entered this field of study a dozen years ago. It was then and is still most satisfying. In the intervening period, John Buxton gave me freedom to develop a course in algorithmic learning at King's College. The students who attended it have been stimulating good company. Saud Al-Mathami, Farid Azhir, Siani Baker, Jimmy Boyce, Pavel Brazdil, Alan Bundy, John Campbell, Arthur Cater, David Chillingworth, Susan Craw, Jim Doran, Simon Fairthorne, David Fraser, Gerald Gazdar, Fausto Giunchiglia, Margaret Harris, Bryan Holdsworth, Philip Johnson-Laird, Satir Kocabas, Yves Kodratoff, Pat Langley, Xue-Mei Li, Edward Little, Teresa Ludermir, Jason McCallion, Lee McCluskey, Gustav Medrano-Cerda, Chris Mellish, Andrew Moore, Katharina Morik, Stephen Muggleton, Peter Niblett, Richard Overill, Stephen Owen, Mark Plumbley, John Pusey, Paul Qualtrough, Hamish Rae, Michael Reade, Steve Reeves, Larry Rendell, Graeme Ritchie, Dermot Roaf, Paul Scott, Aaron Sloman, Alan Smaill, Maarten van Someren, Sam Steel, Ray Streater, Gheorghe Tecuci, Ray Turner, Paul Vitányi, Lincoln Wallen, Gerry Wolff, David Wolstenholme, Richard Young, and Hong Zhu are among many who have helped me. I hope that what follows persuades them that their efforts were well spent.

*London*                                                                                     A.H.
April 1993

# Acknowledgements

The quotation from Gould and Marler on p. 24 is reproduced with permission of *Scientific American*.

Many of the exercises are based on questions from examinations for the degree of B.Sc. at King's College, University of London and are reproduced with permission.

The figures on pp. 121–3 were provided by Farid Azhar and David Fraser. Other illustrations appear in the hand of Matthew Little.

Yu-Wen Tsao kindly drew the frontispiece (p. viii), which is the Chinese word 'to learn'. It was suggested by Ray Streater. Literally, it means 'to fly by oneself'.

# Contents

Contents                                                                 xxi

# How to choose a learning algorithm

The following decision tree is an expanded form of the classification described in
Chapter 1. In keeping with the spirit of the subject, it is heuristic. (That means, it is
not completely reliable.) Numbers in **bold** are section numbers in later chapters.

Is each example described by

- a few real numbers?
  Is the problem-solving task a case of
  - optimizing a dynamical system?
    *Consider* engineering control theory **3.1**.
  - associating outputs with inputs, where outputs depend continuously or
    smoothly on inputs?
    *Consider* a learner which remembers and interpolates **3.2**.
  - heuristic search with weighted rules?
    *Consider* the bucket brigade algorithm **3.3**.
  - game playing, using weighted patterns?
    *Consider* deeper search, or watching the opponent **3.4**.
  - prediction from algebraic laws which should be learned?
    *Consider* Bacon **3.5**.
  - anything else?
    Either rephrase the task or begin research.

- a lot of indistinguishable numbers or bits?
  Does the training set include counter-examples, or does it consist of input/ output
  pairs?
  *Yes*:  Is the training set clean?
  *Yes*:  *Consider* WISARD or a logical neural net **4.3**.
  *No*:  *Consider* a layered perceptron **4.2**.
  *No*:  Does the task involve
  - completing a blurred or partial image?
    *Consider* a Boltzmann machine **4.4**.
  - discovering a description of optimum examples?
    *Consider* a genetic algorithm **4.6, 4.7**.
  - discovering a classification?
    *Consider* a Kohonen net **4.5**.
  - anything else?
    Either rephrase the task or begin research.

- a few distinct features whose values may be noisy?

  Is there enough knowledge to partition the training set into instances which can be regarded as clean and other instances which are suspect?

  *Yes*: *Consider* learning from the clean and noisy subsets separately by different methods.

  *No*:   *Consider* a statistical method **5**.

  Is the data structure used for clustering likely to be
    - a context-free grammar?

      *Consider* Wolff's method **5.2**.
    - a hierarchy?

      Can the learner choose the kind of instance it learns from, at each step?

      *Yes*: *Consider* the Scott–Markovitch method **5.6**.

      *No*:  Can the learning be non-incremental and can each step of the classification be made on the basis of a single feature?

           *Yes*: *Consider* the decision tree algorithm **5.3**.

           *No*:   *Consider* Al-Mathami's method **5.4**.
    - a graph showing interdependencies between features?

      *Consider* the Fung–Crawford method **5.5**.
    - any other:

      *Consider* designing a new clusterer, on the lines
      suggested by Buntine **5.7.2**.

- terms in some strict syntax and is the learner's objective to generalize the forms of the training instances?

  Is the desired form of generalization
  - a first-order predicate formula in a theory not subject to laws?

    *Consider* Plotkin's method **6.2**.
  - an iterative program?

    *Consider* program synthesis **6.3**.
  - a recursive definition?

    *Consider* logic program synthesis **6.4**, or inference by the *FunG/PredG* method **6.5**.
  - any other:

    Either rephrase the task or begin research.

- relations, free from noise?

  Is the learner's objective to invent
  - non-recursive conjunctive conditions or definitions?

    Does the training set consist of
    - classified examples and counter-examples?

      Does the description language come in the form of an upper semi-lattice?

      *Yes*: *Consider* focusing **7.3.1**.

      *No*:  *Consider* the version space method **7.3.2**.
    - only unclassified positive examples?

      *Consider* Vere's method **6.1**.
  - non-recursive disjunctive conditions or definitions?

    *Consider* decision trees **5.3**, or rule ordering **7.6**.
  - recursive conditions?

    *Consider* logic program synthesis **6.4**, or inference by the *FunG/PredG* method **6.5**.

- an attempt at search, using heuristic rules?
  Are all existing rules in the database reliable?
  *Yes*: Is the objective of learning to
  - add rules with composite operators which reduce the depth of search to a goal?
    Can the search space be designed so that
    - all operators are serially decomposable?
      *Consider* macros **7.7.2**.
    - the task is a case of theorem proving?
      *Consider* a deductive method: EBG, lemma generation, or chunking **7.5**.
    - otherwise?
      Should the learned rules be totally reliable?
      *Yes*: *Consider* the STRIPS method **7.7.1**.
      *No*: *Consider* STRIPS **7.7.1**, or chunking **7.5.3**.
  - improve conflict resolution?
    Does the learner have access to an ideal trace?
    *Yes*: *Consider* rule ordering **7.6**.
    *No*: *Consider* the bucket brigade algorithm **3.3**.
  *No*: Is the objective of learning to discover
  - flaws in the database?
    Does the learner have access to
    - ideal traces?
      *Consider* comparing the rule trace with an ideal trace **7.2**.
    - an oracle?
      *Consider* contradiction backtracing **7.2.2**.
    - neither?
      Either rephrase the task or begin research.
  - new rules with more reliable conditions?
    Does the learner have access to an ideal trace?
    *Yes*: *Consider* rule ordering **7.6**.
    *No*: Either rephrase the task or begin research.
  - more reliable conflict resolution?
    Does the learner have access to an ideal trace?
    *Yes*: *Consider* rule ordering **7.6**.
    *No*: *Consider* the bucket brigade algorithm **3.3**.

- any other?
  Either rephrase the task or begin research.

# 1   Characteristics of learning algorithms

## SUMMARY

This chapter surveys the nature of learning. It explains:

- what learning is;

- what kinds of data a system can learn from;

- what aspects of the data affect the process of learning;

- how learning can be viewed as searching;

- what are the main classes of learning algorithms and how each class is best suited to a particular form of data;

- various other features of algorithms which help to classify them;

- how some learning processes are easy, but some depend on probability theory and intelligent guessing.

It also presents some anecdotes about how animals and people learn in real life.

## INTRODUCTION

Learning is the essence of intelligence. A schoolboy who always does exactly what he is told and nothing more is thoroughly dull. Teaching is most rewarding when you can see the pupil improve. The same is true of computers. A program which always does the same thing with the same input, predictably, may be fun to begin with, but it will eventually pall. If it is to remain intriguing then it must improve its performance.

Artificial learning may, we hope, let us do two things. It can show us how to mimic natural intelligent behaviour. This is the commercial motive for studying it. It can also suggest plausible explanations for the mechanisms of the human mind. Some people would probably prefer that such mechanisms remain mysterious, as they find that the mysteries of nature and life are beautiful precisely because they are mysterious. The search for knowledge is to them verging on sacrilege. Pope wrote

> Know then thyself, presume not God to scan,
> The proper study of mankind is man.

He would probably be horrified at today's technical advances. Most of all, he would detest any attempt to explain mind mechanistically. There is a basic fallacy in his attitude. His second line is correct—what matters most to people is people—but all our experience drives us to conclude that people are governed by the principles of science, which Pope regarded as the realm of God. In order to study people, we need to understand more than just people. The proper study of mankind is everything.

The study of mind spans observation and theory and experiment. Observation is the province of psychology. Theory is open to anyone, but is mostly left to philosophers. (Put another way, anyone who attempts theory is called a philosopher.) Experiment on real minds is hard, for a host of practical reasons, so theories are tested as far as possible on computers. In fact, computing offers more than just a test bed. The notion of an algorithm has only blossomed since computers were invented. In the eighteenth century, very sensible men were led astray because they could not conceive of the notion of an algorithm and so they refused to allow that mental issues could possibly be open to analysis. Computers are not just test beds for psychological theories. The computing paradigm also stimulates new theories of how brains work.

There are two approaches to artificial learning. The first is motivated by the study of mental processes and says: *It is the study of the algorithms embodied in the human mind, and how they can be translated into formal languages and programs.*

The second is much more mundane. It arises from practical computing, which ostensibly has nothing to do with psychology: *It is a branch of data processing, concerned with programs which extrapolate from past data and alter their behaviour accordingly.*

I claim that the second is the right approach. Psychology is a valuable motivator, but writing a program is so very unlike any task that a psychologist ever faces that practitioners of the two subjects are likely to mislead each other. A program is best viewed first as an algorithm acting on data and then perhaps later as an embodiment of some attempt at psychological reality.

I shall now make a second claim: *The data are more significant than the algorithm.* The algorithm may be more intriguing, because it is the algorithm which *does* something; but the data are more fundamental. You can have valuable data without an algorithm, but an algorithm with no data is like a duck without a pond (see Fig. 1.1). Furthermore, an approach oriented around data permits more systematic analysis. You can classify algorithms by their space and time complexities, but beyond that, describing them becomes very technical and hard; whereas, data sets have many simple varied features which permit a much more revealing classification. Therefore, we shall first discuss the classification of data sets and then later we shall associate families of algorithms with particular properties of their data.

A typical program will manipulate data in just one domain. It may invoke several files, but all such files can be regarded as parts of a general description of this single data domain. Learning programs are novel because they have at least two domains:

**Fig. 1.1**  An algorithm and its data.

- the **application domain**, which is the basic field of problems which the program solves; and

- the **learning domain** of techniques which the program uses to solve problems in the application domain.

Subtle learners can have several learning domains, such as a domain of problem-solving rules and another of control strategies for applying the rules. At least to start with, we shall keep things as simple as possible. Our examples will present programs with a single domain of learning.

## 1.1   CHARACTERIZATIONS OF TRAINING SETS

The program learns by analysing a finite sequence of data. This is called a **training set**, although perhaps it should be called a **training list**. It often comes from the application domain, though some programs have a separate training domain.

### Closed/open domain

We say that a domain is **closed** if there is only a limited number of features which matter and if in principle we could know what they all are. If we cannot say what

may influence the solution to a practical problem, then we say that the domain is
**open**.

Rubik's cube is a classic case of a closed domain. Solving it is a hard problem
(for some of us), but in principle all the information which could ever bear on a
solution is available from the state of the cube.

Weather forecasting is an open domain. It depends a lot on the present state of
the atmosphere, particularly its gradients of pressure and temperature. It also
depends on ocean currents such as El Niño, but nobody realized that until quite
recently. If you want to predict weather really accurately, then perhaps you should
also take account of sunspots and the roughness of the ground that the wind blows
over. The possible contributing factors are almost endless.

### Clean/noisy data

Learning is relatively easy if there are no mistakes in the training set. One way to
learn chess is to watch a good player and try to understand why his moves are
good. This may work well if the player is Gary Kasparov, because then you can be
sure that the moves really are good. It is a lot harder if the player is Alan
Hutchinson, since in that case a lot of the moves will be silly and the learner will be
faced with the additional problem of deciding which moves it should study and
which it should ignore.

If the training set contains only reliable examples, then it is said to be **clean**;
otherwise it is **noisy**.

### Correction or negative examples/only positive examples

It is reported that children can learn to speak and understand their native language
just by listening to their parents and practising. What is more, when a child utters his
or her first words, the parents usually succeed in understanding. The child is
presented with only good examples of speech and has few failures. He is never told
explicitly that 'We eaten rhubarbs' is not English, but he learns this fact all the same.

By contrast, the average novice cook learns by trial and error just how long to
boil an egg and how runny his batter should be before he pours it into a pan to
make pancakes. Cooks need failures and correction in order to learn, whereas it
has been suggested that when children learn to speak, they don't. The Italians have
a saying: *Sbagliando s'impara* ('One learns by making mistakes'). It is generally
true, but not always.

As always, a little knowledge is a dangerous thing. Before you apply this
classification to any data set, you must think very carefully about your data. For
instance, imagine you are trying to learn to forecast the weather. If you can do this,
then farmers and large grocery chain stores will pay you a lot of money. The
training set might consist of past occurrences of weather patterns. If so, then it
contains only positive instances, because all weather is real. However, suppose you
do not just learn from the weather itself, but also from your own attempts to predict

the weather. Some of these predictions will be wrong and so you can learn from your mistakes. In that case, the training set of your attempts also contains *counter-examples*. Which training set are you really learning from?

Here, then, are three broad characteristics of any training set, considered as a whole. We shall next consider characteristics of each individual datum within the training set.

## 1.2   FORMS OF INDIVIDUAL DATA

In the so-called data-processing industry, a datum is usually a number or a string of characters or a Cobol record. The sort of datum one can learn from does not usually have any of these forms, except perhaps for numbers. The point is significant, because the form of data affects the choice of an appropriate learning algorithm. Here are some broad descriptions of forms.

### *Numeric*

Each datum is a real number (a number with a decimal point in it). For example, when learning to throw a cricket ball, each time it is thrown, how near was it to hitting the wicket?

### *Small sets of distinct attributes*

Each datum is described by stating a few properties that it has. The various kinds of properties are substantially different. For example, to describe an animal, say

(1)   how many legs has it got?

(2)   is it furry?

(3)   how big are its teeth?

and so on.

Notice that the answer to the query 'Is it furry?' is *yes* or *no*; while the answer to 'How many legs has it got?' is a whole number and to 'How big are its teeth?' is a real number. The learning algorithm depends on the kind of answer.

### *Large sets of similar attributes*

The information which the retina of an eye supplies to the brain consists of simple signals passing down lots of almost identical nerves. Each image is depicted by one set of such signals. Somehow, any young sighted animal learns to distinguish that some images are of its mother, while others are of trees, stones, and so on. There is nothing in the value of any one signal which can help the animal to decide what it is looking at. Useful information can only be derived from the whole set.

It is hard for us to imagine learning from data like that, because our senses are limited and we cannot consciously absorb this sort of data.

### Points in a metric space

The term 'metric space' is a bit of jargon which will be explained properly later. The underlying idea is that, if you are given two different examples, then you have a way of telling how different they are from each other. This difference is a number. The set of all conceivable examples, both the examples which occur and those that might occur but in fact don't, is then a metric space.

A class of children forms a metric space, if we say the difference between two children is the distance between where they sit in class. This particular metric is unusually simple. We shall come across more elaborate metrics later.

### Descriptions with strict syntax

There is a right way to draw an engineering design, and anything else is not an engineering design. Every draftsman is taught a set of strict rules about a design's layout and these rules define the syntax of designs.

Designs are a bit unusual because they are two- or three-dimensional. More often, when people talk of syntax, they think of rules such as those which describe a language, either a human language such as English or a programming language like Cobol or Prolog. If you have ever written a program, you will know that if one single character is mistyped, the compiler or interpreter will reject it. Syntax describes what is acceptable.

### Attempts at search

Naturally, we are very interested in learning how to *do* things. One standard approach is to try to do whatever it is and fail, then work out why we failed and try a different tack next time. Trying alternatives, without knowing in advance which will work, is called **searching**. A lot of workers argue that search is the best paradigm for artificial intelligence. It is a good one, though there are others too. Search is a laborious and hazardous process, so we shall consider other safer approaches as well.

## 1.3   SOLUTION SPACES

Any program has an input file (or two or three) and an output file (or so). For a payroll program, the principal input is a list of names and other details of members of staff, and the output is a file of pay slips. For the 'login' program on a shared computer, the input is a user's name and password. Its output is an instruction to the operating system to set up a new task or 'shell' which will fetch and obey this user's instructions.

A typical learning program's principal input is its training set of solved tasks, discussed above. Its output is usually some datum which will be used to solve similar tasks in future. The learner's structure depends on its output, just as on its input. In order to find the right output, the learner is designed so that it can in principle produce any conceivable output of the right form. All such conceivable outputs form a set, its **solution space**. This usually has some sort of internal structure. The learner has the job of exploring this space of all possible outputs, and the way it explores will depend on the structure of this space. (The term 'space' is a general word covering any set with extra structure. A metric space is a set with a notion of distance. The notion of distance is its structure.)

*Example: finding the root of a polynomial*

Suppose your program has to find a number $S$ for which

$$S^5 + 3S^2 - 2 = 0.$$

Its output will be this number, so the place to look for the answer is in the space of all numbers. Now, numbers (that is to say, the real numbers) can be drawn on paper as a line which you can imagine stretches off indefinitely at both ends. That is its structure.

One design for a learner runs thus: observe that when $x = 0$, then

$$x^5 + 3x^2 - 2$$

is less than 0. Its value is actually $-2$. When $x$ is 1, this polynomial is 2, which is greater than 0. As $x$ varies, the polynomial varies continuously, so there must be a number somewhere between 0 and 1 where its value is 0. This is the number $S$. Since the set of numbers is so vast, we can only hope to find an approximation to $S$. An algorithm which does this could perhaps be called a 'discoverer' rather than a 'learner'. It might run thus:

```
x_0 := 0;
x_1 := 1;

WHILE  x_1 - x_0 > 0.0001  DO
    x := (x_0 + x_1)/2;
    IF the value of the polynomial for x is > 0
    THEN
        x_1 : = x
    ELSE
        x_0 : = x;

S := (x_0 + x_1)/2.
```

Figure 1.2 shows the structure of the space that this algorithm searches in.

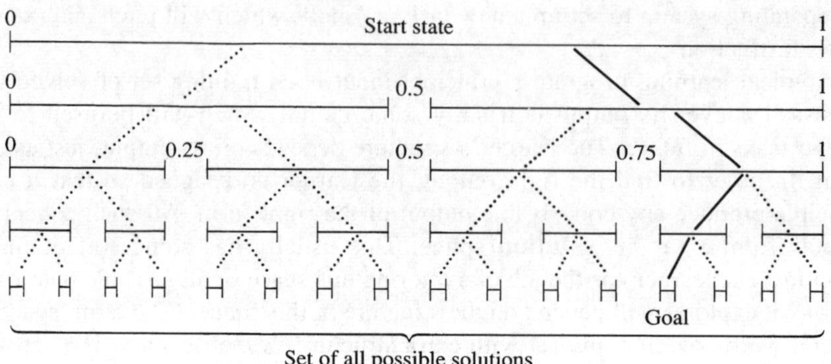



**Fig. 1.2**  Binary chopping.

This algorithm works by a technique called **binary chopping**. At each step, it chops the interval $[x_0, x_1]$ into two pieces and finds which piece contains $S$. Some real numbers can be described exactly as decimals or as fractions. Any other, such as the number $S$, can only be described approximately by describing the ends of an interval containing it. The shorter the interval, the better the approximation. The first two assignments fix an initial interval. Each cycle of the WHILE loop produces a shorter sub-interval. The WHILE loop stops when the approximation is good enough.

If we were not sure at any stage whether $S$ is in $[x_0, x]$ or in $[x, x_1]$ then our algorithm would have to take account of both possibilities. In this task, it just so happens that there is an infallible test for the right sub-interval, and so the algorithm can always choose the best alternative. Its search is very easy. However, it demonstrates the point: it works because the numbers are ordered in a line. Thus, the structure of the algorithm depends on the structure of its output space, the space of numbers.

*Example: learning to recognize an arch*

When you were about four years old, you probably played with wooden blocks, balancing them on top of each other. One of the structures which you built, if you were clever, was an arch. It consisted of two rather precarious columns of blocks with a long block lying horizontally across on top of them, rather like the structures at Stonehenge. A true arch has a neat curved or pointed top. Yours was probably cruder.

In this case, the structure which you built can be described by saying which blocks were in it and which of them rested on which others or which pairs touched. A column, which is simpler than an arch, consists of a few blocks piled one on top of the other. An arch is

(1)  two columns

(2)  which don't touch

(3)  and an extra block

(4)    which rests on top of both columns

Learning this kind of notion is quite tricky. The usual way is to let the learner keep a list of properties which it guesses might form part of the definition of an arch and another list of properties which appear to be absent in all arches. It examines a sequence of arches (the training set) and corrects the two lists in the light of each example. The learner may also use counter-examples—that is, you may offer it a pile of blocks which is not an arch. Counter-examples will help it to construct the second property list. Eventually, all extra examples will match what the learner thinks is an arch. When the program classifies enough consecutive examples correctly, then it can conclude that it knows what an arch is.

The real program will actually be quite a lot more elaborate. One detail that this sketch skated over is the idea of a column. A column can be a single block or a block on top of a smaller column, so the definition of columns is recursive. The learner should really learn that first.

The solution space is the set of all such pairs of lists of properties. It has structure. We say that one list is 'close' to another if they are the same except for a single property. Similarly, pairs of lists are close if they differ in just one list entry.

Such a program has been written. It performs a search through the space of possible description of what an arch might be, as outlined above. It is designed to take small steps in this space—it changes as few properties in its lists as it can, at each example. Thus, it depends on the structure of the space.

## 1.4   LEARNING AND SEARCH

Search is itself a subtle subject. Its basic aspects will be summarized in the next chapter. There are four distinct connections between the processes of learning and searching.

1.   The learner is designed to improve the performance of a problem solver. This problem solver is often a searcher.

(One naturally thinks of the solver and learner as two subroutines in symbiosis within an overall program. This situation is common, but there are cases where they are completely separate, and other cases where the learner and problem solver are the same—the learner learns how to improve its own performance.)

2.   The learner aims to find a good parameter for the problem solver and this parameter is one value in a large space of possible parameters. The learner searches in this space.

There are other 'degenerate' learners which involve no search, but which calculate some parameter by a straightforward algorithm. The first example above, the routine which finds a root of a polynomial, is one such.

The relation between the learner and problem solver is simplest if the solver's search space is fixed and the learner just improves the solver's way of searching it. There are more radical learners which modify the searcher's task.

3.   The parameter produced by the learner may be a **state** of the solver's search space. The problem solver uses this as a new **start state** in future tasks.

The solver's space does not change but, after learning, the solver will begin its subsequent searches from a new point. If the learner is good, this new start will be nearer a solution, so the solver's search will be shorter.

4.   The learner may invent a new way to describe states in the problem space and new ways to express heuristic knowledge about it, which make search in it more reliable.

This process is called adjusting the **bias** of the problem solver. Commonly, such a learner will modify the language used to describe the solver's tasks. If the new language is well chosen, the solver's task may be changed from a searching problem into, say, a problem in pattern matching which is intrinsically easier than search. This kind of learner is searching in a space of languages.

When the learner is itself a searcher, all these techniques can be applied to it too. Thus, one can have a sequence of programs, each a learner improving the performance of its predecessor. Some of the most fascinating learners have such structure, but it is easier to understand the principles of learning if we stay with simple cases.

A program which 'discovers' can be looked on as a sort of general-purpose bias adjuster. Whatever it discovers can be used to abbreviate the description of problem-solving states and knowledge for a whole range of solvers. For instance, the algorithm in the first example will produce the number

   0.76218

This is a much simpler description than

$S,$      where $S \in \mathbb{R}$ and $S^5 + 3S^2 - 2 = 0.$

It would seem odd to refer to '$S$, where …' all the time when we know the value of $S$.

The distinction between learning and discovery is fine. There is an algorithm, useful in problems of diagnosis and prediction, which involves constructing a structure called a **decision tree**. Such decision trees contain a lot of implicit information which can be used for all sorts of analyses. The program which builds the tree is almost always paired with a problem solver which uses the tree for practical diagnostic work, but the two programs are distinct and in principle one program could construct its tree which could then be recorded in some database on the off chance that it might be useful years later to solve tasks. Where does the distinction lie? Should we say that the first program discovers the tree and then, quite separately, the second uses it? Or are the two programs practically inseparable, so that the first learns for the second? I shall dodge the issue. You can make up your own mind whether a particular program discovers or learns.

## 1.5  CLASSIFICATION OF ALGORITHMS

Perhaps the best way to classify learning algorithms is by the kinds of spaces in which they search. That is not what we shall do here, for two reasons. First, the structure of these spaces is sometimes subtle and obscure. Often, two similar algorithms have marginally different search spaces and it is not worth drawing all the distinctions between them. Such a classification would be too fine. Second, there is another simpler classification which is almost as reliable. The structure of input usually dictates a lot of the properties of the search space, hence, learning algorithms with similar input formats often have similar structure. Thus, classification by input data format is nearly the same as by structure of the search space. The most common correspondences are shown in Table 1.1. At this stage, the names in the columns may not all mean very much to you. These notes are intended to explain them.

A word of warning: each of these techniques on its own is rarely enough for a practical learner. Most real systems embody two or more different algorithms, for different aspects of the learning process. Whenever you read a description of some particular learning program, you will find yourself disentangling all these aspects and classifying them individually. The authors of technical papers are not always very helpful in this respect.

Since learning programs are often quite complex, it is not surprising that they can be classified in more than one way. There are several simple distinctions which this scheme does not capture (and which may be more apparent in the schemes devised by Carbonell and Kodratoff, mentioned in the *Further reading* section).

### *Incremental/batch learning*

A problem solver and a learner may be very tightly coupled or hardly connected at all. At one extreme, the learner may produce its output before the problem solver is first called. This is common among statistical correlators. The learner may even be called without any particular problem solver in view. Then its output is desired for general interest. This sort of 'learning' is what Langley calls **discovery**.

At the other extreme, the leaner may be called once each time the solver makes a step in its search, so that the learner analyses that step and the solver can benefit from learning within one search task. This extreme is called **incremental learning**.

**Table 1.1**

| Input data format | Solution Space | Form of algorithm |
|---|---|---|
| Real numbers | Real parameters | Parameter adjustment |
| Attribute–value pairs | States of a neural net | Association and Hill climbing |
| *(Lots of very similar attributes)* | | |
| Attribute–value pairs | Relations between specific attributes | Correlation |
| *(Relatively few attributes with distinct types)* | | |
| Points in a metric space | Possible cluster centres | Clustering |
| Terms in a strict syntax | Patterns with the same syntax | Generalization |
| Attempts at search | All possible sets of heuristic rules | |
| | *Forming composite operators* | Composition; Macros |
| | *Conflict resolution* | adjusting weights/ rule ordering |
| | *Conditions for rules, given:* | |
| | *no underlying theory* | Similarity-based learning (SBL) |
| | *a reliable set of simple rules* | Explanation based learning (EBL) |

### *Forgetting past training instances*

A correlator usually has to keep records of a large set of instances. In incremental learners, this is awkward. Learning is simpler if, once each instance has been examined, it can be forgotten.

### *'Right' answer or optimal answer/convergence to an adequate answer*

When a doctor makes a diagnosis, his conclusion is either right or wrong. Either a patient's cut finger is septic or else it isn't. By contrast, there is no absolutely certain 'correct' dose of penicillin which might cure the sepsis. The doctor will try to strike a balance between too small a dose which will do no good and an overdose which will do as much harm to the patient as the infection. The doctor may have a choice of antibiotics, any of which will probably do, though some are stronger, some are cheaper, and some have less severe side effects.

When medical students study, they learn some 'facts' which are beyond dispute and they also learn more nebulous information and techniques which are subject to debate. In the case of diagnosis, the answer is simply right or wrong. For treatment, many answers are adequate.

As always, beware the naïve approach. Take, for instance, the problem of solving Rubik's cube. In this case, it appears there is a unique right answer: each face should end up with a single colour. However, that is not what you are really trying to find. What you want is a way of rotating the faces until each face is of one colour. It is the sequence of rotations which you seek. Although there is a unique right final state of the cube, there are many possible sequences of rotations which will produce it. Some are shorter than others, so will be better; but when you learn to solve the cube, are you really searching for a shortest solution path or for a path which you can find in shortest time?

### Size of the training set

The more sophisticated algorithms learn a lot from just a few examples (maybe just one). The cruder ones often require large training sets.

### The order of examples in the training set: what is learned

It is a strange fact that seemingly rational people don't always agree. If they were all presented with the same evidence, you would think they would all come to the same conclusion. They don't. Politics is a clear case. Perhaps each of us forms preconceptions from our first exposures to politics and, afterwards, very few of us can form unbiased views.

### The order of examples in the training set: speed of learning

You would think it makes sense to present instances to the learner in some sort of helpful order. When you were learning to spell, your teacher probably first showed you written words like

    dog     cat     pig

which are spelt phonetically, before

    mouse     penguin     pigeon

which contain diphthongs; and he probably left until much later the awkward cases

    though    trough    through

In principle, you might have learned correct spelling even if you saw all these words in any order. However, learning is much easier if you are shown words in the order above. The same is true for many algorithms.

### Does learning ever stop?

This is not as silly as it sounds. A true scholar is always eager to learn, but he does not expect to go on always learning the same sort of thing. After a while, he will have learned how to add and then he will try his hand at multiplication. He has recognized that he has learned all he needs to know about addition.

The same is true of some learning algorithms, even some incremental ones. They can discover when everything that can be learned has been learned. This is a useful feature.

### Is learned information accessible?

Knowledge is not just valuable for straightforward problem-solving in the system which learns it. If a user can extract it and treat the encoded knowledge itself as data, he may find ways to correct and improve the program's knowledge by manipulating it directly. He may even learn from the program.

When learned knowledge is accessible, it also lets us check that the learner is really doing something sensible. It gives users more confidence in their system, and it allows researchers to analyse the system and develop and test theories of how it works. There is some antipathy towards neural nets and genetic algorithms because what they learn is not accessible.

### Underlying predictive theory

To some extent, all algorithms embody some sort of theory, but the theories vary greatly in their assumptions and in predictive power. Often, if you can find an application algorithm with a strong underlying theory, you can use this theory to predict properties of the algorithm's output and even to analyse the likely behaviour of the learner itself. However, strong theories demand strong assumptions and such learners are not usually very widely applicable.

### Extent of dependence on a user

So much for how an algorithm treats its major input and output. It may have other inputs as well. One of these is what the user tells it during the course of its search. Incremental learners commonly take the form of three linked programs which run in a cycle:

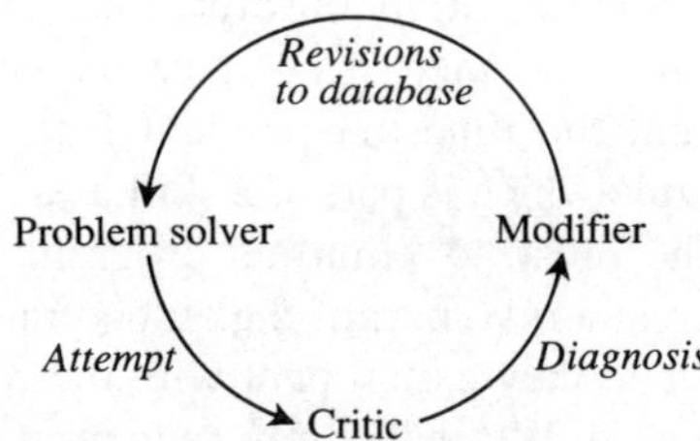



The **problem solver** uses its current database and tries to solve whatever task the user offers it. The **critic** detects whether the answer found was right and, if not, where the fault lies. The **modifier** may change the database, on the basis of output from the critic. Problem solvers are well understood. They are like expert system shells. The modifier is also usually straightforward. The critic is the core of the learning process.

Critics vary immensely, from the trivial

   (*Ask: How should the problem be solved?*)

through the naive

   (*Ask: Did I go wrong and, if so, where?*)

and the constructive query-driven searcher

   (*Repeatedly, ask: Is my diagnosis right? until it is*)

to the theorem prover and generalizer

   (*Ask: Show me a right result and let me work out the method*)

and the totally self-contained analyser. Such learners can be classified by their dependence on guidance from the user.

## 1.6   WHAT IS LEARNED INFORMATION FOR?

What sort of thing can you remember? For a start, you can sometimes remember *events*. Either you said 'Hello' to somebody this morning or else you didn't. If you did, then you did it at a unique place and time. That is characteristic of an event.

Then, you remember *facts* which are not events. You know your own name. You almost certainly cannot remember the event when you first learned that this is your name, but you did learn it nevertheless.

Then, you can remember *skills*, such as how to walk. A skill might perhaps be portrayed as a set of facts—if you put your feet on the ground thus, then you will go forward without falling over—but most people are not conscious of thinking that way when they walk.

Programs are similar. A learning algorithm may store time-dependent information which another problem solver will use as a diary, or it may store facts which are independent of time, or it may record rules which assist a solver in its process of problem-solving. In a computer, all these sorts of data are stored similarly, as bit patterns. Hence, inside a machine, there is no clear distinction between a record of an instantaneous event and the time-independent fact that the event occurred. Similarly, one can treat a rule which is part of a skill as a fact about the skill.

Authors of some of the most adventurous programs deliberately blur these notions. They do indeed treat a rule, part of the embodiment of a skill, as an assertion about that skill, and then they experiment with the rules as if they were facts and so try to improve the skill. We shall study such programs, but writing them is easier if you keep these distinctions in mind.


## 1.7   LEARNING, AESTHETICS, AND COMMON SENSE

Any worthwhile learning process involves a degree of guessing. The first example above, which found the number $S$, does not really deserve to be called learning, but the second does. There are many interesting kinds of structures which can be built out of wooden blocks, and the process of discovering the definition of an arch is not easy.

Science is all about looking for sensible explanations of what we see around us. The notion of an arch is one such sensible explanation. There is a naïve view of the scientific method which runs something like this:
You observe a phenomenon or several similar phenomena. You wish to find an extension of your current theory of things which will explain it and which will preferably also explain and predict some other related phenomena.

You list all the possible alternative theories which could explain what you saw. You ignore explanations which disrupt old established theory too much and explanations which make a lot of assumptions.

You then devise a few experiments, the fewer the better. The different theories should make different predictions for their outcomes, so when you actually do the experiments, just one theory will be compatible with them all. This is the theory you choose.

*Example: why a certain kind of seaweed only grows on the shore*
The possible reasons why the seaweed only grows in the tidal area of the shore, and not on dry land, are

- it has to be kept wet;

- other vegetation which can only grow on dry land would choke it;

- the seaweed cannot tolerate soil structure or some other similar feature of the land;

- rabbits like eating it and rabbits don't like getting their feet wet.

Someone devised a single experiment. He wired off a piece of land just by the shore, so that rabbits could not enter it but otherwise it was indistinguishable from the rest of the dry land. The seaweed became established there, despite its microclimate, soil, and competing vegetation. Hence, the last conjectured theory is correct.

This approach is good as long as it works. In many cases it doesn't work, for two reasons. First, the experiments may not be decisive. Many essential experiments involve statistics, and the most we can ever hope for from them is a degree of confidence in some conclusion. There may not be any absolute test. Second, we may have no way of listing all the possible alternative theories. The theories may even form a continuum.

*Example: the colour of a fruit fly's eyes*

Fruit flies' eyes come in two shades: red and white. Any one fruit fly may have both eyes red or both white. That is the first significant fact about fruit flies. The second is that they are small and harmless and they breed even faster than rabbits, so they provide a good test bed for experiments on inheritance.

In the days before the structure of DNA was understood, a priest called Mendel proposed that inheritance is conveyed by a set of discrete switches. Each feature of a living organism is, he suggested, described by the settings of a particular group of these switches. Its father contributes the father's settings and the mother contributes hers. Mendel did not know what form these switches take, but that does not matter. His idea is that the offspring's switch settings are a random selection from those of its parents.

We guess that some simple features are controlled by just one switch. In the case of a fruit fly, the eye shade is such a feature. In fact, there are many switches which bear on eye shade. However, we can work on the assumption that there is just one.

According to Mendel's theory, each fruit fly carries two copies of the switch, one from each parent. If both its copies are switched to one shade, then it will have eyes of that shade. However, if it has one switch set to 'white' and the other to 'red', then its eyes will actually be red. In the jargon of genetics, the red setting **dominates** the other and the white setting is **recessive**.

To test this theory, we need a population of fruit flies which all have red eyes over many generations, so that we can be reasonably sure that there is no suppressed white switch concealed in any of them. Similarly, we want an established family of pure white-eyed fruit flies. We take one fruit fly from each family and mate them.

According to the theory, all the offspring will have one red switch and one white. One parent can only pass on a red switch setting and the other can only pass on white. Since red dominates, all their eyes will be red. However, if we take any two of these offspring and mate *them*, suppressing any qualms we may have about incest, then any one of the next generation will have an even chance of inheriting red or white from its father and an even chance of inheriting red or white from its mother. Thus, there is

- a 25 per cent chance that both its switches will be red;

- a 50 per cent chance that one of its switches will be red and the other white;

- a 25 per cent chance that both its switches will be white.

Since red dominates, the grandchild can only have white eyes if both switches are white. Thus, the chance of white eyes is 1 in 4 (see Fig. 1.3).

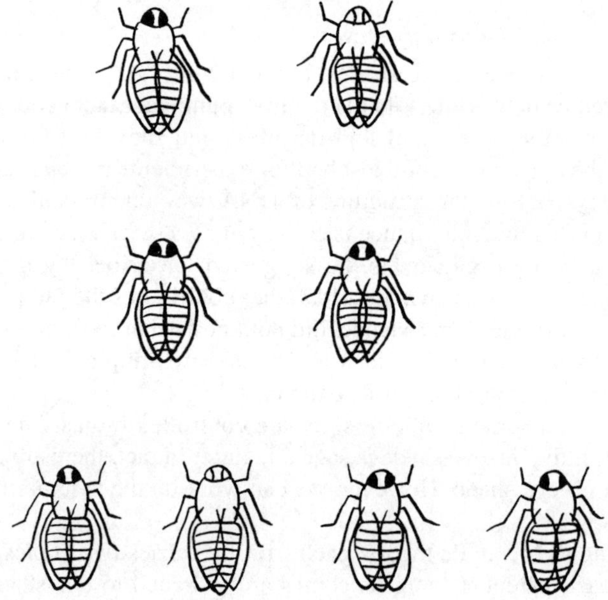

**Fig. 1.3**    Inheritance among fruit flies.

Let us imagine how such an experiment actually appears to a student. His professor tells him the above. He then mates his fruit flies. All the offspring in the first generation have red eyes. He then mates two of these offspring and counts how many offspring in the second generation have each eye shade. Let us suppose his result is:

- 61 offspring with red eyes;

- 23 offspring with white eyes.

At this point, the professor leans over his shoulder and murmurs 'Yes, those results look pretty conclusive.' Of course, they are nothing of the sort. That is common sense. There is no way one can *conclude* from such results that Mendel was right.

So, what is the proper interpretation of the experiment and why do most of us agree with Mendel?

The first point is that the best we can say of this result is that it is *compatible* with the theory. The numbers 61 and 23 are quite close to the ratio 3 : 1, as Mendel would predict. If the student is in a large class who all perform the experiment and they add up all the offspring with red eyes and all those with white, then we can expect that the two sums will be even closer to this ratio, as a proportion of the total number of offspring. If the professor is a little more careful, he will point out that the proportions of offspring with red and white eyes appear to get closer and closer to 3 : 1, the more offspring are counted. That is about all it is worth saying of the experiment.

Secondly, what of the theory? Mendel was not totally original. His idea is an *extension* of an older simpler theory, to wit: offspring inherit traits from their parents. Mendel's contribution quantifies how they inherit.

Without delving into biochemistry and discovering precisely what the switches really are, there is no way this theory can ever be proved. It makes statistical predictions and any test of it can at best be statistical. No doubt one could concoct a spectrum of theories which make infinitesimally differing predictions, but almost all of them would be outlandish. People had confidence in Mendel's suggestion even before they understood DNA because it is *compatible* with experiment and because it is *simple* and *aesthetically satisfactory*.

The foundations of science are simple, yet it is capable of explaining every physical and biological phenomenon which we have yet had the time and resources to investigate. It is a fantastic success. The guiding principle among researchers is that they want to develop its coherence and simplicity. A theory is judged to be good if it both preserves the simple foundations and extends the predictive power of the whole edifice, even if the predictions are statistical and beyond undisputable test.

Researchers in science are in the business of learning. Learning algorithms deserve to be assessed on the same criteria which we use for science. There are all sorts of practical ways of assessing an algorithm. Will it tolerate noisy data? Will it require correction? Over and above these, we also want algorithms which produce acceptable results. In this sense, a result is acceptable if it is compatible with the training set and if it is simple. There are probably lots of ways of describing sets of characteristics common to all the examples of arches in a program's training set. What we want is the simplest characterization. A learning algorithm is good if it is likely to find the simplest explanation.


## 1.8   NATURAL LEARNING

Before we spend effort investigating artificial algorithms, it is worthwhile finding out as much as we can about the natural versions. This section is a short summary of learning among animals other than people. Human learning will be treated in the next section. The examples included here appear in ascending order of subtlety, in so far as subtlety can be measured, but they will not be analysed beyond that. Trying to classify them would be tantamount to trying to present a theory of nature.

Two good introductions to the subject are the book by Halliday and Slater and the paper by Gould and Marler, both mentioned in the *Further reading* section. A lot of the claims made below can be traced to these two references, but sometimes I shall put in snippets of hearsay. Whenever I remember, I shall add a 'reference' such as [*the radio*] after apocrypha, but I may forget to, so this section and the next are definitely *not* authoritative.

The first point is, beware of anthropomorphism. Some seemingly intelligent behaviour probably involves absolutely no intelligence whatsoever.

*Example: mosquitoes*

These creatures are very good at finding their food, us; but the way they do it requires *no memory*. If a mosquito finds itself flying into a region with high humidity and high levels of carbon dioxide, then it will continue in a straight line. If it detects that it is flying out of such a region, then it will turn through random angles. Between these turns, it will fly in straight segments for short periods, and the durations of these periods obey a Poisson distribution. It appears that the mosquito's logic depends on absolutely no memory at all. [*The radio*]

There is also the tale of a goose which saw an egg roll out of its nest and gently pushed it back in again. This appears very clever, but Konrad Lorenz found that the goose would do exactly the same with a beer bottle.

The second point to bear in mind is that experiments with living creatures are often hard to arrange and even harder to interpret. The apparent intelligence of an animal is sometimes limited by the intelligence of the man who observes it. Even when an animal is capable of learning, it may not bother. There is the tale of a flock of wild pigeons who were presented with a peg in a stopper. Whenever any pigeon pecked it, food fell out nearby. When they were all together, just two pigeons learned to peck it. All the others stood and watched these two and then dashed for the food. However, the others were capable of learning. They all did so, when they were separated each with its own peg but so that they could see each other. It seems pigeons are intellectually lazy.

Some small, short-lived insects are capable of both memory and learning of a sort. A butterfly, caught by a botanist while on a flower, avoided that flower the next day. Fruit flies can learn associations between scents and colours or between a scent and an electric shock. A normal fruit fly can remember this sort of associations for hours, after just three training sessions. There are mutant strains of fruit fly which appear normal in all reaspects except that they cannot remember at all or they forget very quickly.

Much learning among animals seems to be of a very few basic kinds:

- imprinting;

- habituation and sensitization;

- classical conditioning;

- instrumental conditioning.

**Imprinting** is the way a lamb or a chick learns to recognize its mother. For a short period in the young creature's life, it will not know its mother and it will following anything which moves and looks and sounds about right. Among all such objects, it will develop a preference for one and stay with that one. The chosen being is usually its mother, but it can be a human or even a metal bucket. Once the young has developed such an imprint, it is much less willing to learn any other.

This form of imprinting is not quite the way that some young birds learn their song. The white-crowned sparrow is born with a crude notion of its species' song. This is enough for it to ignore the songs of a Harris sparrow or a song sparrow. There is a period of about 40 days in the life of the young white-crowned sparrow during which it can learn its true song by listening to adults. During this interval, it will learn all the songs that its parent sings to it and afterwards it cannot learn any more. (A similar phenomenon has been noted in human children. When young children hear nursery rhymes often, they become more facile with language. This works up to about the age of 7 years. A child who has not been taught rhymes remains less fluent, even with extra coaching after that age [*The radio*].)

**Habituation** simply means that if you bump the plank that a snail is walking on, it will withdraw into its shell the first time, but if you do it repeatedly, after a while the snail will take almost no notice. Do we count this as learning? It is a change in behaviour induced by experience, so it is learning of a sort.

**Sensitization** is a little more interesting. A marine worm which is often fed will start crawling out for more food when a light is flashed, but a worm which receives electric shocks will withdraw when the light is flashed. It seems that the worm becomes more ready to act on *any* stimulus, if it has already had some exciting experience. Similarly, if you feed an octopus, then it will become more interested in anything which might be food.

**Conditioning** is the name given to learning by straightforward association. It comes in two varieties. In classical conditioning, the subject (often a rat or a pigeon) sees or hears some brief signal which would normally mean nothing to it and then soon afterwards it is given an emotive experience such as food or a shock. In jargon, the first signal is called the **conditional stimulus** and the experience is the **unconditional stimulus** or **reinforcer**. After several such training sessions, the animal will react to the signal alone. It anticipates the experience, even when the food or shock never happens.

Instrumental conditioning is a bit different. The experimenter does not show a signal to his rat. Instead, he waits for the rat to do something such as press a lever and, soon after the rat presses the lever, the experimenter gives the rat some food. After a few trails, the rat starts pressing the lever uncommonly often. The difference between this and classical conditioning is that in this case the rat has to take the initiative by doing something.

Conditioning is very common among many species. Originally, psychologists imagined that the time lag between stimulus and reward must be quite brief if the animal is to learn anything, and that any creature could associate any stimulus with any reward. All that governed learning, they supposed, was the strengths of the stimulus and reward and the time lag between them and any absolute limitation of the animal. However, this is not so. For a start, rats are very good at remembering when they have been poisoned. They will associate food or drink with an illness which they only feel hours later. Also, rats associate smell with food, and sight and sound with danger, whereas pigeons associate visual stimuli with food.

*Example: tasty water and bright-noisy water*

Water can be made novel for a rat in two ways. It can be flavoured, for example with salt or saccharine, or whenever the rat touches it with its mouth there may be a noise or flashing light. Also, there are two emotive responses which the rat can receive—it can be made to feel sick or it can be given an electric shock. (There are many other 'rewards' which one might offer a rat, but these were the ones used by Garcia and Koelling.)

With these signals and experiences, one can perform four different tests for instrumental conditioning. They are four ways of pairing a signal with an experience. If rats had no predisposition as to what they learn, then all four should produce equally strong effects. However, in practice, the rat associates well:

- tasty water with sickness;

- bright-noisy water with pain;

but it associates hardly at all:

- tasty water with pain;

- bright-noisy water with sickness

Conditioning shows other non-trivial properties.

*Example: recognizing the best signal*

This is an elaborate little experiment. The subject was a rabbit. The reinforcer was a puff of air into its eye, which made it blink. The conditional stimulus was the subtle part. It was composed of a flash of light and a tone. The flash was always the same, but there were two different possible tones, each used on half the occasions.

There were two versions of the experiment. In the first version, the puff followed just half the flashes. The tones were switched randomly, so in practice each tone also preceded half the puffs, but the flash was a better predictor of a puff because it occurred more often. In the second version, the puff followed half the flashes as before, but one of the two tones always preceded a puff and the other tone never preceded one. The outcome was that the rabbit did the sensible thing. In the first version, it learned to blink after any flash, lest there was a puff, but it ignored both tones since they supplied no extra information. In the second version, it ignored the flash and the tone which never preceded a puff and always blinked at the tone which it learned to associate with a puff.

This and similar tests show that animals do not learn arbitrary associations. They recognize which associations are most significant. They learn those and ignore all others. A rat will associate sickness with the thing it last ate or drank, even if something else which it drank earlier correlates just as well.

As mentioned before, interpreting experiments like these can be hard. One well-known case involves a pigeon which is rewarded with food or drink when it pecks a disc. Naturally, it learns to peck the disc. This would seen to be straightforward instrumental conditioning. However, careful study reveals that the pigeon does not peck the disc the same way for food and for drink. If the reward is food, then it pecks with its beak open and its eyes shut, which is how it eats. When the reward is a drink, it pecks with its eyes open and its beak nearly closed, as if drinking. It seems that the bird is not simply associating the action of pecking with gaining the reward. Perhaps rather, it learns to think of the disc as if the disc itself were the reward. This is not what the straightforward theory of instrumental conditioning predicts.

Of course, conditioning does not exhaust animals' learning skills. Rats are celebrated for their mastery of mazes, and psychologists have wondered how they do it. The obvious possibilities are:

- the rat associates each corner with some single feature and navigates by it;

- the rat recognizes each corner by its composite background and is not dependent on any one feature;

- the rat learns to go left or right by rote;

- the rat has a mental map and it is aware all the time whereabouts it is in the whole maze.

It appears that the first possibility is wrong. The rat can still find its way when any one feature is moved. Rats seem to prefer to navigate by a general synthesis of the scene they are in. However, they are also capable of learning how to navigate when there are no landmarks, so they can either learn left/right turns by rote or they have mental maps. If one wanted to find out which, one would have to put the rat down not at the entrance to the maze but at a random point inside it. If the rat navigates by rote, it will be lost and remain lost. If it has a mental map, then if it is clever, it will recognize the structure of the walls and corners and begin to navigate.

Rats and other animals have mental maps. This was shown by an analogous experiment with bees. As they left the hive for a bank of flowers some distance away, they were caught by the experimenter, who carried them in a dark box to a point off their route and let them go. With few exceptions, the bees headed from there directly towards the flowers. If they navigated simply by remembering a sequence of landmarks along each route, then they would have had to return to the hive first. It appears that bees do have mental maps.

Another simple experiment with bees reveals a bit about the structure of their memory. A bee may recognize a flower by its colour or scent or shape. The experimenter presented his bees with an artificial imitation 'flower' of a particular hue, shape, and scent and taught them that they could gain sugar solution from it. Then he offered the bees a choice between several differing 'flowers', none of them quite like the original. The bees went first for the one with the original scent. If there was no scent cue, then they opted for one with the original colour. Only if neither scent nor colour guided them did they go by shape. This makes sense, since a flower may be battered out of shape and yet still have the right colour and scent, whereas if it loses its scent then it is probably no longer making nectar.

This experiment was extended a stage further by Bogdany. The bees first learned to feed from a peppermint-scented blue triangle. This was changed to an orange-scented blue triangle. Gould and Marler write:

The foragers learned the new odor in one visit, but they completely forgot the colour and shape, even though these characteristics had not been changed. On the other hand, when bees were trained to an odorless blue triangle and then presented with a peppermint-scented blue triangle, they learned the new odor without forgetting the colour and shape. Apparently the [bee's] appointment book has an entry for each cue; the entries are structured in such a way that blanks can always be filled in but that if even one item is changed, the entire entry is erased.

We humans like to imagine that we are superior to other creatures because we can handle abstract notions (such as the concept of learning). The question arises, can other animals do this too? To some extent, some can. Pigeons reared indoors can form the general notion of a tree, from pictures of some species of trees. If a rat knows its way through a maze which contains a black box and a white box and then it is given a shock in a different white box, it will avoid the white box in the maze. It has also been suggested that rats have a notion of positions in order: first, second, third, ....

*Example: the notion of repetition*

A rhesus monkey was taught that, given two pots with different lids, it would find a nut or a raisin under one of the lids but not the other. The positions of the pots were varied, but one of the lids always covered a reward and the other lid always covered an empty pot. It took the monkey some time to learn this.

The lesson was repeated with many pairs of lids. After a hundred or so, whenever the monkey was faced with a novel pair of lids, its first trial was random, but the second time, it always got its reward immediately. It had learned *how to learn* which was the right lid.

This form of learning can be explained as a kind of conditioning. The conditional stimulus is the outcome of the first trial with new lids: it finds the food or it doesn't. The reinforcer is its strategy at the next trial. Thus, what the monkey learns is that:

- if the first trial succeeds, then the *same* lid will hide food next time:

- if the first trial fails, then the *other* lid hides food next time.

This is a form of association. The advanced step is that the monkey formulates the conditional stimulus and the reinforcer for itself, and these are abstractions such as *same*ness. Dolphins can do this too. There are also reports that baboons can imagine how other baboons are thinking, and plan accordingly.

*Example: social skills*

A baby baboon watched an adult dig for bulbs. Every time the adult found one, the baby howled as if it was being attacked. Its mother rushed to the rescue and drove off the other adult, whereupon the baby ate the bulb.

A young male startled a baby, which ran off crying. The whole tribe rushed up to chase away the adolescent, but he promptly gazed around the horizon, as if looking for an enemy. The tribe did likewise and left him in peace.

## 1.9   HUMAN LEARNING

The most spectacular human faculty is the ability to learn and use language. A normal healthy well-educated adult has a vocabulary of between 20 000 and 80 000 words. (The figure depends somewhat on what you call a word.) Most of this is learned in about 15 years. Hence a child learns words at the rate of at least three or four a day on average, and maybe several times that, for many years.

When a child masters a new word, it seems that his memory of it goes through a few distinct phases. Words are not learned by rote. Each word must have an associated meaning. In the child's memory, there has to be some notion for the word to associate with. In the first phase, this notion is formed.

*Example: words for colours*

Two experimenters, Carey and Bartlett, established that a group of 3-year olds did not know the name of the colour which we call olive. Then they painted a cafeteria tray olive and another one blue. They casually asked each child 'Hand me the chromium tray. Not the blue one, the chromium one.' (They might equally well have said 'Hand me the olive tray.' It is not clear why they chose to call it 'chromium'.)

A week later, the children could not recall the word 'chromium' but they could remember that an olive tray was not called 'green' or 'brown'. They appear to have learned that there is a colour, olive, although they had not at that stage learned a name for it.

In a second phase, the child learns the word itself and learns its semantic category. That is, he learns that a word such as *olive* is a colour and that a *delphinium* is a flower, although he may not recall which colour or which flower. The final phase, which is often never completed, consists of associating the word with its exact meaning. Often the child's notion may be wrong to begin with—too restricted or too broad—and needs correcting.

Looking up a word in a dictionary is not a very effective way to learn its meaning. Learning a new notion is easier from examples. What is more, it helps if the first few examples provide a narrow view of the notion and then later examples exhibit it in broader contexts.

By the time a child is at school and reading, the context is often just a phrase in a book, which is not much to go by. The child's first few words are the ones taught by his mother and for these he has more help. A mother's speech is organized, simple, and redundant (that is, she says the same thing two or three times in different ways). However, learning depends on more than just this.

A mother may ask her child questions and repeat and expand on what her child says; she often discusses objects and properties which are at the child's focus of attention; she uses specific names for these objects, rather than pronouns or general terms; and when she changes the topic of conversation, she often reinforces the new topic with movement of the eyes and other gestures. A child's learnt language skill appears to depend on all these factors. Furthermore, most of a child's first ten words are used in unique contexts, although a few seem to be referential. Thus, as one would expect, a child's learning depends a lot on semantics and social interaction.

There is little point in learning something unless you can remember it. What is the nature of memory?

Tell someone the name of a digit and he will probably remember it for a few minutes. Tell him ten such digits in succession and he will probably succeed in remembering the first few and maybe the last, but he is much more likely to forget the ones in between. This simple little experiment is one of many which Alan Baddeley describes vividly in his book *Your memory—a user's guide*. It suggests that there are at least two different sorts of memory involved. One, so-called **short-term memory**, stores the first few digits but then becomes saturated. It behaves like a queue in a von Neumann machine, but a queue of limited length. Anything added after it is full will be lost. However, the very last digit may well be remembered too. The memory which holds this is called **sensory memory**. It stores the very latest word the listener heard. It is like a single cell in a computer's memory, since it can only store one item.

Unless he has some particular reason for remembering, the subject will soon forget all the digits and even the fact that you recited them to him. He is not using the **long-term memory** which stores the information which he would put in his autobiography.

Whenever you want to discover how something works and you cannot take it apart, it pays to see

(1)   how quickly it works;

(2)   what aspects improve or hinder it;

(3)   when and how it fails to work.

Different algorithms have different limitations. So far, we have come across three such modes of failure in memory: short-term human memory, which behaves like a queue, sensory memory, which behaves like a single cell, and bees' memory for a flower, which is like a record which may be erased all at once.

Alan Baddeley's book is full of simple but revealing experiments of this sort which you can practice on yourself. Besides that, quite apart from its material, the book as a whole is a beautiful demonstration of how to do experimental science. Here is a summary selection.

### Ebbinghaus total time hypothesis

Ebbinghaus practiced learning sequences of nonsense syllables. Then the next day, after he had forgotten them, he re-learned them. He found that the time needed to re-learn was

$t - k \times$ the time spent learning on the first day

where $t$ and $k$ were constants.

### Distribution of practice

If the syllables were not completely learned on the first day, then it pays to spend about equal periods during each learning session. He learned equally effectively if he spent seven and a half minutes on both days or if he spent one minute on the first day and 20 minutes on the second.

More recently, the British Post Office decided that some of its staff should learn to type. They tried four different training schedules with different groups:

- 1 session of 1 hour per day;

- 1 session of 2 hours per day;

- 2 sessions of 1 hour per day; and

- 2 sessions of 2 hours per day.

Measured by total hours of practice, the first group learned quickest and the fourth learned slowest. Also, the first group remembered the skill better several months later. However, the first group were dissatisfied because the others were able to type after fewer days.

### Micro-distribution

Assuming you are English, how do you learn French? One way is to look at a word, try to remember it, and then check it again a minute or so later. There is a trade-off: the longer the gap between looking and checking, the better you will reinforce your recall of the word, but if the gap is too long, then when you check the word, you will not be able to recall it at all. It is best to check the word just before you would probably otherwise forget it. If you recall something four times or more, you have a good chance of remembering it permanently.

### Motivation

A Swedish psychologist, Lars Goren Nilsson, asked three groups to remember
groups of words. One group was given no extra motivation. The second group was
offered a prize for the best recaller, but they are only told about it after they had
studied the words. The third was told about the prize before they began learning.
All three groups could remember about the same number of words.

It seems that motivation of itself does not help learning. What it does is to
encourage the learner to spend more time concentrating, which is what matters.

### Repetition and learning

A local radio station warned its listeners more than 1000 times that it was about to
change its wavelength.

- Almost all its listeners remembered that there would be a change.

- Most could remember when it would occur.

- Very few could recall the new wavelength.

It seems that repetition alone is useless. The listeners only learned what they
thought mattered to them. Most of them never thought in terms of wavelengths
when they tuned their radios.

### Meaning helps memory

It is easier to remember strings of 'nonsense' syllables when they have associations. It is also easier to remember concrete nouns than abstract nouns.

Meaning plays a big role in many aspects of memory. It is easier to make sense
of a message if you know in advance what it is about. This makes life hard for us
programmers, because we have a better idea how to store sounds and visual images
than meanings.

### Cultural bias and memory

Bartlett, who believed that there is more to memory than strings of random
syllables, told a story to his subjects. It was a myth from among North American
Indians, but the subjects were Europeans. When they tried to recall it, they often
replaced the original Indian ideas with analogous European ones. Bartlett believed
that we struggle to impose meaning on what we see and recall.

### Organization and visual images aid memory

If you want to learn a group of words, it is easier if they are divided in a
hierarchy, like a thesaurus. Similarly, if you are trying to remember words, then

it helps to form pictures involving them. For example, you will probably remember

    rabbit       steeple

once you imagine a rabbit clinging to the top of a steeple (but not if you just visualize the rabbit on the grass at the foot of the steeple). An exceptional character called Shereshevskii could remember long meaningless formulae by inventing stories around them. Perhaps chess grand masters understand their games so well because they are good at recognizing patterns of pieces and classifying these patterns.

### Cues and structure

Suppose you are asked to think of the name of a fruit beginning with the letter O. This is easy. It is harder to think of a fruit whose name ends with the letter R, and it is also harder to think of the name of something if you are first told that its name begins with a particular letter and then only afterwards that it is a fruit.

As well as experiments on remembering, Baddeley describes some on forgetting.

### Logarithmic forgetting curve

Ebbinghaus found that the number of syllables he could recall decreased logarithmically with time.

### Recall and recognition

One can still recognize old faces decades later, although one cannot recall them. There are odd exceptions, when recall survives. The inhabitants of a fishing village could recall a scandal which took place there 70 years earlier.

### Prompting aids recognition

It is easier to hear a word through background noise if you have already just heard that word.

### Interference

There are two simple theories of forgetting: either memory *fades*; or subsequent learning *interferes* with old memories.

A team of rugby footballers were asked to recall the names of teams they had played. Different members had played in different games, so the time elapsed since (say) the player's fourth last game varied from player to player. The time elapsed since a game did not affect memory. The critical factor was the number of intervening games. Hence, this kind of forgetting is due to interference.

One can do a similar test on cockroaches. If a cockroach enters a cone lined with tissue paper, it will stay inert as if asleep. When it comes out again, it can remember better than other cockroaches which have been living an active life.

### Interference by kind of material

McGeoch and MacDonald measured how different kinds of material produce interference. Their subjects learned lists of adjectives. Then, some of them rested, some learned strings of numbers, some learned other random adjectives, and the last group learned other adjectives with similar meanings. The first group could recall their adjectives best and the last group fared worst. The more similar the interfering material is, the more it will interfere.

### Proactive inhibition

Earlier memories sometimes supplant more recent ones.

### Emotional factors in forgetting

This is the title of Chapter 5 of Baddeley's book. Freud suggested that people sometimes **repress** memories with unpleasant associations. **Arousal** makes it harder to recall immediately after learning, but in the long term, it may be easier to remember something with emotional overtones. Exceptional crises may produce **amnesia**, when the sufferer forgets his own life history.

People who use words a lot sometimes say that mental processing is done largely in terms of words. Perhaps they do most of their conscious processing that way. However, it is not always so. Under suitable circumstances, one can recognize that a picture of, say, a robin shows a bird before one can remember the word 'robin'. Language is based on perception, not the other way round.

Computing has provided some useful analogies for how memory is structured. A common notion is that concepts like

    substance     food     fruit     orange

form hierarchies. Hierarchies help various aspect of data processing. For example, they allow guesses at defaults—we guess that Aristotle had feet. They also suggest that some tasks will be easy because they only involve a short search in a hierarchy and others will be harder because they require deeper search. This is borne out of experiments. However, simple computational hierarchies do not give the whole answer. In a computer, it is equally easy to answer the questions

- is a robin a bird? and

- is a penguin a bird?

Most people find the second question harder. In human memory, categories such as 'arch' and 'bird' do not have neat definitions consisting of pairs of lists of attributes. They are blurred.

## FURTHER READING

There are some detailed studies of the workings of individual neurones, and also of whole regions of the brain, in Rolls (1987).

The classification scheme for algorithms used here owes a debt to Yves Kodratoff and to Carbonell (1989). Shalin, Wisniewski, Levi, and Scott (1990) provide another analysis. Craw, Sleeman, Graner, Rissakis, and Sharma (1992) have designed an expert system for selecting a learning algorithmn, and Sleeman (1994) expands on some of the lessons they learned. Their system uses considerably more information than just the format of the data.

The story of DNA is related by John Gribbin (1985). Bryan Shorrocks (1972) gives a full account of what was known 20 years ago about the genetic make-up of fruit flies.

Halliday and Slater (1983) and Gould and Marler (1987) both give careful summaries of much thorough experimental evidence. It is not immediately obvious how an inventor of computational learning algorithms can benefit from these biological studies, but we would be foolish to ignore Nature's achievements. The *New Scientist* carried accounts of learning among wild pigeons (29 October 1987) and butterflies (17 September 1987), and it reported the many talents of sheep (12 November 1994). This journal has also carried an account of the social life of baboons.

For a thorough introduction to language learning see, for instance, Margaret Harris and Max Coltheart (1986) and their group's other studies: M. Harris, D. Jones, S. Brookes, and J. Grant (1986) and M. Harris, M. Barrett, D. Jones, and S. Brookes (1988). The anecdote about the olive tray comes from G. A. Miller and P. M. Gildea (1987). Alan Baddeley's book, *Your memory—A user's guide* (1983), is a classic which you can give to almost anyone for a Christmas present.

## EXERCISES

1.  Write down the characteristics of a training set for the program which invents a definition of an arch. Suggest a format for each example in its training set.

2.  Write a summary of experimental evidence which suggests that

    (a)  some useful behaviour in animals does not involve any memory;

    (b)  the ability to retain memory is inherited;

    (c)  in some species, learning is easy in domains which will aid the species' survival and not in other domains;

    (d)  some natural memories permit associations between similar unrelated events, and mental maps.

3.  On a remote group of islands, the local tribes have strict social hierarchies. These
    people greet each other by rubbing noses together. Any newcomer in these parts is
    welcome if the first person he rubs noses with is male, has at least one wife, and is
    related to the local head man. Otherwise he has to run away.

    The first explorer who comes there has to learn this rule by trial and error.
    Fortunately, he can attempt the procedure several times with different tribes on
    different islands. Describe the characteristics of his training set and suggest the
    sort of algorithm he might use for learning.

4.  Consider the following training sets and learning situations:

    (a)  thawing snow flakes, when learning their crystal form;

    (b)  plots of cheap romantic novels, when learning social conventions;

    (c)  economic history, when learning how to predict recession;

    (d)  examples of electrical circuits from a textbook, when learning how to design
         amplifiers, filters, and oscillators;

    (e)  your own attempts to design and build electrical circuits, when learning how
         to design amplifiers, filters, and oscillators.

    In each case,

    (i)    characterize the training domain and the forms of data;

    (ii)   suggest a suitable language in which the data might be expressed, so that an
           algorithm could learn from them;

    (iii)  suggest a suitable form for the information which the learner is trying to
           learn;

    (iv)   is there likely to be an appropriate incremental learning algorithm?

5.  Give an example of a clean training domain which contains counter-examples and
    of a noisy training domain consisting only of positive examples.

    Consider the problem of learning how to forecast the weather given a training set
    consisting of either

    (a)  examples of weather charts on successive days; or

    (b)  charts of the weather on successive days and your own attempts to predict the
         weather on each day except the first from the previous day's chart.

    In each case, (a) and (b), what kinds of learning algorithms might be suitable? Give
    reasons for your choices. When you have studied such algorithms in more detail
    from later chapters, list their strengths and limitations.

6.  A certain learning system consists of

    •  a theorem prover (its problem solver), and

    •  a database of rules about how to discover proofs, and

    •  a learner which can modify and invent such rules.

    The input to the theorem prover consists of

- a set of hypotheses, and

- a conjecture, and

- the database.

Its output is either

*yes*        if it finds a proof of the conjecture; or
*no*         if it disproves the conjecture; or
*maybe*      if it finds that the hypotheses are consistent with both the conjecture
             and its negation; or
*fail*       if it runs out of space or time.

The inputs to the learner are

- the three inputs to the problem solver, and

- the problem solver's output, and

- the search tree which the problem solver constructed.

From the point of view of learning, what are the main characteristics of the learner's training set as a whole, and of individual training instances? What sorts of learning algorithms might be appropriate? What are the likely characteristics of such algorithms?

# 2 Some basic ideas

## SUMMARY

Herein are summaries of some of the techniques which will be used later. They include:

- logic and graphs;

- search spaces, with some classic examples;

- the notion of a description language and how it affects search;

- different varieties of search: forward or goal-directed, search for a goal or for a path to a goal, and the distinctions between monotonic, reversible, and irreversible search;

- some basic search algorithms;

- heuristic guidance of search;

- metric spaces;

- experimental backing;

- complexity.


## INTRODUCTION

The algorithms we shall investigate will call for quite careful description. Plain text could get us a long way, but words are an imprecise medium, and pictures and other symbolic descriptions often help, so we shall need a few mathematical ideas and notations. Several good references are listed in the Further reading section, but you need not worry about them now.

You will find this material easier to read if you know the names of some letters of the Greek alphabet. Also, do not be surprised if you see conventional symbols used in odd ways. For instance, the symbol '<' may represent any partial order, not just the standard order on numbers. The term 'iff' is often used instead of 'if and only if '.


## 2.1 LOGIC

This is not the place to explain logic, but there are a few bits of terminology which occur often in this subject and which are not entirely standard.

- An **atomic formula** is one of the form

  $p\,(t_1, t_2, \ldots t_k)$

where each $t_j$ is a term and $p$ is a predicate symbol which expects $k$ arguments (that is, $p$ has **arity** $k$).

- A **literal** may be either an atomic formula or a negated atomic formula:

  $\neg p(t_1, t_2, \ldots t_k)$.

- A **(disjunctive) clause** is a sentence composed only with the connectives

  $\vee$     and     $\neg$

  in which all variables are universally quantified. The quantifiers are usually omitted. Thus, the sentence

  $\forall x\ (\exists y\ \text{likes}(y, x) \Rightarrow \text{happy}(x))$

  can be rewritten as

  $\forall x\ \forall y\ (\text{likes}(y, x) \Rightarrow \text{happy}(x))$

  and so it is equivalent to the clause

  $\neg \text{likes}(y, x) \vee \text{happy}(x)$.

  Both can be read as 'Anyone who is liked is happy'.


## 2.2   GRAPHS

The graphs we are concerned with are quite unlike the ones which physicists and engineers draw on graph paper. To a professional, a **graph** consists of a set $V$, called the set of vertices or nodes, and another set $E$ of arcs or edges:

$G = (V, E)$.

An edge is an unordered pair of nodes. True pedants add the proviso that an edge cannot also be a node. Imagine what the graph would look like if it was!

Most graphs which you will come across in the references are finite. We shall require infinite ones too, because almost all interesting search spaces are infinite. There is always a trivial algorithm for searching in a finite space—just list all nodes and examine them in turn.

That is the simplest case. Next, we come to the notion of a **directed graph**, sometimes called a digraph. This is a set of nodes $V$, as before, but now an edge is an *ordered pair* of nodes:

$G = (V, E)$      where $E \subseteq V \times V$.

In pictures, a graph consists of points and lines as in Fig. 2.1, while in a digraph, Fig. 2.2, the lines have arrows on them. The arrow points from the first node $a$ of the pair $(a, b)$ towards the second, $b$.

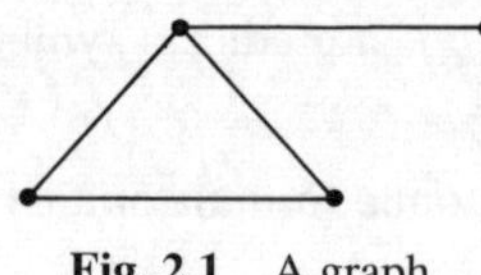

**Fig. 2.1**   A graph.

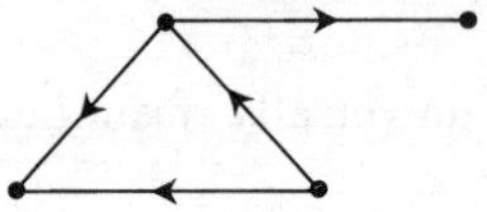

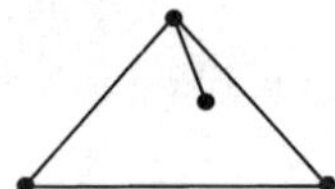

**Fig. 2.2**   A directed graph.

Note, by the way, that a picture of a graph, like Fig. 2.1, is not itself the graph. It is an *embedding* of the graph in the plane of the page. One graph has lots of possible pictures. Fig. 2.3 is another picture of the same graph as Fig. 2.1.

**Fig. 2.3**   A different picture of the graph in Fig. 2.1.

In any graph, each node is different from all the other nodes. If you like, each node has its own name. One often meets graphs in which the arcs have names too. This is called a **labelled graph**. The name attached to an arc is called its label. A label can be more than just a sequence of characters. We shall come across labelled graphs in which the labels are elaborate operators. When arcs have labels, there can be more than one arc between two nodes. Of course, two arcs with the same ends must have different labels.

A **path** in a graph is a sequence of distinct nodes

$$a \quad b \quad c \quad \ldots \quad d \quad e$$

with arcs between them from $a$ to $b$, $b$ to $c$, $\ldots$ $d$ to $e$. Its **length** is the number of arcs. If $e$ is the same as $a$ and the length is at least 1, then it is a **cycle**. A graph is **connected** if any two nodes in it are the ends of a path. The same ideas apply in digraphs, except that all the edges must point in the same direction.

To a true graph theorist, a **tree** is a connected graph with no cycles. However, nobody thinks of a tree like that. Every tree which occurs in computing has a distinguished node, called its **root**. When we have decided on a root, then the tree automatically becomes directed: each arc is given an arrow pointing away from the root, as in Fig. 2.4. Then, to each node $e$, we can associate its set of **children** which are the far ends of the arcs pointing away from $e$. A **leaf** is a node with no children. Any node with children is said to be **internal**.

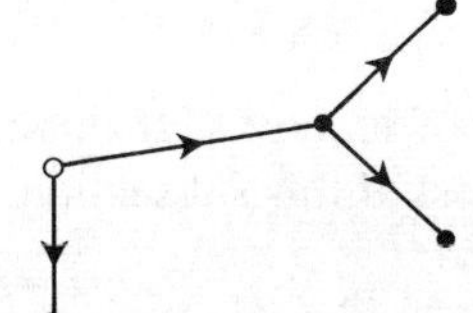

**Fig. 2.4**   A rooted directed tree.

The **depth** of a (rooted) tree is the length of a longest path from the root to a leaf. The depth of any node is the length of the path from the root to it. Thus, the root itself is at depth 0.

The typical number of children of each internal node in a tree is called its **branching ratio**. One often comes across trees in which every node is either a leaf or else has just two children. Such a tree is called a **binary** tree and, of course, its branching ratio is 2. Often, there is no fixed number of children per node. In that case, the 'branching ratio' is a suitable number which makes the calculations work the way we want. For instance, it is often convenient to be able to estimate how many nodes there are in the whole tree. If all leaves are at roughly the same depth $d$ and the number of children per internal node is fixed, say $n$, then this is

$$1 + n + n^2 + n^3 + \ldots + n^d.$$

If the number of children varies, then the branching ratio $n$ can be taken to be the number which makes this sum give the right answer.

In fact, computing people add yet another layer to this stack of definitions, though they very rarely make it explicit. In practical cases in computing, in a tree, the children of each node are often not just a set. They are ordered, so they form a list. Any picture of such a tree, with ordered lists of children, conveys a much more thorough image than does a picture of an average graph, because the picture can exhibit the order in each list of children. The root is at the top, as in Fig. 2.5 (not at the bottom as in botany). Each node's children appear below their parent, in order from left to right. This is standard convention.

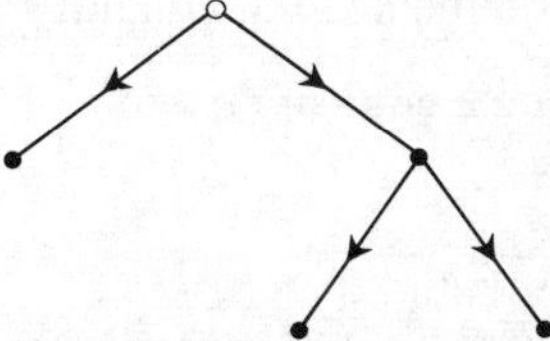

**Fig. 2.5**   A rooted directed tree with ordered children.

Since almost everyone who writes about computing does not usually bother to make all these distinctions, neither shall I. Whenever you come across a graph or a tree, it will be up to you to recognize what kind it is.

### 2.2.1  COVERING TREES

Fig. 2.6 shows a graph and part of its cover. If this picture makes some sense to you, then you need not read the rest of this subsection.

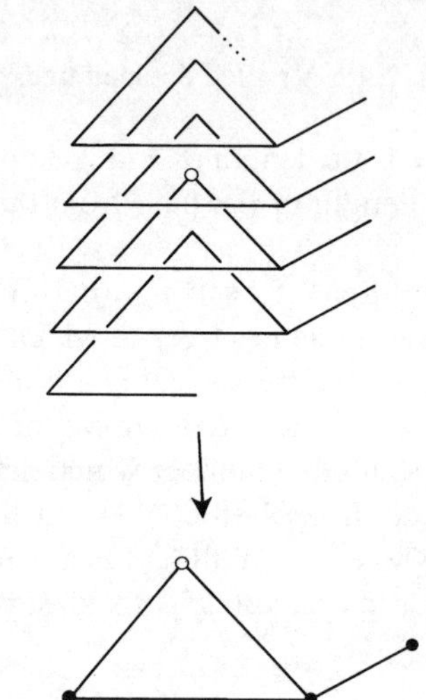

**Fig. 2.6**   The cover of a rooted graph.

Suppose you are presented with any connected graph $G$ with vertices $V$, edges $E$, and a root $r$. (Even when $G$ is not a tree, we can still choose a root for it.) Unless it is a tree, there will be some cycle in it. There is a standard way of constructing a tree, called the (universal) **cover** of $G$, or 'gee twiddle',

$$\widetilde{G} = (\widetilde{V}, \widetilde{E}, \widetilde{r})$$

with the following pleasant property: There is a projection from the tree $\widetilde{G}$ onto $G$. That is,

- every node of $\widetilde{G}$ is 'above' some node of $G$; and

- every edge of $\widetilde{G}$ is above some edge of $G$; and

- $\widetilde{G}$'s root $\widetilde{r}$ is above $r$.

Furthermore, $\widetilde{G}$ is the largest possible such tree.

- Every node and every edge of $G$ is underneath a matching object in $\widetilde{G}$;

- if you were an ant crawling around in $\widetilde{G}$, it would seem at each node as if you were on $G$. The only way to tell the difference is that in $G$, if you go far enough in the right direction, then you arrive back where you began, whereas in $\widetilde{G}$, the only way to return to the start is by retracing your steps.

All such covering trees look the same. They are isomorphic graphs and they go by a common name: $\tilde{G}$. For the technically minded, there is a standard way of constructing one such $\tilde{G}$ from $G$. It involves the notion of a **walk**. A walk is like a path, except that it may cross itself:

$$(a \quad b \quad c \quad b \quad \ldots \quad a \quad d \quad e).$$

If the walk does not have any triple of successive nodes like

$$\ldots \quad b \quad c \quad b \quad \ldots$$

so it never goes from one node to another and then straight back again, then let us say the walk is **reduced**. You can always convert any walk to a reduced walk by deleting pairs of nodes:

$$\ldots \quad f \quad g \quad b \quad c \quad b \quad h \quad \ldots$$

changes to

$$\ldots \quad f \quad g \quad b \quad h \quad \ldots$$

Given any node $a$ in $G$, the nodes of $\tilde{G}$ above $a$ are *reduced walks* in $G$ from $r$ to $a$:

$$(r \quad s \quad t \quad \ldots \quad u \quad a).$$

This may seem a curious idea. The point is, mathematical notation gives us complete liberty, unless it explicitly states otherwise. You can make *any* set into the set of nodes of a graph. The set chosen here, the reduced walks from $r$ in $G$, fulfils all we require of $\tilde{V}$.

The root $\tilde{r}$ of the cover is represented by $(r)$, which is a walk of length 0.

If there is an arc in $G$ from $a$ to another node $b$, then

$$(r \quad s \quad t \quad \ldots \quad u \quad a \quad b)$$

is a walk in $G$ so, when reduced if necessary, it yields a node of $\tilde{G}$. The arc $(a, b)$ in $G$ is covered by the arc

$$((r \quad s \quad t \quad \ldots \quad u \quad a), (r \quad s \quad t \quad \ldots \quad u \quad a \quad b))$$

in $\tilde{G}$.

If you want an idea of what is going on, look at Fig. 2.7. This shows a graph $G$, which is actually a tree, and its cover $\tilde{G}$. Since $G$ is already a tree, it and $\tilde{G}$ are isomorphic, but the nodes of $\tilde{G}$ are unlike those of $G$.

Similarly, any connected digraph with a root has a covering tree. However, all the paths in the construction are directed, that is, all the arrows on a path point the same way, away from the root. This means that there are fewer paths, so the cover of a digraph may be smaller than the cover of the undirected graph. In fact, the cover of a digraph may even be smaller than the original digraph. The cover of a digraph may be finite, while if you leave out all its arrows, its undirected cover is infinite. Fig. 2.8 shows an example.

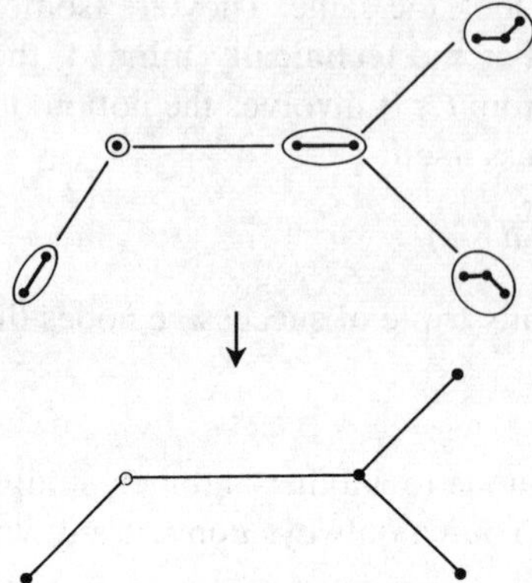

**Fig. 2.7**   The cover of a rooted tree, showing the distinct nodes in the cover.

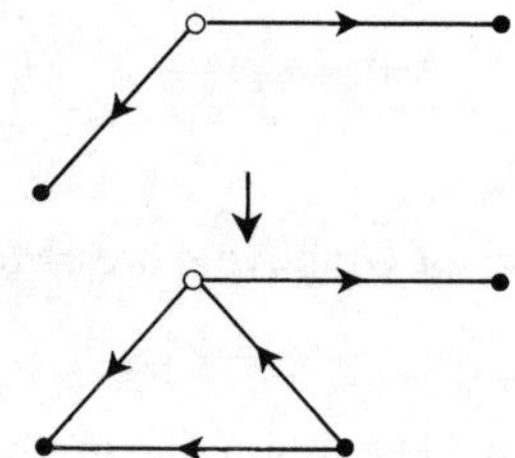

**Fig. 2.8**   The cover of a rooted digraph.

As an exercise, you may care to draw a picture of a labelled directed rooted graph $G$ and then draw a picture of part of its cover. The cover is a labelled tree. Each edge in the cover will be labelled with the same label as the edge it covers.

## 2.3   SEARCH

Searching consists of trying to find a way to change the present state of affairs into some desired state. This desired state is called a **goal state**. The change is made in a finite sequence of discrete steps. An **operator** is a process or action which makes one step. It generates a new state from an old one.

*Example: binary chopping, again*

When you perform this algorithm, the states are pairs of numbers

$$(x_0, x_1)$$

where $x_0 < x_1$. Given any such state, there are two operators which act on it. The first produces a new state

$$(x_0, x)$$

and the other produces

$(x, x_1)$

where $x = (x_0+x_1)/2$. The goal is any state $(x_0, x_1)$ in which $x_1-x_0 \leq 0.0001$.

*Example: learning to recognize an arch, again*

In this case, the important aspect of a state is what is in the two sets of properties. Remember, the first set contains properties which are common to arches and the other holds properties which arches don't have.

One step in the search consists of examining a pile of wooden blocks and modifying the contents of the sets accordingly. The operator adds a property to one of the two sets.

The goal is a state in which the set of properties common to all arches includes the assertions:

- there are two columns;

- there is an extra block;

- the extra block rests on both columns;

and the other set of properties contains the assertion:

- the two columns touch.

Search is hard because there are (usually) many paths from the start state. The art of search lies in choosing a short sequence of operators which will lead the search to a goal quickly. Choice of operator is often made using **rules**. There are various forms of rule. The standard form is

IF *conditions*
THEN *action*.

This means that, if the *conditions* are all true in some state, then the operator called *action* can be applied to the state and, what is more, this operator may help the search.

*Example: the 'blocks world'*

There is a well-known setting for search called the 'blocks world'. In it, there are three wooden blocks labelled **A**, **B**, and **C**, and a **table**. Any state is described by saying what is on what. Initially, we can imagine that **A** is on top of **B** which is

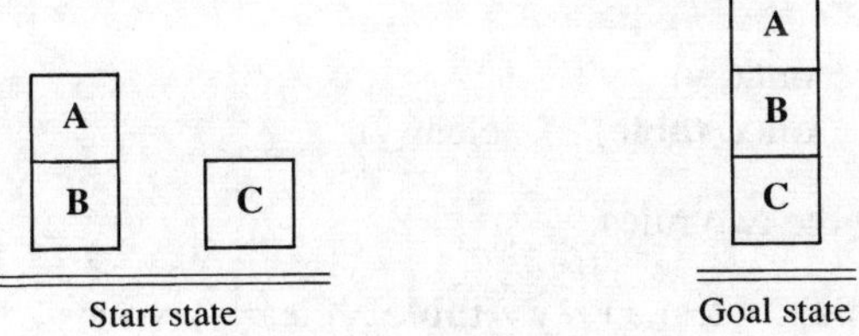


**Fig. 2.9**   The blocks world.

on the **table**, and **C** is also on the **table**. The goal is the state in which **A** is on **B** and **B** is on **C**, on the **table**, as illustrated in Fig. 2.9.

Thus, there are four constants in the blocks world:

**A   B   C   table**

and one relation which is specific to this task:

$\text{on}(x, y)$.

Using just it, the start state is described as

$\text{on}(\mathbf{A}, \mathbf{B}) \wedge \text{on}(\mathbf{B}, \textbf{table}) \wedge \text{on}(\mathbf{C}, \textbf{table}) \wedge$

$\forall x \, \neg \text{on}(x, \mathbf{A}) \wedge \forall x \, \neg \text{on}(x, \mathbf{C})$

and the goal is

$\text{on}(\mathbf{A}, \mathbf{B}) \wedge \text{on}(\mathbf{B}, \mathbf{C}) \wedge \text{on}(\mathbf{C}, \textbf{table}) \wedge \forall x \, \neg \text{on}(x, \mathbf{A})$.

In practice, the blocks world is more conveniently described if we introduce a second predicate:

$\text{clear}(x)$    which, by definition, means  $\forall y \, \neg \text{on}(y, x)$.

Using it, the start state is represented as the statement

$\text{on}(\mathbf{A}, \mathbf{B}) \wedge \text{on}(\mathbf{B}, \textbf{table}) \wedge \text{on}(\mathbf{C}, \textbf{table}) \wedge \text{clear}(\mathbf{A}) \wedge \text{clear}(\mathbf{C})$

and the goal state is

$\text{on}(\mathbf{A}, \mathbf{B}) \wedge \text{on}(\mathbf{B}, \mathbf{C}) \wedge \text{on}(\mathbf{C}, \textbf{table}) \wedge \text{clear}(\mathbf{A})$.

The natural rules for this task have conditions which also include the predicate

$\neq$

which checks that two things are different. One does not usually bother to mention it, but we shall make it explicit this time.

There are two operators. The first, *stack(x, y)*, takes a block *x* off the **table** and puts it on another block called *y*. The second, *unstack(x)*, does the reverse. Thus, the operator *stack* forms a new state from the current one by

    deleting the facts    $\text{on}(x, \textbf{table})$        $\text{clear}(y)$
    adding the fact        $\text{on}(x, y)$,

and unstack works by

    deleting the fact      $\text{on}(x, y)$
    adding the facts      $\text{on}(x, \textbf{table})$        $\text{clear}(y)$.

They are applied by the two rules:

    IF        $\text{on}(x, \textbf{table}) \wedge \text{clear}(x) \wedge y \neq \textbf{table} \wedge \text{clear}(y) \wedge x \neq y$
    THEN    $\text{stack}(x, y)$.

IF      on($x$, $y$) $\wedge$ clear($x$) $\wedge$ $y \neq$ **table**
THEN   unstack($x$).

In certain states, a rule can be applied in more than one way. A particular choice for the variables $x$, $y$, and $z$ in a rule gives an **instance** of the rule. In the start state of the blocks world, this does not happen. The two rules can each be applied in only one way. For instance, the second rule can only be applied when $x$ is **A** and $y$ is **B**. It produces the state

on(**A**, **table**) $\wedge$ on(**B**, **table**) $\wedge$ on(**C**, **table**).

In this new state, the rule for *unstack* cannot be applied at all, but the rule for *stack* has six instances:

- $x$ is any one of the three blocks;

- $y$ is any block other than $x$.

In this example, the shortest path to the goal is via the instance

$x$ is **B**      $y$ is **C**

which produces the state

on(**A**, **table**) $\wedge$ on(**B**, **C**) $\wedge$ on(**C**, **table**).

*Example: finding a forced win in noughts and crosses*

In this game, the board is a $3 \times 3$ array of squares, like

|   | a | b | c |
|---|---|---|---|
| 1 |   | ✗ |   |
| 2 |   | ○ |   |
| 3 |   |   |   |

The individual squares will be called '$a_1$', '$b_2$', and so on. The structure of the board can be described by eight facts stating which triples of squares form lines:

linear $(a_1, a_2, a_3)$, ... linear $(a_3, b_2, c_1)$, ...

(The order of arguments in linear doesn't matter.) The other predicates which we use to describe the game are

empty($x$)          which signifies that the square $x$ is empty;
filled($x$, $C$)        where $x$ is a square and $C$ is a player;
to-move($C$)       which is true if it is $C$'s turn to play;
win($C$)            which is short for the formula

$$\exists x\, \exists y\, \exists z \quad (\text{linear}(x, y, z) \wedge$$
$$\text{filled}\,(x, C) \wedge \text{filled}\,(y, C) \wedge \text{filled}\,(z, C)).$$

If you are playing with $\bigcirc$ then the goal is win($\bigcirc$).

Once the board's structure has been given, the position in this picture is completely described by the facts

empty($a_1$)      empty($a_2$)      empty($a_3$)   empty($b_3$)       empty($c_1$)
empty($c_2$)      empty($c_3$)      filled($b_1$, $\times$)      filled($b_2$, $\bigcirc$)

and, if it is $\bigcirc$'s turn to play,

to-move($\bigcirc$).

In the game, there is just one operation, called play $(x, C)$. It deletes the facts

empty($x$)      and      to-move($C$),

which are this operation's necessary conditions and adds

filled($x$, $C$)      and      to-move($D$)      where $D \neq C$.

The task of this example is not to play the game but to discover forced sequences of moves. In any such forced sequence, moves will occur in pairs: first player $\bigcirc$ plays and then player $\times$ plays. Hence, an operation in this task's search space consists of a *pair* of *play* operations. An operation

play($x$, $\bigcirc$); play($y$, $\times$)

of the search deletes the facts

empty($x$)      and      empty($y$)

and adds

filled($x$, $\bigcirc$)      and      filled($y$, $\times$).

Here are two rules which may help to plan a forced win

IF    to-move($\bigcirc$) $\wedge$ linear($x$, $y$, $z$) $\wedge$ filled($x$, $\bigcirc$) $\wedge$ empty($y$) $\wedge$ empty($z$)
THEN
      play($y$, $\bigcirc$); play($z$, $\times$).

IF    to-move($\bigcirc$) $\wedge$ linear($x$, $y$, $z$) $\wedge$ linear($x$, $y'$, $z'$) $\wedge$
      $\{y, z\} \neq \{y', z'\}$ $\wedge$
      empty($x$) $\wedge$
      filled($y$, $\bigcirc$) $\wedge$ empty($z$) $\wedge$
      filled($y'$, $\bigcirc$) $\wedge$ empty($z'$)
THEN
      any move beginning    play($x$, $\bigcirc$)    will lead to    win($\bigcirc$).

Strictly speaking, this second is not a proper rule because it does not end with a single well-defined action. You can consider it as an abbreviation for the two rules with this condition and the actions

$$\text{play}(x, \bigcirc); \text{play}(z, \times)$$

and

$$\text{play}(x, \bigcirc); \text{play}(z', \times).$$

A **search space** is a labelled directed rooted graph. It depicts the present situation, the desired situation, all ways of obtaining all possible situations, and the possible steps between them. Each possible situation is a 'state'.

- Its *nodes* are the possible states.

- Its *edges* are labelled with the operators between states.

- An *edge* is directed *from* the old state *to* the new state which its operator generates.

In this book, an edge will often be identified with the operator labelling it. In addition, a search space has

- a **goal**

- a **start state**.

The start state is the unique state where search begins. It is the graph's root. The goal is a characterization of the set of goal states. It may be

- the particular goal state, if there is just one, or

- a predicate which defines goal states.

This predicate is sometimes given in terms of an entity which occurs in goal states and not in others. The artificial intelligence community sometimes uses terminology a bit loosely.

*Example: the eight tiles problem*

You can buy versions of this in toy shops. It consists of a plastic $3 \times 3$ square containing eight little tiles and one empty space. The tiles have grooves and ridges on them so that they do not fall out. There is one empty position. Let us suppose that the tiles have numbers 1, 2, 3, ..., 8 and that the goal is the configuration

```
1  2  3
8     4
7  6  5
```

where the empty space is in the middle. If, when you first pick it up, the tiles are in the positions

```
1   3   4
8       2
7   6   5
```

then one path to the goal proceeds as in Fig. 2.10.

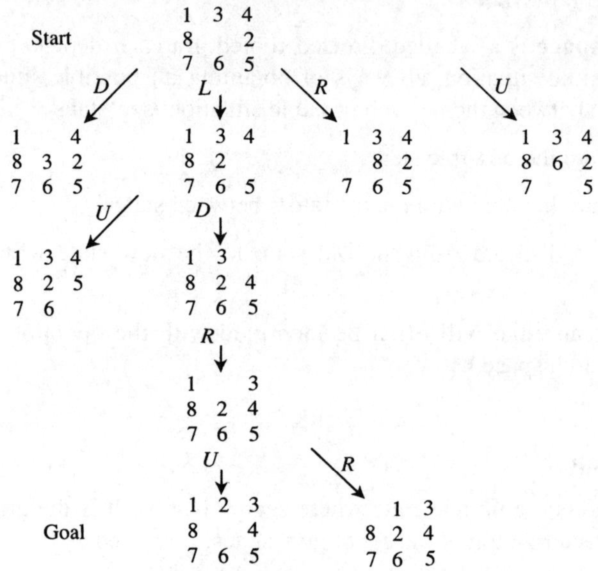

**Fig. 2.10**   Part of the search space of the eight tiles problem.

Some of the search space's side branches have also been shown. The whole search space is big—much too big to be drawn on one page. This task has four operators: a tile may be slid Up, Left, Right or Down. The edges in the search space are marked accordingly.

In binary chopping, a state is a pair of numbers. A goal is any pair close enough together. By contrast, in the blocks world, there is a unique goal.

There are various algorithms for searching in different cases. Some of them will be described soon, but first, as always, it behoves us to study their data. The principal datum input to any search algorithm is a description of its search space.

*Examples: characteristics of search spaces*

   *1.  Binary chopping*

      •  The start state is the pair (0, 1).

- A goal is any pair of numbers $(x, y)$ for which

$$x \le S \le y \qquad \text{and} \qquad y - x \le 0.0001.$$

2. *Inventing the notion of an arch*

   - The start state is the pair of two empty sets $(\{\ \}, \{\ \})$

   - A goal is any pair of sets $(L_1, L_2)$ for which every property in $L_1$ is fulfilled by every arch, and anything which is not an arch has at least one property in $L_2$.

   - An operation adds a single property to one of the two sets.

3. *The eight queens problem* A state of this task consists of a chess board, with up to eight pieces on it so that no two pieces are on a common row or file or diagonal.

   - The start state is the empty board, with no piece on it.

   - An operation consists of adding one extra piece. Naturally, it must not be put on a square which is on the same row or file or diagonal as another piece.

   - A goal is any state with eight pieces on the board.

4. *The eight tiles problem*

   - The start state is whatever state the toy is in when you find it.

   - An operation involves sliding a tile into the empty position. This changes the positions of the moved tile and the empty space.

   - The tiles are numbered or painted, and the goal is a particular arrangement of them. There is just one goal.

5. *The crates problem* A warehouse is nearly full of heavy crates. One at the back has to be delivered, so it must be moved to the front door, but others are in the way. There are two views of this task, the real one and the naïve one.

   - A naïve state is some arrangement of crates.

   - A naïve operation consists of pushing one crate into an adjacent free area. It is raining, so no crate may ever be moved outside. Also, stacking crates on top of each other is not allowed.

   - A naïve goal is any state in which the crate to be delivered is by the door.

   - The naïve start state is the state of the crates when you arrive.

This description does not depict the real problem. What is really wanted is not to get the important crate to the door, but to get it there without expending more effort than necessary. Hence, the real goal is a **path of least effort** through the naïve space.

- A real state consists of a naïve state and a sequence of naïve operations which produces it from the naïve start state.

- A real operation is almost the same as a naïve one. It involves pushing one crate into an adjacent space, thus producing a new naïve state, and also extending the sequence of naïve operations by this extra push.

- A real goal is some state consisting of a naïve goal, with the crate to be delivered at the door, and a sequence of pushes getting it there which involves least effort.

- The real start state is the naïve start state and the empty list of pushes.

These spaces have quite different characteristics.

1. The search space is a tree. No two different paths from the start will ever arrive at the same state.

2. The search space is not necessarily a tree, because properties may be added to the two sets in various orders depending on the order of training instances. However, once a property has been added to either set, it is never removed. Hence, the space does not have cycles.

3. As for 2, the search space is not a tree. Pieces can be put on the board in any order, so one position can arise via several sequences of operators. Once a piece is on the board, it is never removed, so the search space does not have cycles.

4. This space has cycles. It is easy to slide tiles around and find they are back in some position they were in before.

5. The naïve space has cycles. The real search space does not because a real operation extends the list of naïve operations which move crates. (This is not much consolation, if ever you find you have pushed all the crates in circles.)

The real search space for the crates task is almost the cover of the naïve one. It would be the cover, if we were clever enough never to push a crate and then decide to push it straight back to where it was before.

A similar situation arises in any task which involves finding an *easiest* solution. There is a naïve search space, whose states are the situations which you could produce and the real space whose states record ways of obtaining naïve states.

Here are some more examples, with summaries of their properties.

6. *Composing crossword puzzles*

- a state is a partly filled form of a crossword;
- an operation consists of entering a new word;
- the start state is the empty form;
- a goal is any form with all positions for letters filled in.

The task space is a finite digraph without cycles.

- Objective: find a goal.
- Branching ratio: about 100.
- Depth: constant (depending on the crossword).
- Ratio of goals to all leaves: maybe $\leq 1/10000$.

7. *Theorem proving* There are various approaches. For the sake of argument, consider the tableau method which is described in the book by Wilfrid Hodges mentioned in the *Further reading* section.

- A state is a tree. Each path from the root to a leaf includes all hypotheses, the negation of the statement to be proved, and various consequences of them.
- An operation consists of choosing a composite statement in some branch and extending the branch by expanding this statement into simpler ones.
- A goal is any tree in which every branch contains two contradictory statements.
- The start state is the minimal tree containing just the hypotheses and the negation of the statement to be proved.

The task space is a digraph without cycles.

- Objective: find a goal.
- Branching ratio: finite (may be unbounded).
- Depth: may be infinite, depending on the task.
- Ratio of goals to all leaves: may be very small or zero.

8. *Exploring the world* Five hundred years ago, the map of the known world had large gaps in it. Many learned people still thought that the world was flat. Each time a new expedition returned from foreign parts, a little more of the map could be filled in.

- A state is a state of the map of the known world.

- An operation consists of mounting an expedition and then filling in a bit more of the map when it returns.

- The goal is the completed map.

- The start state is the current state of the map, just before Columbus first sailed west.

The task space is a digraph without cycles.

- Objective: find a goal.

- Branching ratio: $\geq 100$ (as there were some hundred possible directions for a new expedition).

- Depth: $\geq 100$.

- Unique goal.

### 2.3.1 DESCRIPTION LANGUAGES

The language in which a task is posed can have a crucial bearing on its solution. In most of the examples above, this language is not specified. Really, it should be made explicit. Different languages lead to different task formulations and half the art of searching lies in how you pose the task.

*Example: the eight queens problem in more detail*

Here are three different ways to represent a state.

*Representation 1: as a set of coordinates of filled squares.* In this representation, the state shown in Fig. 2.11 appears as

$$\{(1, 8), (2, 6), (3, 4), (4, 2)\}.$$

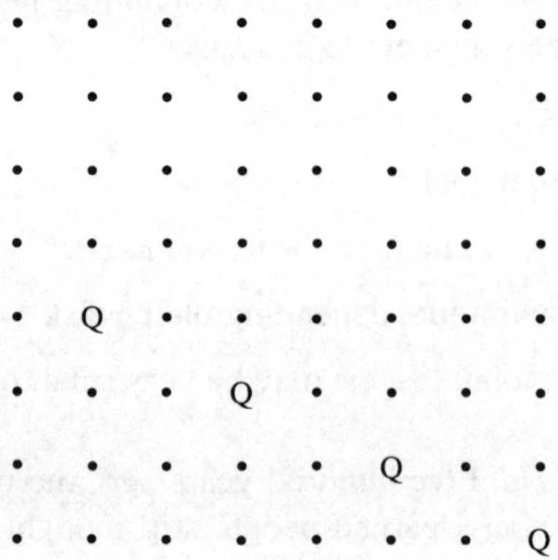

**Fig. 2.11**

The start state, the empty board, is represented as

{  }.

*Representation 2: as a list with one entry for each row of the board.* If there is a queen on the board on the square in the $i$th row and $j$th column, then the $i$th entry in the list is the number $j$. If there is no queen in the $i$th row, then this list entry is 0. Thus, the empty board is

[0, 0, 0, 0, 0, 0, 0, 0].

In this representation, the board in Fig. 2.11 appears as

[8, 6, 4, 2, 0, 0, 0, 0].

*Representation 3:* This also portrays the board as a list, but it can only represent boards which have queens in the bottom $i$ rows for some $i$ and none in higher rows. Thus, the empty board is

[  ]

and the board in Fig. 2.11 is represented as

[8, 6, 4, 2].

This representation cannot depict the state shown in Fig. 2.12.

**Fig. 2.12**

*Example: the 'blocks world' again*

The obvious way to describe this task is with

constants:      **A   B   C   table**

relations:      on   =

connectives:    $\wedge$   $\vee$   $\Rightarrow$   $\neg$

variables:        $x$   $y$   ...

quantifiers:      $\forall$

This is the apparatus which we first assumed, in the task's original definition. Any state, and the condition for any operation, can be written in this language. In practice, though, states and conditions can be described in a simpler language containing only the symbols

constants:        **A   B   C   table**

relations:        on    clear   $\neq$

connectives:      $\wedge$

variables:        $x$   $y$   ...

If we select a more suitable set of relations, then we can get by with fewer connectives. Also, since all variables are assumed universally quantified, there is never any need for an explicit quantifier.

This second language is less expressive than the first. All the ideas expressible in the second can be reproduced in the first, but, for instance, the notions of disjunction ($\vee$) and implication ($\Rightarrow$) and negation ($\neg$) which occur in the first cannot be expressed in the second. Doing without negation is a particularly helpful feature, for a reason which will be explained below.

The process of choosing a language for a task is sometimes called 'adjusting the task's bias'. Usually, we shall take the language for granted. Adjusting bias will be discussed in Chapter 6 and again in Chapter 8. Notice that the language discussed here is only intended to describe the individual states and the goal of the task and the conditions for applying operators. It is not adequate for describing the task's operators, nor does it suffice for a description of the whole search space. Thus, in the crates problem, the naïve states can be described just by stating which crate is where, but the states of the real search space involve subtler relations.

It is often a good idea to use an unexpressive language. The reason is simple. The more expressive the language, the more possible states can be described in it and, hence, the larger will be the space of states that a solver may have to search through for a goal. Also, an unexpressive language permits few operators, so the search's branching ratio is small.

*Example: eight queens again*

Representation 1 allows us to depict *any* set of pairs of coordinates. There are $2^{64}$ possible states which can be described in this representation. This is a vast set of states. Most of them are illegal because two queens conflict. In addition, the branching ratio at the start is 64 and it never falls below 57. Searching this space directly would be totally impractical.

Representation 2 can depict $9^8$ possible states (eight entries in the list, each of which can take one of nine possible values). It can depict all legal states, so all the excluded ones are illegal. It is a great improvement over representation 1. However, a state with $k$ unoccupied rows has $8k$ children, so the branching ratio starts at 64 and remains high for most of the search. Searching in this space would be difficult.

Representation 3 can depict

$$8^0 + 8^1 + 8^2 + 8^3 + \cdots 8^8 \;\; = \;\; 19\,173\,961$$

states. There are some legal arrangements of queens on the board which this representation cannot depict. However, it can depict all goals (because in any goal there is a queen in every row). Also, this representation suggests a particular limited set of just eight operators: given any incomplete state, the $j$th operator adds a queen in the $j$th column. In this representation, the search space is a tree with branching ratio no more than 8. When illegal states are excluded, the branching ratio is less. Search in this space is easy on a small computer, and is even practical manually.

Half the art of search lies in choosing a language which suffices to describe the task and an adequate solution but very little more. If you can find such a language, then there will be very few alternatives and so your search will probably not go astray. Your task may cease to be a search task at all.

There are circumstances when it pays to use an expressive language. Sometimes a task can be posed in a simple language but the best search path leads beyond the space describable in that language. However, these cases are hard. (It is tempting to say that they are rare, but the truth is that they only seem rare because we avoid them.) As a matter of principle, it is better to use an unexpressive language.

Heuristic information (that is, suggestions about how best to search) is often more easily exploited when the language is enlarged, but that does not necessarily mean that it has to be made more expressive. For instance, the predicates '*clear*' and '$\neq$' were used in the conditions for the operations in the blocks world, but adding them did not extend the descriptive power of the language. They just make these conditions appear simpler.

### Predicates and functions

You may have noticed that the description we have given of the blocks world is quite unlike all the descriptions of the eight queens task. A state of the blocks world is portrayed as an **assertion**, formed with predicate symbols. By contrast, in each of the three representations of the eight queens problem, each state is a **term**. Terms are constants or are constructed using function symbols. For instance, the empty list

$$[ \;\; ]$$

is a constant. Any longer list

$$[a, b, c, \ldots]$$

is a term built from the list's head (in this case, $a$) and its tail (the rest: $[b, c, ...]$). The list notation does not make any function symbol explicit; but in Prolog, for instance, one can write such a list as

$. (a, [b, c, ...])$

The '.' is actually a function symbol.

Usually, there is no great difficulty in translating from a description using predicates to one using functions or the other way round. In Section 6.5, we shall see that the choice of one style of description, rather than the other, can have a substantial effect on learning. For the time being, just bear in mind that there are these two styles.

### 2.3.2 REVERSIBLE AND IRREVERSIBLE SEARCH

In most of the examples above, every operator can be reversed. The eight little plastic tiles can always be pushed back into their original state (if you can remember the sequence of moves you have made so far). A queen can be put on the chess board and then taken off again. If you really want to, you can push the crates around and then push them all back again and nothing will have changed except the dust on the floor and your state of health. You can erase words from a partly designed crossword and you can prune a partly completed tableau by just forgetting some of it.

Exploring the world is different. Once part of map of the world is known and that knowledge has been disseminated, there is no way that it can be censored. Once Columbus had found his new continent and the discovery was common knowledge, anyone else mounting an expedition westward would have a rough idea what he would find. The common pool of knowledge grows inexorably. Extensions of the map are irreversible.

Note: saying that an operator is reversible is *not* the same as saying that it is an invertible function on the set of states. For instance, all four operators $U, D, L$, and $R$ of the eight tiles problem are reversible, but they are *partial* functions on the set of states and so they do not have inverses.

*Example*

Another real life example of irreversible search occurs in the kitchen.

9.  *Cooking*

    - A state is an arrangement of cooking utensils and ingredients in a kitchen.

    - The operations of cooking include moving things, heating, mixing ingredients, cutting, and breaking eggs.

    - The goal is supper.

- The start state is the state of your kitchen when you arrive home after work.

Moving something is a reversible operation, but mixing, cutting, and breaking are irreversible. Heating is usually irreversible, because foods change their textures irreversibly when heated.

**Planning** involves simulating a difficult or irreversible search. A planner is given some such task, where any venture into search would be rash. It constructs a second more tractable search space. There are two ways whereby the planning search may be easier than the real one:

- the operators in the planner's simulation may be reversible; or

- the planner may be able to remember several states. We shall soon see that some good search algorithms involve remembering many states.

Thus, if one search path turns out to be wrong, the planner can recover and try a different path. A **plan** is a path in the simulation from its start to a goal.

If the plan is faithful to the original task, then

- there will be one operator for each original operator;

- the search graph in the simulation will be isomorphic to that of the original task.

Planning need not be faithful.

When cooking, a plan can be invented in the cook's head. He imagines the sequence of states which he will have to construct on the way to supper. When he has invented such a plan, if he is wise, he will check that it really is faithful.

Suppose that the goal is a mushroom omelette. The initial plan might consist of the operations:

(1)   break eggs into a bowl;

(2)   whisk eggs;

(3)   heat fat in frying pan;

(4)   pour whisked eggs into hot fat in the pan;

(5)   chop mushrooms;

(6)   mix the chopped mushrooms with the eggs; and so on.

This plan is not faithful to reality. As soon as the eggs meet the hot fat, they will harden and the mushrooms could never be mixed in with them. Also, by the time operation (5) is complete, the eggs will be burning. A more nearly faithful plan would be

(1)   break eggs into a bowl;

(2)   whisk eggs;

(3)   chop mushrooms;

(4)   mix the chopped mushrooms with the eggs;

(5)   heat fat in frying pan;

(6)   pour whisked eggs and mushrooms into hot fat in the pan; and so on.

Planning is valuable for solving irreversible search, as in cooking. It is also useful when some operators in real life are reversible but 'expensive' and they can be simulated by cheaper ones. Pushing heavy crates expends a lot of energy, so planning would be appropriate in that task too.

A fourth case where planning helps occurs when the real task involves a deep subtle search and there is a simpler simulation of it. The search space of the simulation is formed by partitioning the real search space's set of states into disjoint subsets. Each subset is a single 'state' of the simulation. For example, when cooking, the whole solution can be abbreviated into three major steps:

(1)   prepare ingredients;

(2)   cook ingredients;

(3)   lay table and serve.

If this abbreviated plan is convincing, then it can be refined to a more nearly faithful one. The process of constructing a sequence of plans like this, each a refinement of the one before, is called **hierarchical search** or **stepwise refinement**. It is often quicker than direct search in the original task's space.

### 2.3.3   GOAL DIRECTED SEARCH

Study the search space which is drawn is Fig. 2.13a. This directed graph has a significant property which is common in real tasks: at almost all nodes, the node has at least as many children as parents and several nodes have more children than parents. Also, there is only one goal state.

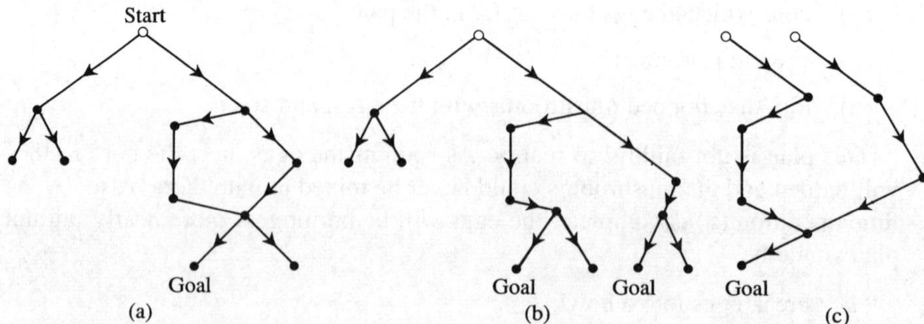


**Fig. 2.13**   (a) The object space; (b) forward search space; (c) goal-directed search space.

Let us suppose that

- on average, each state has more children than parents;

- all operators are reversible;

- there is just one unique goal state;

- the task specification includes an adequate description of the goal.

When this happens, we can convert the search of Fig. 2.13a into another simpler search as follows.

- The states of the new space are the same as those of the original one.

- Its start state is the (unique) goal of the original.

- Its goal is the start state of the original.

- Operators of the new space are the *reverses* of operators of the original.

The new space has as many nodes as the original, but often a lot of its nodes will not be accessible. In Fig. 2.13a, there is no directed path along reversed edges from the node marked 'Goal' to any other leaf of the space. Hence, the relevant part of the new space is smaller, so search is easier.

Figure 2.13b shows the space which a naïve search algorithm will explore for this task. It is the cover of the original space. A naïve algorithm may generate some state twice by two different paths and not detect that this state occurs twice. Figure 2.13c shows the space which the same algorithm would explore if it were given the backwards version of the task. Observe that Fig. 2.13b has fourteen nodes but Fig. 2.13c has only nine.

An algorithm which performs this trick before it starts searching is said to be **goal directed** or a **backwards** searcher. There are cases when backwards search is no use. In the examples,

1, 2, and 3:  the specifications of these tasks do not tell us what a goal state is. The objective is to find out. There are many goals.

4:  in the eight tiles problem, there is just one goal state and we know what it is and the objective is to construct a path to it. However, every state has just as many parents as children, so searching backwards won't help.

5:  as in 4, each state has many parents. Also, there are many naïve goal states, differing in the positions of all the other crates.

However, in the case of cooking, goal directed search can help. If you can make up your mind what you want to eat then there is a unique goal. The real task does not have reversible operators, but any planning simulation of it will.

Goal directed search is useful much more often than this discussion suggests. The eight queens and eight tiles tasks are not typical of real life search situations. They were adopted by academics for the purpose of studying search. They are good little theoretical exercises because they are easily described and algorithms for solving them are easily programmed, but they have a lot of symmetry, which is most uncommon in real life.

### 2.3.4  MONOTONIC OPERATORS AND THEOREM PROVING

In this section, we shall assume that any task is described by assertions formed with predicates. The notion of monotonicity does not mean anything if states are portrayed as terms.

There is a class of search tasks, called **monotonic search**, in which each operator establishes some new truth compatible with all other previously known truths. An operator which adds some new knowledge, and which does not undo any fact already established, is said to be **monotonic**. The two operators of the blocks world are not monotonic. *Stack*$(x, y)$ deletes the facts

on($x$, **table**)        and        clear($y$)

and *unstack*$(x)$ deletes the fact

on($x, y$)

for some block $y$.

*Example: letter arithmetic*

There is just one way to replace all the letters by digits in the multiplication

$$
\begin{array}{r}
\mathrm{ABA} \\
\mathrm{CC} \times \\
\hline
\mathrm{DAE} \\
\mathrm{DAEF} \\
\hline
\mathrm{DGHE}
\end{array}
$$

so that the calculation is correct and no two letters stand for the same digit.

This task begins from the start state

$$
\begin{aligned}
&\mathrm{A}\neq\mathrm{B} \;\wedge\; \mathrm{A}\neq\mathrm{C} \;\wedge\; \ldots \mathrm{A}\neq\mathrm{H} \;\wedge\; \mathrm{B}\neq\mathrm{C} \;\wedge\; \ldots \mathrm{B}\neq\mathrm{H} \;\wedge\; \ldots \mathrm{G}\neq\mathrm{H} \;\wedge\; \\
&\{\mathrm{A, B, C, D, E, F, G, H}\} \subseteq \{0, 1, 2, 3, 4, 5, 6, 7, 8, 9\} \;\wedge\; \\
&\mathrm{AC}\ \mathrm{mod}\ 10 = \mathrm{E} \;\wedge\; \\
&(\mathrm{BC} + \mathrm{AC}\ \mathrm{div}\ 10)\ \mathrm{mod}\ 10 = \mathrm{A} \;\wedge\; \\
&\mathrm{AC} + (\mathrm{BC} + \mathrm{AC}\ \mathrm{div}\ 10)\ \mathrm{div}\ 10 = \mathrm{D} \;\wedge\; \ldots
\end{aligned}
$$

Each child of the start state consists of the above assertions and one additional one, such as

E < D.

Each state is a conjunction of facts extending its parent. The goal is a state containing eight assertions of the forms

A = $d_A$, B = $d_B$, ... H = $d_H$

where the symbols $d$ are all digits. Each operator adds a single new fact which is a direct consequence of what is known already.

Problems in the form of this example are called **constraint satisfaction** tasks. This particular example has a unique solution, but a constraint satisfaction task can have many goal states which all fulfil the task's goal.

Monotonic search is usually efficient in space, because the searcher does not have to construct each state's representation from scratch. Apart from the start state, which must be presented in full, any state can be described as its parent plus a few more details. Hence, in monotonic search, one can often keep many states easily. Since constructing new states is simple, monotonic search can sometimes be time efficient too.

Almost all monotonic operators are reversible: if the operator adds a new truth, then its reverse just forgets that truth again. Exploring the world is intriguing because its operator is monotonic but not reversible (among other reasons).

Note: some authors describe monotonic search slightly differently. They say that search is monotonic if, whenever two operators can be applied to a state, either can still be applied to the state produced by the other. The definition used here is closer to the notion of monotonicity in logic.

Monotonicity is a sensitive property. It depends critically on the description language. Consider, for instance, exploring the world. Describing a map with words and symbols, such as appear on this page, is not very easy, but you can imagine that each entry on the map is recorded as a fact of the form

here-found ($x$, $y$, **thing**)

where $x$ and $y$ are coordinates, longitude and latitude, and **thing** is any feature such as a mountain or river or city. A state of knowledge is a set of facts of this form and so the search can be described adequately in a language containing just the names of things which might be found, all possible coordinates, and the relation 'here-found'. When this is the chosen language, exploring the world is indeed monotonic.

However, suppose that we extend the language with negation, '¬'. In the extended language, the fact

¬here-found (…, …, America)

is true in the start state but it is contradicted in some later states. Hence, depending on what facts we choose to assert about states, this task in the extended language may not be monotonic.

Search is never monotonic if the description of a state can include negations of assertions which are true in its children. For this reason, the description languages used in artificial intelligence do not usually include '¬'. It would be too tedious, each time a new state is formed, to contradict all the assertions of ignorance in its parent. If a negated conjunct is really necessary in some condition, then the standard trick is to introduce a new predicate which stands for it. The two predicates '*clear*' and '≠' of the blocks world serve just this purpose.

**Theorem proving** is a special case of monotonic search. Its distinguishing feature is that it uses only a very few basic operators—typically just one, called **modus ponens,** or some equivalent—which are very well understood and reliable. All the searcher's other knowledge is in the form of rules about how best to apply this operator and of facts about states which this operator can be applied to. The searcher may go astray because either some supposed fact is wrong or it applies modus ponens in an irrelevant way, but it will never arrive at any result which is not a strict consequence of its start state.

Theorem proving is a particularly easy form of search. All its operators are both monotonic and reversible. Furthermore, the states of a theorem prover are sets of axioms and proved assertions, and these are easily remembered, so it is easy to keep many states simultaneously.

The major difference between theorem proving and other forms of monotonic search is just the choice of basic operators. From the point of view of searching, this difference is not very great. When it comes to learning, there is a more significant difference because a general monotonic learner may be permitted to invent new basic operators and these may be ill understood and unreliable. However, in most of what follows, we shall treat theorem proving and other cases of monotonic search as if they were the same.

If you know enough about a task, you can always translate it into an exercise in theorem proving. In the translation, each assertion is about the space which the original task searches.

- A state of the new task is a set of assertions.

- An operator is a deduction step.

Hence, a child node consists of all the assertions of its parent and one extra consequence of them.

- The start state is the specification of the original task's search space.

- A goal is a set of assertions which includes one of the form 'A certain sequence of operations will lead to a goal of the original task'.

If ever the theorem prover discovers one of its goals, then back in the original task, you can run off this sequence of operations without any search at all. The limitation of theorem proving is that the description of the original task may be inadequate. For instance, it is no use trying to prove that a certain sequence of steps will lead to supper if you don't know what is in your larder.

### 2.3.5 SUMMARY PROPERTIES OF SEARCH TASKS

Of any search task, one can ask the following questions:

- How big is the search space?

- Is the search space a tree? If not, does it have directed cycles? (A space without directed cycles is a **partial order**.)

- What is its maximum branching ratio?

- What is its depth?

- Is there a unique goal? If not, among all states, roughly what is the proportion of goals?

- Is the search for a naïve goal or for a path to a naïve goal?

- Are all operators reversible?

- Are all operators monotonic?

(We assume some standard representation of the task by predicates.)

For the examples we have seen so far, the answers to these questions are listed below.

### Binary chopping (to some given precision)

- How big is the search space? *Finite; depends on the precision.*

- Form of the search space? *Tree.*

- What is its maximum branching ratio? 2.

- What is its depth? *Small* (proportional to the log of the precision).

- Proportion of goals? *Unique goal.*

- Is the search for a goal or for a path? *A goal.*

- Are all operators reversible? *Yes.*

- Are all operators monotonic? *Yes.*

### Learning to recognize an arch

- How big is the search space? *Small.*

- Form of the search space? *Partial order.*

- What is its maximum branching ratio? *Low.*

- What is its depth? *Low.*

- Proportion of goals? *Unique goal, we hope*.

- Is the search for a goal or for a path? *A goal*.

- Are all operators reversible? *Yes*.

- Are all operators monotonic? *Yes*.

### The 'blocks world'

- How big is the search space? 13 *states*.

- Form of the search space? *Has cycles*.

- What is its maximum branching ratio? 6.

- What is its depth? *Low*.

- Proportion of goals? *Unique goal*.

- Is the search for a goal or for a path? *A path*.

- Are all operators reversible? *Yes*.

- Are all operators monotonic? *No*.

### Finding a forced win in noughts and crosses

- How big is the search space? *Small*.

- Form of the search space? *Partial order*.

- What is its maximum branching ratio? *Low*.

- What is its depth? *Low*.

- Proportion of goals? *May be high*.

- Is the search for a goal or for a path? *May be either*.

- Are all operators reversible? *If searching for a goal: no; if searching for a path: yes*.

- Are all operators monotonic? *No*.

### The eight tiles problem

- How big is the search space? 181 440 *states*.

- Form of the search space? *Has cycles*.

- What is its maximum branching ratio? 4.

- What is its depth? *Large*.

- Proportion of goals? *Unique goal*.

- Is the search for a goal or for a path? *A goal*.

- Are all operators reversible? *Yes*.

- Are all operators monotonic? *No*.

### The eight queens problem (in representation 3)

- How big is the search space? *Less than* 4 000 000.

- Form of the search space? *Tree*.

- What is its maximum branching ratio? 8.

- What is its depth? 8.

- Proportion of goals? *Perhaps about* 1 *in* 40 000.

- Is the search for a goal or for a path? *A goal*.

- Are all operators reversible? *Yes*.

- Are all operators monotonic? *No*.

### The crates problem

- How big is the search space? *Finite but large*.

- Form of the search space? *The naïve space has cycles*.

- What is its maximum branching ratio? *Perhaps about* 20 *depending on the number of crates*.

- What is its depth? *Perhaps about* 10.

- Proportion of goals? *Perhaps about* 1 *in* 10 000.

- Is the search for a goal or for a path? *A path*.

- Are all operators reversible? *Yes* (*in the naïve space*).

- Are all operators monotonic? *No*.

### Composing crossword puzzles

- How big is the search space? *Finite but large*.

- Form of the search space? *Partial order*.

- What is its maximum branching ratio? *Large*.

- What is its depth? *Perhaps about* 50.
- Proportion of goals? *Low.*
- Is the search for a goal or for a path? *A goal.*
- Are all operators reversible? *Yes.*
- Are all operators monotonic? *No.*

### Theorem proving

- How big is the search space? *May be infinite.*
- Form of the search space? *A partial order.*
- What is its maximum branching ratio? *Depends on the example.*
- What is its depth? *May be infinite.*
- Proportion of goals? *Usually negligible.*
- Is the search for a goal or for a path? *A path.*
- Are all operators reversible? *Yes.*
- Are all operators monotonic? *Yes.*

### Exploring the world

- How big is the search space? *Finite but large.*
- Form of the search space? *Partial order.*
- What is its maximum branching ratio? *About* 100.
- What is its depth? *Large.*
- Proportion of goals? *Unique goal.*
- Is the search for a goal or for a path? *A goal.*
- Are all operators reversible? *No.*
- Are all operators monotonic? *Yes.*

### Cooking

- How big is the search space? *Finite but large.*
- Form of the search space? *May have cycles.*
- What is its maximum branching ratio? *Perhaps about* 20.
- What is its depth? *Large.*

- Proportion of goals? *Low.*

- Is the search for a goal or for a path? *A goal; but the planning version is for a path.*

- Are all operators reversible? *No.*

- Are all operators monotonic? *No.*

### Letter arithmetic

- How big is the search space? *Perhaps a few million states.*

- Form of the search space? *Partial order.*

- What is its maximum branching ratio? *Perhaps about* 10.

- What is its depth? *Perhaps about* 20.

- Proportion of goals? *Unique goal.*

- Is the search for a goal or for a path? *A goal.*

- Are all operators reversible? *Yes.*

- Are all operators monotonic? *Yes.*

### 2.3.6 SOME PRIMITIVE SEARCH ALGORITHMS

Be warned: most of the algorithms described in the next few sections are rarely used in practical systems. The search algorithms commonly used in the real world are almost invariably the very simplest. They are instances of a method called **hill climbing**, which will be explained shortly. For all that, it is wise to know a little more about search than just that method, so here are a few relatively elaborate algorithms which are also sometimes useful.

The algorithms listed in this section are 'blind': they cannot take advantage of any prior knowledge of the search space. They all involve storing several states. Hence, they would be no good for making supper directly, though they could construct a plan for making supper.

### Breadth-first search

This algorithm works when the objective is to find a goal in the search space, rather than a path to the goal. It keeps a list, called *OPEN*, of nodes which the algorithm has found and which are not goals but whose children might be goals. It also has a variable called $N$ whose value is the current node of interest.

```
OPEN : = [   ]
N : = start
WHILE N ≠ fail and N is not a goal
    FOR each child C of N
        add C to the back of OPEN

    If any such child C is a goal
    THEN
        N : = C
    ELSE
    IF OPEN = [   ]
    THEN
        N : = fail
    ELSE
        N : = the head of OPEN
        OPEN : = the tail of OPEN.
```

When this stops, either $N$ is a goal or else it is *fail* and the task has no solution. The algorithm examines every shallow state, which can be constructed by a short path from *start*, before it examines deeper states. Hence, if all goals are deep or the branching ration is large, it will be very slow. It also takes up a lot of memory for *OPEN*, remembering all states which might lead to a solution.

The process of constructing all $N$'s children is called **expanding** $N$. Note that the order of children in the FOR loop is not specified. This order will depend on the representation of the task.


### *Depth-first search*

This is a variant of breadth-first search. It differs in just one step: instead of

    add $C$ to the back of *OPEN*

it reads

    add $C$ to the front of *OPEN*

Depth-first search searches all descendants of one state before considering the state's siblings. If the first state among siblings always leads to a goal, it will be much more efficient than the breadth-first version, but otherwise it may be worse. In fact, if the search space has unlimited depth, then depth first search may *never stop* whereas breadth-first search would succeed.

Depth first search depends critically on the order in which the children $C$ of $N$ are added to the front of *OPEN*. Thus, it depends on the *order* of children in the *FOR* loop:

    For each child $C$ of $N$

It is very useful if, at each node on a path to a goal, the child leading to the goal comes first (or at least, none of the children preceding it leads to a large subtree). If you can design the task's representation so that this is so, then depth-first search is better.

### Breadth-first search with back pointers

This is the corresponding algorithm which finds a path to a goal. It differs from naïve breadth-first search, much as the naïve version of the crates problem differs from the real version. Each node is given a pointer to its parent. Thus a node of this algorithm's search space is a pair:

$(N, p)$

where

> $N$ is a node of the naïve task
> $p$ is a pointer to the parent of $(N, p)$.

The algorithm has a new variable, $S$, whose value is such a pair. It goes:

```
OPEN : = [   ]
N : = start
S : = (N, p) where p points to S itself.

WHILE N ≠ fail and N is not a goal
   FOR each child, C of N
        p : = a new pointer to S
        add (C, p) to the back of OPEN
   IF any such child C is a goal
   THEN
        N : = C
   ELSE
   IF OPEN = [   ]
   THEN
        N : = fail
   ELSE
        S : = the head of OPEN.
        N : = the naïve node in S
        OPEN : = the tail of OPEN.
```

When this stops, either $N$ is *fail* or else the pointer $p$ in the current $S$ points back along the sequence of operators which leads to $N$. There is a version of this which corresponds to the basic depth-first algorithm.

### Backtracking depth-first search

This is a modification of depth-first search which avoids finding all children of $N$ straight away. It is more economical with space. The nodes of its search space are a little more elaborate than the task's basic states. Each node consists of:

- a state of the original task's search; and

- a list of operators which can be applied to this state.

Thus, the algorithm searches a space in which each node is a pair:
   $(N, LOp)$
In one cycle, the algorithm takes one operator out of the list $LOp$ and applies it to $N$, so producing just one child.

*Avoiding duplicate states*  This is a simple modification of the basic algorithms. A node is only added to the *OPEN* list if it has never been created before. The algorithm keeps a second list, called *CLOSED*, of nodes which have been expanded and removed from *OPEN*. Whenever a child is created, it is only added to *OPEN* if it is not already in *OPEN* or *CLOSED*.

### Back marking

Sometimes a searcher can detect that a large branch of its search space does not contain a goal. For instance, in the eight queens problem, the position shown on Fig. 2.14 cannot be extended to a goal because no queen can ever be put in the left file. However, a naïve searcher might explore children and children's children of this state, such as in Fig. 2.15, before discovering this fact. As soon as the back marking algorithm finds a proof that some state $N$ does not lead to a goal, it tests to see if the same proof applies to ancestors $A$ of $N$. When the same proof works for $A$, it prunes $A$ and all its other descendants too.

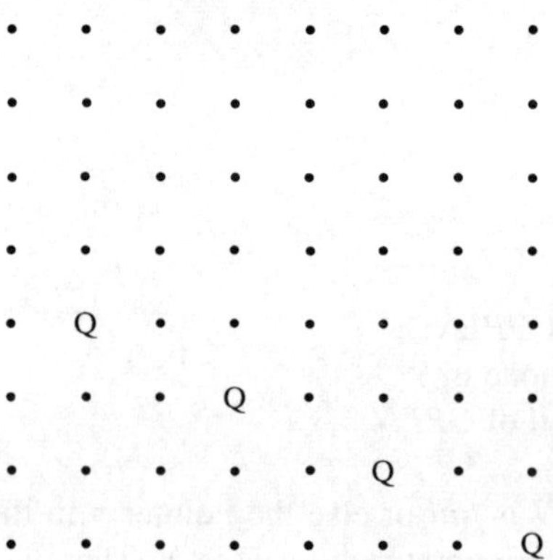

**Fig. 2.14**

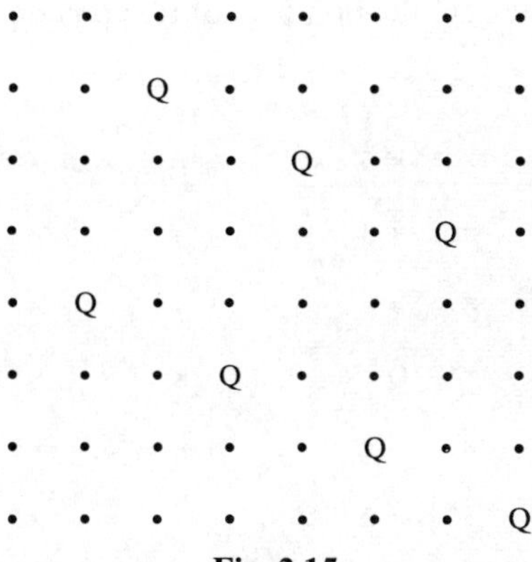

**Fig. 2.15**

### 2.3.7 HEURISTIC FUNCTIONS

Breadth-first search lies employs no knowledge about the task. Depth-first search can employ a little knowledge, if you can represent the search space so that the searcher always tries the best child first. However, the chances are that, if you can really represent the search space optimally before calling the algorithm, then there is no need for search at all. You could solve the task without it.

The key to clever search lies in choosing the most likely state in *OPEN*. If the algorithm can detect a short path to a goal and always choose the $N$ from *OPEN* which is on this path, then it will be efficient. **Heuristic** evidence is any guiding information which suggests how to choose a good node $N$ from *OPEN*. A heuristic rule may be helpful, but it may be wrong. A good **heuristic function** is a function $f$ on states whose values tend to be large on nodes on paths to goals. The rule of thumb is that, whenever you choose a next node, the one you choose has as large a value $f(N)$ as possible. There are good and bad heuristic functions. A function is bad if it is large on nodes which lead nowhere. Part of the art of search lies in designing good heuristic functions.

*Examples of heuristic functions*

   3a.   *Eight queens: number of children*

Suppose $N$ is a state of the eight queens problem and $r$ is any row on the board without a piece in it. If $N$ can be extended to a goal, then it must be possible to put a piece in row $r$ somewhere. Say $n_r$ is the number of positions in this row where one might put a piece, according to the rules. Then $f_a(N)$ is the smallest value of $n_r$, for all such rows $r$.

   For instance, suppose $N_1$ is the left of the two positions shown in Fig. 2.16, and $N_2$ is the one on the right. In them,

   Q represents a piece;
   × represents a square where a piece cannot be put;
   ○ represents a square where one could put a piece.

The numbers beside rows are the numbers of free positions.

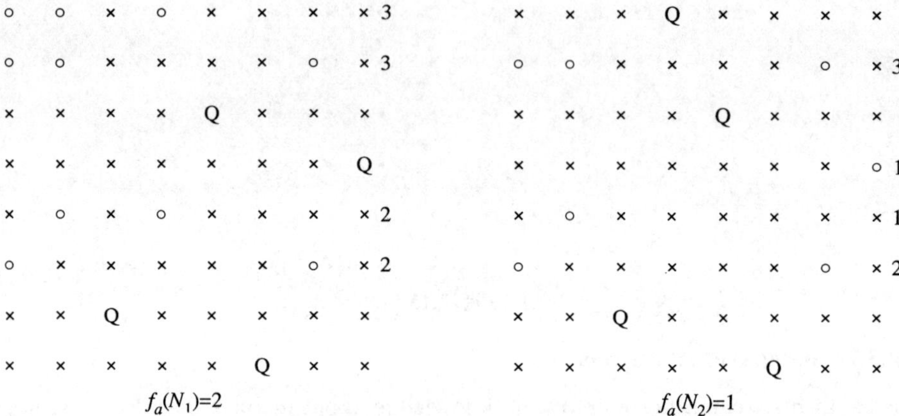



**Fig. 2.16**   Heuristic values of states in the eight queens problem.

3b.   *Eight queens: minimum number of extensions in any row*

Suppose

- $c_N$ is the number of rows in $N$ which already contain a piece;

- For each unoccupied row, $r$, say, $n_r$ is the number of positions in this row where one might still put a piece.

Let $f_b(N) = c_N^2 \times f_a(N)$. It is easy to estimate the worths of a few significant nodes listed in Table 2.1.

**Table 2.1**

|  | $f_a(N)$ | $f_b(N)$ |
|---|---|---|
| Node $N$: |  |  |
| *start* | 8 | 0 |
| any child of *start* | $\geq 5$ | $\leq 7$ |
| any grandchild of *start* | $\geq 2$ | $\leq 24$ |
| any grandparent of goal | $\leq 2$ | $\geq 36$ |
| any parent of a goal | 1 | 49 |

$f_b$ is a better heuristic function than $f_a$ because it assigns relatively large values to states near goals.

The values of a heuristic function need not be numbers. Here is another even better function for the eight queens problem whose values are list of numbers.

3c. *Eight queens: ordered lists of numbers of vacancies*

If $N$ is any state, $f_c(N)$ is a list with as many entries as $N$ has empty rows. The entry in the list corresponding to row $r$ is the number of positions in this row where one could put a piece. The subtlety is that $f_c(N)$ is arranged in increasing order. Thus,

$$f_c(N_1) = [2, 2, 3, 3] \quad f_c(N_2) = [1, 1, 2, 3].$$

If $M$ and $N$ are any two positions, then say

$$f_c(M) > f_c(N)$$

if either

- all entries in $f_c(M)$ are strictly positive, and

- $f_c(M)$ is shorter than $f_c(N)$;

or

- $f_c(M)$ and $f_c(N)$ have the same length, and

- the first few entries in $f_c(M)$ and $f_c(N)$ are equal and the next entry of $f_c(M)$ is larger than that of $f_c(N)$.

This is a novel use of the symbol '>'. Generally, people think of inequality as a test which is performed on numbers. Here we are following Humpty Dumpty's maxim: a word (or symbol) can mean whatever we want it to mean. In this case,

$$f_c(N_1) > f_c(N_2).$$

The state $N_1$ is more promising than $N_2$ because for each row in $N_2$, there is a row in $N_1$ with at least as many free positions. The novel definition of '>' was chosen to reflect this fact.

Of all three heuristic functions, $f_c$ is almost certainly the best, because it contains most information. Both $f_b(N)$ and $f_a(N)$ can be derived from $f_c(N)$.

### 2.3.8 SOME HEURISTIC SEARCH ALGORITHMS

#### *Best-first search*

In this algorithm, the collection called *OPEN* of untested nodes is a set, not a list. The algorithm is supplied with a heuristic function, called $f$, and it chooses nodes from *OPEN* in order of their $f$ values. It is designed to find a goal as quickly as possible.

```
OPEN : = {   }
N : = start

WHILE N ≠ fail and N is not a goal
    FOR each child C of N
        add C to OPEN
    IF any such child C is a goal
    THEN
        N : = C
    ELSE
    IF OPEN = {   }
    THEN
        N : = fail
    ELSE
        N : = the node in OPEN whose f value is greatest
        OPEN : = the rest of OPEN.
```

(Traditionally, the new node $N$ chosen from *OPEN* is the one with least $f$ value. The version given here suits the above examples.)


### The $A^*$ algorithm

This is the heuristic analogue of uninformed search with back pointers. Given a few reasonable assumptions about $f$, it is guaranteed to find a best path from the start to a goal. It has to extend its search one stage beyond the point where it finds the best path, in order to check that it is the best.

$A^*$ is based on the assumption that each step in the search has a cost, and the best path is one which costs least. Say

$$cost(N, c)$$

is the cost of stepping from a node $N$ to one of its children, $C$. Since the aim is to find a path, the space which the algorithm explores has nodes consisting of pairs

$$S = (N, p)$$

where $N$ is a naïve state, and $p$ is a pointer to another such pair. Each such node $S$ is at the end of a path from the start, and this path can be traced back from $S$ by following pointers. The cost of this path is a function $g$ of $S$. If $C$ is a child of $N$, and

$$T = (C, q) \quad \text{where } q \text{ points to } S$$

then

$$g(T) = g(S) + cost(N, C).$$

The heuristic component of the algorithm is given by some estimate $h(N)$ which is a guess at the cost of a cheapest path from $N$ to a goal. Thus, $h(N) = 0$ whenever $N$

is a goal, and $h(start)$ is an estimate of the cost of a cheapest solution. As well as the set *OPEN* of known nodes which might lead to goals and where children have yet to be explored, the algorithm keeps a set called *CLOSED* of other known nodes whose children have been constructed.

```
OPEN : = {  }
CLOSED : = {  }
N : = start
S : = (start, p) where p points to S itself.

WHILE N ≠ fail and N is not a goal
   add S to CLOSED

   FOR each child C of N
       IF C is not the naïve state of any node in OPEN or CLOSED
       THEN
               add a new node (C, q) to OPEN
               where q points to S
       ELSE
               {say OPEN ∪ CLOSED contains some state T = (C, q)}
               IF g(T) ≤ g(S) + cost(N, C)
               THEN
                       do nothing
               ELSE
                       discard T
                       T' : = (C, q') where q' points to S
                       put T' on OPEN

   IF OPEN ≠ {  }
   THEN
       S : = the entry in OPEN, and
       N : = the naïve state in S, for which
             g(S) + h(N) is least
   ELSE
       N : = fail.
```

This algorithm is guaranteed to find a cheapest path to a goal, if

- there is a goal somewhere and a path in the search space from *start* to it; and

- for every naïve state $N$, $h(N)$ is not more than the least cost of any path from $N$ to a goal; and

- there is a number $\delta > 0$ which is less than the cost of any step:

    $cost(N, C) \geq \delta$   for every node $N$ and child $C$.

Recall the warning which you read above: all these fancy search algorithms which you have just read about are not really much use for algorithmic learning. They are good educative examples of the search process, but it can be argued that they have little to do with practical intelligence. These algorithms are most suited to tasks with small branching ratios and reversible operators and with simple states so that the algorithm can remember several states at a time. Real life is harder.

### Hill Climbing

This algorithm is so simple that it is often not taken seriously. It sets out to find a state $N$ with the greatest possible worth, $f(N)$. Such a state is treated as a goal. Hill climbing is so called because it is like what one would do to find the top of a hill in a Scottish mist: keep going up.

$N : = start$

WHILE $N$ has a child $C$ for which $f(C) > f(N)$

$N : =$ the child $C$ of $N$ for which $f(C)$ is largest.

The result is the final value of $N$. Note that the value $f(C)$ is assumed to be the *true* worth of the state $C$, at least when $C$ is a goal.

Hill climbing can be adapted to a range of situations. The other algorithms are all designed as if the children of each state form a discrete set with no structure. When the set of children of $N$ is infinite, it will not be feasible to calculate $f$ for them all. It can happen that the children of $N$ form a continuum, $C_t$ where $t \in \mathbb{R}$; and $N = C_0$, and $f(C_t)$ is a smooth function of $t$. Then the natural approach is to find the slope $f'$ of $f$ at $N$. The algorithm's step then becomes

$N : = C_\delta$

where $\delta$ is a suitable positive number if $f' > 0$; $\delta < 0$ if $f' < 0$. The magnitude of $\delta$ is arbitrary. Perhaps it could be large if the second derivative of $f$ is small and vice versa.

Hill climbing has several limitations. Anyone who has done much climbing in Scotland will be familiar with the first two.

### The foothill problem

There may be several states $N$ with greater worth than any of their children. The algorithm may arrive at the wrong one and miss the true goal with maximum worth. This error is like climbing below a col and coming up on the wrong side of it. The true goal is a summit to the right of the col, but the climber may begin a bit too far to the left and go up the other way.

### The plateau problem

A node and its children may all have almost identical $f$ values, so the algorithm lacks sensible guidance. This is analogous to wandering around in a misty marsh in

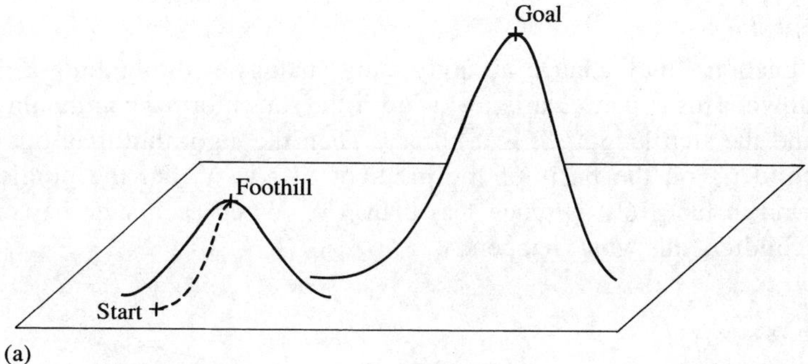


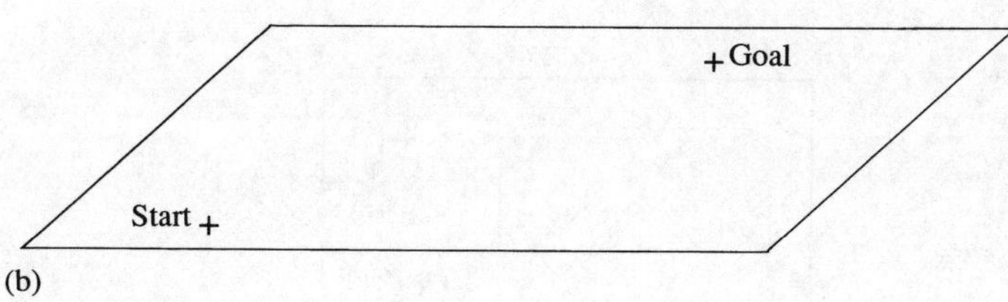


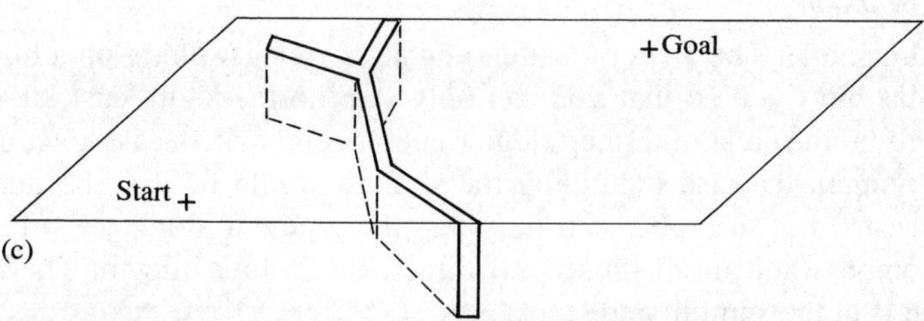


**Fig. 2.17** Problems with hill climbing: (a) the foothill problem; (b) the plateau problem; (c) the canyon problem.

a river valley surrounded by invisible summits. The other algorithms presented above do not suffer so seriously from this problem because they can all examine alternative paths for several steps.

### The canyon problem

The worth function $f$ may change abruptly. Any analogy with climbing a gentle hill breaks down. This is particularly awkward if the states form a continuum and $f$ is smooth and the step length, $\delta$, is arbitrary. Then the algorithm may opt for a particular child $C_\delta$ on the basis of the gradient of $f$ at $N$, but the gradient is misleading and in fact $f(C_\delta)$ is much less than $f(N)$. When each state has only a finite set of children, this won't happen.

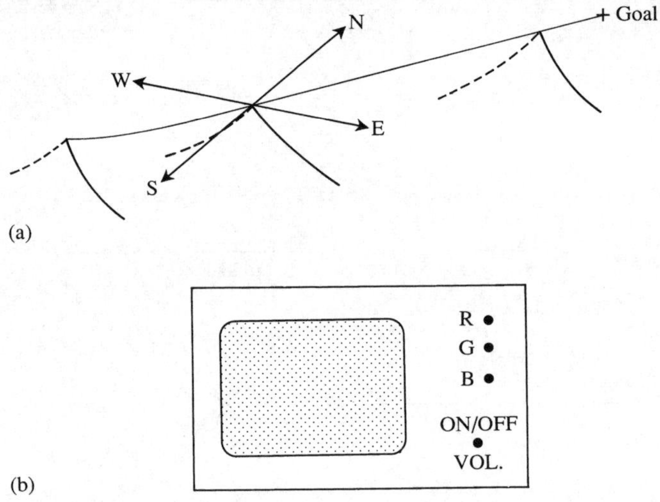



**Fig. 2.18**   (a) The knife edge problem; (b) a manifestation of the knife edge problem.

### The knife edge problem

This is a bit more subtle. The analogy is that you are a robot walking up a hill and your designer has built you so that you can only step north, south, east, or west. You find yourself on the crest of a steep ridge running gently north east towards the summit. For a human, the task of finding the summit would be easy because he could see north east, but the robot can only test the gradient along the principal points of the compass and in all those four directions it slopes down. The robot concludes that it is at the summit and stops.

Another similar case occurs no doubt every evening in thousands of homes. A television set has three colour controls: red, green, and blue. Twiddling any one of them makes the colour worse, but if all three or perhaps just two were adjusted simultaneously then the colour of the picture would improve.

Despite that, there are many cases when hill climbing is the best or perhaps the only practical approach:

- when the children of each node form a continuum (see above);

- when each state is complex and so only one state can be remembered at a time;

- when any two states have a common descendant goal, so choosing a child which is not optimal does not matter too much. This often happens in theorem proving;

- when the task's operators are irreversible. In that case, there is no point in remembering alternative branches of the search, because when one has been chosen, the others are no longer attainable.

This last situation is common in real life.

*Example*

10. *Politics*
Once something has been said, it cannot be unsaid.


### Hill climbing with planning

The usual strategy in real irreversible searches is the following. Choose some function $f$ of the operators of the search task. $f$ is meant to be an estimate of the benefit gained by applying an operator. This $f$ is used only to detect when the algorithm has reached a summit or plateau. It is not used to choose between alternatives. To do that, at each step, the algorithm constructs a planning sub-task. It simulates the bit of the search space which is rooted at the current state, and plans an acceptable next move.

The major limitation of this algorithm is the quality of the simulation in the planning stage. If not much is known about the properties of operators which could act on $N$ and its children, then the plan may be misleading.

$N := start$

WHILE any operator $Op$ applies to $N$, and $f(Op) > 0$
    Set up a subsidiary planning task to simulate a search
    for a path starting at $N$, leading to a better state.

    Explore this planning task until it has found

    - either a path to some state $C$ which is substantially
      better than $N$

    - or a child $C$ of $N$ such that every path to any
      better state begins through $C$

    $N := C.$

In real life, all the clever algorithms with *OPEN* lists are verging on the irrelevant. Hill climbing with planning is used almost universally. Even the planning step is commonly done by pattern matching or some similar technique which does not involve keeping alternatives.

### Heuristic rules

So far, we have taken the operators for granted. In practice, finding which operators apply to some node is often a substantial part of the searching process. Commonly, each operator is incorporated into a *rule*. This is a data structure with:

- a condition;

- an action;

and maybe

- a name;

- other details.

The *action* is the operator, which is applicable to some states $N$ but not to all. The *condition* is a predicate which is true of just those states $N$ that the action can be applied to. Rules of this form are often called *IF ... THEN ...* rules: if the condition is true of a state, then the action can be applied.

There is a simple searcher which uses such rules. Say the set of all rules available to the searcher is called *KB*, for 'knowledge base'. The searcher goes:

```
N : = start
TRACE : = [  ]

WHILE N is not a goal and N ≠ fail
   CS : = conflict-set(N, KB)

   IF CS = {  }
   THEN
      N : = fail
   ELSE
      R : = best-rule(CS, Goal-conditions, N)
      {say R has  action  Op
               name   Name}
      add Name to the back of TRACE
      N := do-action (Op, N).
```

The *TRACE* is a record of which rules were applied to find a goal. It is often useful for later learning. The three major sub-functions are:

- *conflict-set*. This finds which rules have conditions satisfied by $N$.

- *best-rule*. This makes a heuristic choice of one rule $R$ out of the conflict set, *CS*. (The work 'heuristic' means that the choice is based on incomplete or unreliable information.) It is this rule's action which generates the child of $N$ for the next cycle.

- *do-action*. This applies the action $Op$ of the rule $R$ to the current state $N$ and produces the next state of the search.

The 'other details' in the rule, which were left unspecified above, are to help *best-rule* to make a good choice of rule. *Best-rule* can also make its choice depend on the node $N$ itself and on the properties which characterize a goal, *Goal-conditions*. The process of choosing the best rule is called **conflict resolution**.

Many learning systems have a problem-solving component of this form. Learning is relatively easy with such systems, because the searcher's knowledge is accessible in the rules.

### *Generate and test*

There is a particular form of *best-rule* which run thus:

> $Children := \{do\text{-}action\,(Op', N) \mid \exists\, R' \in CS\,(Op'\text{ is the action of }R')\,\}$
> $C :=$ the state in *Children* which most nearly satisfies *Goal-conditions*
> $R :=$ a rule in $CS$ whose action $Op$ produces $C$ from $N$.

This method of search, which creates all children of $N$ and opts for a best one, is called **generate and test**. It is a sort of hill climbing. Of course, if the searcher is really written thus, then it creates the chosen next state twice, which is inefficient, so one would never actually write a searcher thus. The optimized code goes:

```
N := start
TRACE := []

WHILE N is not a goal and N ≠ fail
    CS := conflict-set (N, KB)

    IF CS = {}
    THEN
        N := fail
    ELSE
        Children := {do-action (Op', N) | ∃ R' ∈ CS (Op' is the action of R') }
        N := the state in Children which most nearly satisfies Goal-conditions.
```

Generate and test only makes sense when we have some means of measuring how nearly each child $C$ satisfies *Goal-conditions*. Often, there is such a measure. Also, this form of search is only possible when the searcher can generate and then retract states or store several states simultaneously. so it is not suitable with irreversible operators.

## 2.4   METRIC SPACES

A metric space is a set, commonly called $X$ or perhaps $S$. The elements of $X$ are commonly called 'points', since they are supposed to have no size. Instead, there is a notion of **distance** between points. This distance is a function

$$d : X \times X \to \mathbb{R}$$

which satisfies the following properties. For any points $x$ and $y$ in $X$,

(1)   $d(x, y) \geq 0$

(2)   $d(x, y) = d(y, x)$;

(3)   $d(x, y) = 0 \iff x = y$;

(4)   $d(x, y) \leq d(x, z) + d(z, y)$   for any $z$ in $X$.

The last condition, (4) is called the **triangle inequality**. If you draw a triangle which corners $x$, $y$ and $z$, then it says that the length of the side between $x$ and $y$ is no more than the sum of the lengths of the other two sides.

*Example: normal distance—the Euclidean metric*

If you choose a particular position, such as the doorknob on your front door, you can describe any position in your house by saying how far it is

(1)   behind the front door (its $x_1$ coordinate);

(2)   to the right of the line perpendicular to the front door (its $x_2$ coordinate);

(3)   above the doorknob.

Thus you can describe it by the three numbers

$(x_1, x_2, x_3)$.

The set $X$ is the set of all positions in your house.

If $(y_1, y_2, y_3)$ is a second such point, then the Euclidean distance between $x$ and $y$ is

$$d(x, y) \;=\; \sqrt{(x_1 - y_1)^2 + (x_2 - y_2)^2 + (x_3 - y_3)^2}\,.$$

This function $d$ is just the usual distance between any two places. It satisfies the five conditions of a metric.

*Example: Manhattan distance, in the eight tiles problem*

This is a metric on the space of states in the game studied above, with eight tiles which have to be moved to the goal position:

$$\begin{array}{ccc} 1 & 2 & 3 \\ 8 & & 4 \\ 7 & 6 & 5 \end{array}$$

Consider the positions

$$a: \begin{array}{ccc} 5 & 3 & 6 \\ 1 & & 4 \\ 7 & 8 & 2 \end{array} \quad \text{and} \quad b: \begin{array}{ccc} 7 & 1 & 6 \\ 8 & & 4 \\ 3 & 5 & 2 \end{array} \quad \text{and} \quad c: \begin{array}{ccc} 5 & 3 & 6 \\ 1 & & 4 \\ 7 & 8 & 2 \end{array}$$

If tile 4 were out of the way, as in $c$, then position $b$ could be obtained from $a$ by sliding the tiles 7, 8, 3, 5, and 1 in a loop. If they are moved in that order then it would take twelve operations to move from $a$ to $b$. The Manhattan distance $Md(a, b)$ is 12.

In general, for any state $x$, the position of each tile $t$ can be described by its horizontal and vertical coordinates: $(t_{xh}, t_{xv})$. Thus, in $a$ the coordinates of tile 5 are $(1, 3)$, while in $b$, 5 is at coordinates $(2, 1)$. The Manhattan distance $Md (x, y)$ between any two states $x$ and $y$ is

$$\sum^{8} \left| t_{xh} - t_{yh} \right| + \left| t_{xv} - t_{yv} \right|.$$

It is the number of operations needed to move from $x$ to $y$ if no tile got in the way of any other. (You should check that this function Md really is a metric.)

*Example: Venn diagram metric*

Take any set $S$, and let $X$ be the set of finite subsets of $S$, so a point $x$ in $X$ is a finite set

$$\{s_1, s_2, s_3, \ldots s_N\}$$

of elements of $S$. The number of elements in $x$ is sometimes written $|x|$. In this case, it is the number $N$.

Say $y$ is another such finite subset. The Venn diagram distance between $x$ and $y$ is the number of elements of $S$ which are in either $x$ or $y$, but not in both:

$$Vd (x, y) = |x| + |y| - 2 |x \cap y|.$$

See Fig. 2.19.

We should check that this really is a metric. You can convince yourself that it satisfies the conditions (1)–(4). To check the triangle inequality, we need to know that

$$Vd (x, z) + Vd (z, y) - Vd (x, y)$$

is not less than 0. This is most easily seen by drawing a picture like Fig. 2.20.

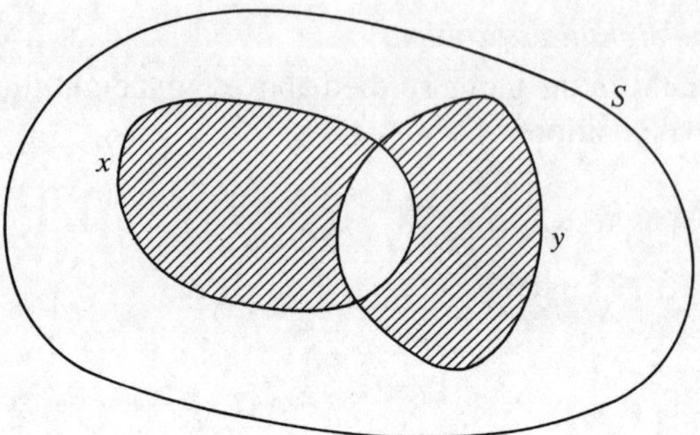

$$\mathrm{Vol}(x,y) = |x| \cdot \qquad \mathrm{Vd}(x,y) = |x| + |y|\ 2\,|x \cap y|$$

**Fig. 2.19**   The Venn diagram metric.

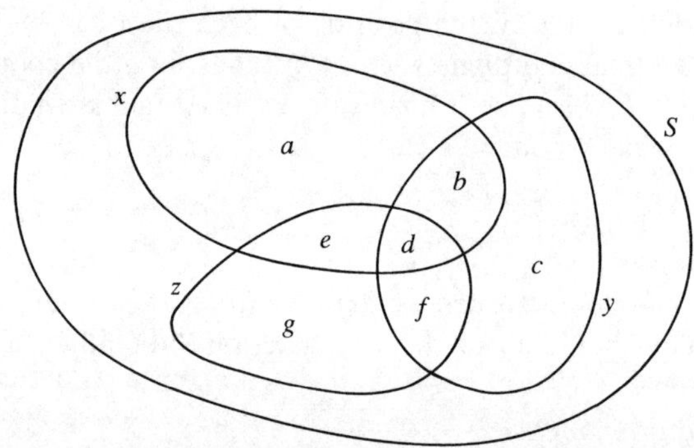

$$\mathrm{Vd}(x,z) + \mathrm{Vd}(y,z) - \mathrm{Vd}(x,y) = (a+b+f+g) + (b+c+e+g) - (a+e+c+f) = 2b+2g \geqslant 0$$

**Fig. 2.20**   The triangle inequality for the Venn diagram metric.

The expression above is twice the number of points which are

(1)   in $z$, but not in $x$ or $y$;

or

(2)   in $x$ and $y$, but not in $z$.

Thus, it cannot be negative.

*Example: the Hamming distance*

This is a metric between $N$-tuples of 0s and 1s. Here, $N$ is a number. An $N$-tuple is a sequence of $N$ digits

$$(0 \quad 1 \quad 1 \quad 1 \quad 0 \quad 0 \quad 1 \quad 0 \quad 0 \quad 0 \quad 0 \quad 0).$$

The set $X$ is the set of all such sequences. There are $2^N$ points in $X$. If $N$ is 3, then you can think of a cube, drawn in the corner of the three axes of Cartesian geometry. A point is a corner of this cube. The Hamming distance between two tuples

$$\mathrm{Hd}\,(x, y)$$

is the number of positions where the two tuples have different entries. For example, if

$$
\begin{aligned}
x &= (0 \quad 1 \quad 1 \quad 0 \quad 0 \quad 0 \quad 1 \quad 0 \quad 1 \quad 1) \\
y &= (0 \quad 1 \quad 0 \quad 0 \quad 1 \quad 0 \quad 1 \quad 1 \quad 0 \quad 1) \\
  &\phantom{= (0 \quad 1 \quad } {*} \phantom{\quad 0 } {*} \phantom{\quad 0 \quad 0 } {*} \quad {*}
\end{aligned}
$$

then

$$\mathrm{Hd}\,(x, y) = 4.$$

You should check that this function Hd also satisfies the conditions of a metric.

We shall use the Venn diagram metric and the Hamming distance later.

## 2.5    EXPERIMENTING WITH CHANCE

There is very little that we can be absolute certain about. Descartes said: '*I think, therefore I am.*' The point of this statement is not that it is true. It is just about the *only* fact which is unquestionably true. Anything else, for instance the 'fact' that Socrates talked with Plato, depends (from Socrates' point of view) on the assumption that there really was someone called Plato who often visited him. Socrates might have dreamed about Plato. For all we in this century know, Plato may have dreamed all the conversations with Socrates. You may have dreamed all the 'experiences' in which you ever heard mention of Socrates and Plato. If you find this hard to swallow, try to work out what is reality in the more fanciful novels of Daphne du Maurier.

**Ontology** is the branch of philosophy concerned with existence. How can you be certain that anything exists, apart from you yourself? What does it mean to say that something exists? It is an aspect of **metaphysics**, which is the study of what we are justified in believing from first principles. By its nature, metaphysics almost never provides us with certainties, so rather than bother with it, most people 'believe' what to them seems most plausible. Since Plato (dreamed that he) saw Socrates most days and this was among the most vivid and consistent of all his experiences (dreams?) and since Socrates said things which he, Plato, had not thought of before and which sometimes surprised him, Plato found it most consistent and convenient to 'believe' that Socrates really did exist.

Probability is all about justifying beliefs. Our belief in Mendel's theory of genetics is based on experiments, with fruit flies and similar organisms, whose interpretation depends on probabilities. Almost all learning is similarly based. We observe a set of events and decide that they have one explanation which is vastly more probable than all others, so we come to believe that this explanation is correct and we use it to make predictions in future. When you have seen an egg broken into a frying pan of hot fat so that it turns from a clear fluid into a rubbery opaque solid, you are inclined to believe that this is quite a good way to cook an egg. When you have seen it done twice and, furthermore, your mother does it with complete confidence and expects you to eat the result, then you are sure. You have *learned* about frying eggs. You simply do not admit the possibility that your mother is being whimsical and that these two eggs were exceptional. It would be just too unlikely. However, if you were completely rational, you would admit the possibility–and your mother would go frantic.

Let us analyse the fruit flies experiment on page 17 more carefully. Some results fit the prediction very well (for example, if the numbers of offspring with red eyes is just three times the number with white eyes), some fit it quite well (for example, if 61 have eyes and 23 have white eyes), and some would fit it very badly (for example if all but one had white eyes).

To make the numbers a bit easier, let us suppose that there are just three offspring. If *n* have red eyes, then (assuming Mendel's model of inheritance) we can tabulate in Table 2.2 how may ways this can happen.

**Table 2.2**

| $n$ | Number of ways this can happen |
|---|---|
| 0 | 1  (they are all white) |
| 1 | 3  (Because any one of the three off spring might be the one with red eyes) |
| 2 | 3  (3 ways to choose the first with red eyes, 2 ways to choose the other, but then, only half $3 \times 2$, because the order of the two doesn't matter.) |
| 3 | 1 |
| Total: | 8 |

However, these are not really the numbers which matter. What we should consider are all the alternative ways that a fruit fly can inherit individual switch settings. There are four possible outcomes: It gains

- red from the mother and red from the father;

- red from the mother and white from the father;

- white from the mother and red from the father;

- white from the mother and white from the father.

According to Mendel's theory, all these situations are equally likely. Let us tabulate in Table 2.3 how many ways each can occur, among three offspring.

**Table 2.3**

| rr | rw | wr | ww | Number of ways this can happen |
|----|----|----|----|--------------------------------|
| 0 | 0 | 0 | 3 | 1 |
| 0 | 0 | 1 | 2 | 3 |
| 0 | 0 | 2 | 1 | 3 -- |
| 0 | 0 | 3 | 0 | 1 |
| 0 | 1 | 0 | 2 | 3 |
| 0 | 1 | 1 | 1 | 6 -- |
| 0 | 1 | 2 | 0 | 3 |
| 0 | 2 | 0 | 1 | 3 -- |
| 0 | 2 | 1 | 0 | 3 |
| 0 | 3 | 0 | 0 | 1 |
| 1 | 0 | 0 | 2 | 3 |
| 1 | 0 | 1 | 1 | 6 -- |
| 1 | 0 | 2 | 0 | 3 |
| 1 | 1 | 0 | 1 | 6 -- |
| 1 | 1 | 1 | 0 | 6 |
| 1 | 2 | 0 | 0 | 3 |
| 2 | 0 | 0 | 1 | 3 -- |
| 2 | 0 | 1 | 0 | 3 |
| 2 | 1 | 0 | 0 | 3 |
| 3 | 0 | 0 | 0 | 1 |

Total:  64

Out of these 64 possible outcomes, the 27 marked '--' have just one fruit fly with both switches set to white. Suppose the student actually found that two have red eyes and one has white. According to the theory, the chance of this result is (27/64).

The idea of confidence runs thus: rather than look at the probability of getting this precise result, we find the probability of getting a *less likely* observed result. The probabilities of all possible observable results are listed in Table 2.4.

**Table 2.4**

| Number with red eyes | Probability |
|----------------------|-------------|
| 0 | 1/64 |
| 1 | 9/64 |
| 2 | 27/64 |
| 3 | 27/64 |

The chance of getting a less likely result is 10/64 or about 15 per cent. This is reasonable. In a typical experiment, the chance of getting a less likely result is no more than 50 per cent.

This number, the chance of getting a less likely result, is called the **backing** which the experiment offers the theory. If an experiment's possible outcomes are $O_1$, $O_2$, $O_3$, ... and a theory predicts that each outcome $O_j$ may occur with probability $P_j$ and the outcome actually observed is $O_n$, then this observation gives the theory a backing of

$$\underbrace{\sum P_j}_{j \,:\, P_j < P_n}$$

By various tricks which save us from the full load of naïve combinatorics, one can show that the student's original result (61 with red eyes, 23 with white) gives Mendel's explanation a backing of nearly 100 per cent; so the professor was right.

In fact, the calculation above hides a significant assumption: we have taken it for granted that there are precisely three offspring. If you allow that there might be any number of offspring, the backing is not quite this value.


## 2.6   COMPLEXITY OF DESCRIPTIONS

The student's experiment provides good backing for Mendel's theory. The questions arise: Does the same experiment provide equally good backing for any other theory? If so, why do we opt for Mendel's? The answer to the first is, Yes. There are lots of theories (for example, that fruit flies have a probability of 61/84 of inheriting red eyes rather than white). The reason for opting for Mendel's is that it not only has good experimental backing, it is also simple. This simplicity can be quantified (in principle). It is done thus.

We choose once and for all some language for describing theories. English might do, if you can decide just what is English, but theoreticians prefer something more precise. Each candidate theory is described in this language. We then measure the complexity of each description and choose the one with adequate experimental backing and the simplest description. Solomonoff and Kolmogorov independently invented a systematic way of measuring complexity.

Choose any programming language. If $D$ is a description of some theory, then observe that $D$ is string of characters. Its complexity $K(D)$ is the length of any shortest program which prints $D$ and then stops. Kolmogorov showed that, if you choose a different programming language, the differences in the value $K(D)$ is less that some constant which depends on the languages but not on $D$. More details will appear in Chapter 8.

Even this is an unduly myopic attitude. We do not just consider Mendel's theory in isolation. Science is a unity and the test of simplicity should really be applied to the whole of science. We do not believe that fruit flies have a chance of 61/84 of red eyes because on its own this is a silly idea; but also, it does not fit with the rest of our accumulated knowledge. Mendel's idea fits.


## FURTHER READING

Many relevant techniques are described in detail in the compendium edited by Bibel and Jorrand (1987).

There are plenty of good books on the propositional and predicate calculi. If these topics are new to you and you feel the urge to investigate them, try the ones by Binmore (1980) and Hodges (1977). If you are more adventurous, look at Johnstone (1987) or Mendelson (1987).

There are several good books on graph theory too. However, the kind of graph theory which mathematicians investigate is almost totally irrelevant for computing and even the basic definitions which computing people use are not always the standard graph theoretic ones. The proper approach appears in Bollabas (1979) and Wilson (1985) and various other references.

Rich and Knight (1991) present a good introduction to the theory of search, including a discussion of the various features characterizing search spaces. For a thorough introduction to the subject of search algorithms, you can study the books by Nils J. Nilsson (1980) or Judea Pearl (1984).

Experimental backing is an aspect of probability theory and statistics. For more details, see any one of many good books, such as the one by N. C. Barford (1967).

The original references on complexity are Kolmogorov (1965) and Solomonoff (1964, 1978). More references and a discussion of Kolmogorov's contribution appear in Razborov (1990). Applications to machine learning are discussed at greater length by Li and Vitányi (1993).


## EXERCISES

1. Suppose that $A$ is a finite set of symbols, such as the Roman alphabet. The set of all finite sequences of symbols from $A$ is called $A^*$. For instance, if $A$ is the Roman alphabet, then every word on this page is in $A^*$. If A is the set $\{0, 1\}$ then $A^*$ consists of all finite binary sequences. One sequence which occurs in $A^*$, whatever $A$, is the **null sequence** of no symbols. It is usually called $\varepsilon$.

   If $x$ and $y$ are sequences in $A^*$ then say

   $$\mathrm{lcp}(x, y)$$

   is the longest common prefix of $x$ and $y$. It is the longest sequence which is an initial segment of both $x$ and $y$. For example:

   - if $x$ = bigger and $y$ = biggest  then $\mathrm{lcp}(x, y)$ = bigge

   - if $x$ = big     and $y$ = biggest  then $\mathrm{lcp}(x, y)$ = big

   - if $x$ = little   and $y$ = biggest  then $\mathrm{lcp}(x, y)$ = $\varepsilon$.

Suppose that we are given some order on $A$, such as

$$A < B < C < \dots Z < a < b < c < \dots z$$

or

$$0 < 1.$$

It is then possible to invent two different relations on $A^*$.

> *Dictionary order*, $\leq_D$:
> $x \leq_D y$ if either $\mathrm{lcp}(x, y) = x$
> or   in $x$, the symbol after $\mathrm{lcp}(x, y)$ is $a$,
>        and in $y$, the symbol after $\mathrm{lcp}(x, y)$ is $b$,
>        and $a < b$.
>
> *Lexical order*, $\leq_L$ :
> $x \leq_L y$ if either $x$ is shorter than $y$,
>        or   $x$ and $y$ have the same length,
>        and $x \leq_D y$.

Show that both $\leq_D$ and $\leq_L$ are orders on $A^*$; that is, for any $x$ and $y$,

- either $x \leq y$ or $y \leq x$ or both; and

- $x \leq y$ and $y \leq x$   iff   $x = y$.

Show that any non-empty subset of $A^*$ has a least member under the order $\leq_L$ (that is, $\leq_L$ is a **well-order**). Show that $\leq_D$ is not a well-order.

2.   (a)   Draw pictures of all graphs with just four vertices. (There are 11 of them.)

   (b)   Draw pictures of all directed graphs with just four vertices and three edges. (There are 10 of them.)

3.   For the 'blocks world', write down a sequence of instances of operators which leads from the start to the goal. Also write down the sequence of states that this operator sequence produces. Describe each state with the predicates 'on' and 'clear', without using negation. Check that the first operator can be applied to the start state and that each later operator can be applied to the state produced by its predecessor.

4.   In the 'blocks world', certain conditions should always be true:

- any block can have at most one block on it:

  $$\forall x \, (x = \textbf{table} \lor \forall y \, \forall z \, (\mathrm{on} \, (y, x) \land \mathrm{on} \, (z, x) \Rightarrow y = z));$$

- nothing can be on itself:

  $$\forall x \, \neg \mathrm{on} \, (x, x);$$

- the four entities **A**, **B**, **C** and **table** are all distinct.

Show that if these three conditions are true to start with, then they will still be true in all subsequent states.

5. Consider the following tasks:

   (a) the eight queens problem;

   (b) the eight tiles problem;

   (c) cooking supper.

   In each case,

   (i) describe a language for describing states of the task. Using this language,

   (ii) describe the task's goal and its start state;

   (iii) describe the task's operators. Each operator is a partial function from states to states;

   (iv) for each operator, describe any conditions on a state which must be satisfied if the operator can be applied to it.

   For the eight queens problem, carry out (i)–(iv) for two representations, such as representations 1 and 3 in the example of Section 2.3.1.

   The descriptions which you have provided probably involve operators with side effects. In each case, describe a task whose search space's graph is isomorphic but in which all operators are monotonic.

6. The description of the blocks world, in Section 2.3, was in terms of predicates. Each state is describe by an assertion. Invent another language for describing this task, with functions rather than predicates, in which each state is described by a term.

   All the three descriptions of the eight queens problem described its states as terms constructed with functions, not predicates. Invent a language with predicates but no functions, and show how to describe a state of the eight queens problem in that language.

7. Write down a summary characterization of the task of politics.

8. Choose some particular programming language, such as Pascal or C. Using it,

   (a) design data structures which represent the search space of the eight queens problem. Recall that the search space includes the operations, so the data structure will include ways to identify the operations;

   (b) encode the operations as procedures or functions. For each operation, add a condition which is true of a state iff the operation is applicable to the state;

   (c) encode the depth-first search algorithm, and apply it to the eight queens problem in the form you have created in steps (a) and (b).

   If Pascal is your chosen language, this exercise is quite hard because Pascal does not have a simple way to refer to procedures and functions from within data structures. Hence, the data structure for the search space cannot actually include the operations. The exercise becomes a lot easier if you cheat and encode the operations as part of the search algorithm.

Repeat this exercise with different languages:

(d) Prolog, with the operations coded separately from the search space (this is easier than the Pascal version because Prolog can represent data structures more succinctly than can Pascal);

(e) Prolog, with operations encoded as part of the search space and invoked by Prolog's built-in 'call' predicate;

(f) a purely functional language in which functions are first-order objects, such as Miranda or pure Lisp.

In (e) and (f), it is possible to design the data structure for the search space so that it includes all the operations. It is then possible to write the search algorithm itself so that it is completely independent of the particular search task.

9. Repeat exercise 8 for the eight tiles problem and the best-first heuristic search algorithm, using the Manhattan distance as the basis of a heuristic function. Note that the output of the search is a path, not a state of the naïve search space.

10. Discuss the ontology of

(a) your bicycle;

(b) machine learning;

(c) Father Christmas.

11. Imagine that the experiment with fruit flies is performed three times, with the following results:

(a) 3 in the third generation; 2 have red eyes and 1 has white.

(b) 6 in the third generation; 4 have red eyes and 2 have white.

(c) 9 in the third generation; 6 have red eyes and 3 have white.

In each case, calculate the backing that the result gives to the hypothesis that the genetic model is correct.

Find a readable book on statistics and study the $\chi$-squared test. This test is often used as an approximate way to estimate backing. For each of the three sets of results, find the estimate which the $\chi$-squared test gives for the backing.

# 3 Learning algorithms with numeric input

## SUMMARY

This chapter describes learning algorithms which handle individual real numbers. The particular examples discussed are:

- control theory for differential equations;

- remembering and extrapolating;

- adjusting heuristic weights, for conflict resolution and also for evaluating a position during game playing;

- discovery of algebraic relations between numeric parameters.

In the process, we shall discover an instance of a phenomenon called the credit assignment problem, and we shall see that the classification scheme set out in Chapter 1 does not always work perfectly.

## INTRODUCTION

Recall that we shall classify algorithms by their forms of input. The first class consists of those which handle real numbers.

Real numbers are both very versatile and very simple, so it is not surprising that learning processes use them in many different ways. Very broadly, the algorithms summarized here fall into three groups: those that **optimize** a parameter, those that **record** numeric observations and **extrapolate** from them, and those that use real numbers as **weights**, to choose between alternatives. Weights also occur in many neural nets, which will be described in Chapter 4. The essential difference between the weights in neural nets and the numbers used here is that, in a neural net, each datum conveys information by virtue of its position among other similar data, whereas each number in the algorithms of this chapter conveys a distinct piece of information.

## 3.1 OPTIMIZING METHODS

The very simplest example of this kind is the binary chopper, presented in Chapter 1, which finds a root of a polynomial.

Some of the very earliest 'clever' programs were elaborations of this method. Chemical engineering companies found that their very large complex expensive process plants could be controlled better by computer than by human operators.

Typically, such a plant's behaviour can be described by a differential equation which depends on coefficients. The optimum values of these coefficients characterize the plant. When engineers build the plant and install the computer control, they do not know the precise values of these coefficients, so they start the process running with approximate values and write control programs which improve the values. This is a simple form of learning.

If the plant's behaviour depends continuously on a single coefficient, then binary chopping is a safe standard approach. If it is known that behaviour depends smoothly and the second derivative of the dependence is never zero and its first derivative is bounded away from zero, then the Newton–Raphson method is better. More commonly, the plant can only be characterized by several coefficients which must all be optimized simultaneously. The science of adjusting these coefficients goes by the name of **control theory**.


## 3.2  OBSERVING AND EXTRAPOLATING

The crudest form of learning imaginable consists of just remembering. The learner notes what has happened in the past and assumes that in like circumstances, the same will happen in future. This algorithm, if it deserves the title, becomes valuable if it is used together with one simple assumption: we assume that in *almost* identical circumstances, *almost* the same will happen again. In jargon, we say that effects depend on causes **continuously**. It will take us even further if we assume that the dependence of effects on causes is **smooth**.

*Example: the birth of a robot*

A new robot is screwed down on to its workbench. It has a single arm with seven joints in it. Three joints let it move its gripper to any point in three-dimensional space and then two more let it point its gripper in any direction, the sixth lets it twist its gripper round the gripper's axis and the last opens and closes the gripper. Each joint can move through an angle within fixed limits, say up to 180 degrees.

As yet, the robot has no way of knowing whereabouts its gripper will be, when its joints are set at particular angles. Even more to the point, how can it move its gripper to a given position? What joint angles will get it there?

Two ordinary TV cameras are fixed so that they both scan the robot's work area, from different perspectives. The robot holds a small torch in its gripper. Thus, the gripper's position can be chosen by selecting settings for joint angles and measured by reading the torch's images on the TV pictures.

Any position in the robot's reach can be described by its two images in the cameras' fields of view. We can specify a point by saying what its coordinates are in the first field of view—say $(x_1, y_1)$—and its coordinates in the second— $(x_2, y_2)$. Of course the field of view is three-dimensional, so in principle we need only state three of these four coordinates, but there is no point in discarding information and so we may as well keep all four.

*Problem solving* Given a point in the robot's work area, what settings for its joint angles will move the gripper to this point? The point is described by the four image coordinates $(x_1,y_1,x_2,y_2)$, called $I$. A person can provide $I$ by holding a torch at the desired point. The problem solver should find a corresponding 7-tuple of joint angles,

$$(j_1, j_2, j_3, j_4, j_5, j_6, j_7)$$

which we shall call $J$.

*Learning* Given a set of solutions to past problems, find information which will help the problem solver in future.

Observe that this task can be solved by just altering the first three joints. There is no need to twist the gripper or alter the direction the torch points in, and the gripper should most certainly not be opened. Hence, $j_4$, $j_5$, $j_6$, and $j_7$ are irrelevant.

*Simple method, depending on continuity* The robot keeps a set, called $R$, of all the solutions it has found. Each entry in $R$ is a pair $(I, J)$ of a point's image coordinates on the TV pictures and of joint settings which take the gripper to this point.

*Problem solving* Given some image coordinates $I$, the robot looks up in $R$ for the pair $(I', J')$ with $I'$ closest to $I$. It moves all joint angles to $J'$. If this leaves the gripper too far from the desired point, the robot tries adjusting each joint a little in turn to see which adjustment moves the torch's image most quickly towards the point $I$. It selects the most significant joint and keeps moving this one until the torch's image is as close as it can get to $I$, then it tries adjusting another joint, and so on, until the gripper is close enough. This process of fine adjustment is a form of hill climbing.

*Learning* This consists just of adding a new pair $(I, J)$ to the set $R$ every time the robot moves to a new position.

*Elaborate solution, depending on smoothness* This keeps the same set $R$ as the simple solution. It also keeps a similar set $R_2$, explained below.

*Problem solving* This is almost exactly as in the simple solution. The only difference is that the problem solver searches in the combined set $R \cup R_2$ for a pair $(I', J')$ with $I'$ closest to $I$.

*Learning* In this second approach, learning takes three forms.

1. As before, whenever the robot moves its gripper to a new position, it records the joint settings $J$ and the torch's image coordinates $I$ as a new entry $(I, J)$ in its set $R$.

2. In order to gain reasonable competence by the simple approach, the robot has to record more than 1000 entries in the set $R$. This takes a long time. To overcome the delay, the robot's software 'dreams' approximate solution pairs $(I', J')$ and uses them like observed entries. It finds these approximate solutions by linear interpolation from the first few genuine observations. The interpolated points are recorded in the set $R_2$. The details appear below.

3. The 'dreamed' entries in $R_2$ are inevitably a bit inaccurate. When there is time, the robot's joints are switched on again and for each entry $(I', J')$ in $R_2$, the robot's arm angles are set to $J'$. Then the cameras record the torch's true image coordinates $I$ and the true observed pair $(I, J')$ is added to the set $R$. The 'dreamed' entry $(I', J')$ is removed from $R_2$.

Here are the details of the 'dreaming' stage. After the robot has accumulated several observations $(I, J)$ in $R$, the joint motors are switched off and the software chooses new joint angles $J'$ at random or perhaps it chooses $J'$ in a poorly explored region. The software finds the four entries

$$(I_0, J_0) \quad (I_1, J_1) \quad (I_2, J_2) \quad (I_3, J_3)$$

in $R$ for which the joint settings $J_k$ are closest to $J'$. (In practice, it may be best to choose the four settings $J_k$ first and then $J'$ be any point between them.) Unless the points $J_0, \ldots J_3$ all happen by accident to be linearly dependent, there will be four numbers $w_0, w_1, w_2, w_3$ for which

$$J' = w_0 J_0 + w_1 J_1 + w_2 J_2 + w_3 J_3$$

and

$$w_0 + w_1 + w_2 + w_3 = 1.$$

These relations define the numbers $w_k$ uniquely. (We choose four points $J_k$ because the space of $J$s is three-dimensional. The $J_k$s form the corners of a tetrahedron.) The dreamed approximate solution $I'$ is the point

$$w_0 I_0 + w_1 I_1 + w_2 I_2 + w_3 I_3.$$

This 4-tuple $I'$ of coordinates in the TV images may not actually correspond to any real point in the three-dimensional space of possible positions of the torch, but that does not matter.

Clocksin and Moore, who performed this experiment, found that the robot could indeed learn its way around its workspace in a fairly short time (between a quarter and half an hour). The details of their experiment were a little more subtle. The pairs $(I, J)$ were not stored as a simple list, but as a $k$-D tree, so that search for closest points is faster.

Moore says that the cameras were accidentally jogged a little during their tests, without much effect. The software included a second $k$-D matrix which recorded errors in the first matrix. An error found at one position was assumed to apply over a large region of the workspace.

The old alternative approach to robot control involved some very complicated equations describing their kinematics. These equations are hard to solve, and even if one can find a solution, it may be inaccurate. The Clocksin–Moore approach to problem solving and learning has nothing to do with a robot or visual images. The software behind the robot need not embody any knowledge of the robot at all. It is basically a way to associate tuples of real numbers

$$I \leftrightarrow J.$$

### Strengths of the method

- It is very general. It will work for any robot with any number of joints. It is also independent of the positions of the TV cameras. The same logic will work in any situation where one set of real parameters depends smoothly on another set.

- It uses all the available information from experiments.

- It does not depend on any description of the robot's structure. Such descriptions are complex and only approximate.

- Learning is fast enough for realistic applications.

- Performance is also fast enough.

- Learning is incremental.

- The algorithm implemented with this particular robot could re-learn after its TV cameras were jogged. Thus, obsolete learned information can be forgotten or overridden.

### Limitations

There may be more than one joint setting which brings the gripper to a desired point. The relation $I \leftrightarrow J$ is one-to-many. The algorithm may not find the 'best' answer. Clocksin and Moore chose joint angles which were most 'comfortable' for their robot.

## 3.3   CONTROLLING SEARCH WITH WEIGHTS

Optimization of a few parameters is not what people think of when they speak of learning. However, simple adjustments to numbers can produce worthwhile learning behaviour. A prime example occurs during search with heuristic rules, in the choice of a rule from a conflict set.

Recall the version of hill climbing which depends on rules, in Chapter 2. Suppose we use it to decide how to travel from home to somewhere. The possible operations are

walk;   bicycle;   drive by car;   take a train;   fly.

Taking the train is the pleasantest method, so we invent a rule:

IF the destination is near a station
THEN take a train
NAME *train rule*
WEIGHT 10.

The last entry, the rule's WEIGHT, should be large for reliable rules and small for unreliable ones.

It often happens that the initial choice of weights is poor and the program should learn improved weights. An extremely simple algorithm which does this runs thus. Suppose there are altogether $N$ rules.

Start with given weights.
Search with these weights.

IF the search reaches a goal by a path of length $k$
THEN
    add $1/k$ to the weight of each rule
      whose name appears in the *Trace*;
    subtract $1/(N-k)$ from
      the weight of every other rule
ELSE
    subtract $1/k$ from the weight of each rule
      whose name appears in the *Trace*;
    add $1/(N-k)$ to the weight of every other rule.

### Strengths of this algorithm

- It is simple.

- If it is run twice on similar data, it will probably work better the second time.

### Limitations

- It only learns at the end of a search. A truly incremental algorithm would improve its performance during a single search, so that information learned early can be used later in the same search.

- Weights will grow indefinitely.

- It reduces the weights of rules which may never have occurred in any conflict set. Some of these rules may be good in other situations, and deserve high weights. If they are not employed, then their weights will fall too low.

The last flaw is an instance of the **credit assignment** problem. Although a search may be successful, we do not know which key rule led to its success. There may be just one rule whose weight deserves to be raised, but this algorithm raises all weights of rules in the trace indiscriminately and suppresses all others. Similarly, if the search fails, this algorithm penalizes all the chosen rules, when perhaps only one is flawed. We do not know which aspect of the search was significant.

### 3.3.1  THE BUCKET BRIGADE ALGORITHM

This algorithm is an alternative way of learning weights. It is incremental and it only credits and penalizes rules which were actually involved in the search. It gets its name from an analogy with a chain of firefighters passing buckets of water. Each time a bucket is passed from hand to hand, a little water is spilt. The algorithm passes credit between rules, similarly. Each time credit is passed from rule to rule, the rule which passes it keeps back a little for itself. It works as follows.

We fix, once and for all, two numbers: $C$, the value of a reward, which is any positive number, and $K$ which is between 0 and 1. Any rule chosen from the conflict set has its weight reduced by the factor $K$. If it is the last rule of a successful search, so its action leads immediately to a goal, then the reward $C$ is added to its weight. Any other rule used during search is rewarded by the amount of weight which the next rule loses.

Thus, suppose a search reaches a goal in three steps, by rules $R_1$, $R_2$, and $R_3$. To start with, their weights are $W_1$, $W_2$, and $W_3$. After the search,

$R_1$ will have weight $(1-K)W_1 + KW_2$
$R_2$ will have weight $(1-K)W_2 + KW_3$
$R_3$ will have weight $(1-K)W_3 + C$.

In a similar search which fails, $R_1$ and $R_2$ will end up with the same final weights, but $R_3$ is not rewarded:

$R_1$ will have weight $(1-K)W_1 + KW_2$
$R_2$ will have weight $(1-K)W_2 + KW_3$
$R_3$ will have weight $(1-K)W_3$.

Usually, all rules begin with the same or nearly equal weights. After one search, all have nearly unchanged weights except the last rule, $R_3$. After two such searches along the same path, $R_2$ will have a substantially changed weight. The three rules will then have weights:

$R_1$ will have weight $(1-K)^2W_1 + 2(1-K)KW_2 + K^2W_3$
$R_2$ will have weight $(1-K)^2W_2 + 2(1-K)KW_3 + KC$
$R_3$ will have weight $(1-K)^2W_3 + (1-K)C + C$.

The code of the hill climbing rule-based searcher can be modified easily, so that it incorporates this form of learning. In this version of the algorithm, we shall forget about names and the trace. The learner is incremental. In fact, its code is incorporated into the searcher, so that the two are rather hard to tell apart. The algorithm includes an extra variable called *LR*, for 'Last Rule', which is set to the rule which generated the current state $N$.

$N := start$
$LR := a\ dummy\ value$

WHILE $N$ is not a goal and $N \neq fail$
   $CS :=$ the conflict set
      IF $CS = \{\ \ \ \}$
   THEN
      $N := fail$
   ELSE
      $R :=$ a rule in $CS$ with greatest weight
      *Weight of LR* :=
         *Weight of LR* $+ K \times$ *Weight of R*
      *Weight of R* :=
         $(1 - K) \times$ *Weight of R*
      $LR := R$
      $N :=$ *Result of R's action applied to N*

IF $N$ is a goal
THEN
   *Weight of LR* := *Weight of LR* $+ C$.

### Strengths of the bucket brigade algorithm

- It is incremental, even to the extent that learning can take place during a search.

- It does not require any memory. Past examples are not saved. In fact, it does not even require memory for the trace. Some learning algorithms do.

- It accepts noisy data. If one search misleads it and sets the weights a bit wrong, then later learning efforts can correct the mistake.

- The weights remain bounded.

### Limitations

- It is slow. If a search path to a goal takes $n$ steps, then the first rule on the path will gain no credit until the $n$th traversal of the path.

- It is even slower at learning from failures at search. A poor rule will not be penalized until the $N$th attempt at search, where $N$ is the number of states in the entire search subspace accessible through this rule. Until then, the poor rule may even gain weight.

- Even when the first rule eventually gains credit, the amount of credit it gains is only

$$K^n C$$

where $n$ is the depth of the goal. This is not much, if $n$ is large.

- It is sensitive to the order of examples.

- It never stops learning.

- It is only good for learning weights and weights are not always a reliable way to resolve conflicts. If a rule is good in some circumstances and bad in others, it will sometimes be missed when it should be used if its weight is low, but it may be misapplied if its weight is high. What we really want is an algorithm which learns improved conditions for a rule. We shall meet some such algorithms, but not yet.

### 3.3.2 THE HIERARCHICAL BUCKET BRIGADE ALGORITHM

This next version of the bucket brigade algorithm seems to be the best way to solve the credit assignment problem when conflicts between rules are resolved with simple weights.

Several flaws in the simple bucket brigade algorithm arise only if the search path is long. When paths are short, every rule on the path will gain an improved weight after just a few searches and, for a good rule, the gain in weight will be significant even if the rule is first on the path. The hierarchical algorithm is designed to keep all paths short.

The idea is this. The basic search algorithm is as before, except that actions are of two sorts.

(1) Basic actions are the primitive operations which change state in the search space.

(2) Hierarchical actions are themselves search problems, which can be solved just like the main search. Whenever conflict resolution selects a rule whose action is hierarchical, it sets up a new search problem and solves it by calling *search* recursively. When an action is hierarchical, of course its rule $R_1$ will gain weight $KW_2$ from the following rule $R_2$, just as before. The trick is, this same gain in weight is passed as the credit $C$ to the recursive call of search. Thus, at each lower level of the hierarchy, the last applied rule also gains credit proportional to the weight of $R_2$.

*Example*

Let us suppose that:

- the whole problem is solved by rules $R_1$ and $R_2$, which both have hierarchical actions;

- $R_1$ has constituents $R_{1,1}$ and $R_{1,2}$, which are both basic;

- $R_2$ has constituents $R_{2,1}$ and $R_{2,2}$;

- $R_{2,2}$ is basic;

- $R_{2,1}$ is hierarchical, with constituents $R_{2,1,1}$ and $R_{2,1,2}$ which are both basic.

To start with, all rules have weight 16. To make the numbers simple, suppose that $C = 8$ and $K = 1/4$. After successive successful searches, the various rules' weights are shown in Table 3.1.

**Table 3.1**

| | | | | |
|---|---|---|---|---|
| $R_1$: 16 | | $R_2$: 20 | | |
| $R_{1,1}$: 16 | $R_{1,2}$: 16 | $R_{2,1}$: 16 | | $R_{2,2}$: 20 |
| | | $R_{2,1,1}$: 16 | $R_{2,1,2}$: 16 | |
| $R_1$: 17 | | $R_2$: 23 | | |
| $R_{1,1}$: 16 | $R_{1,2}$: 17 | $R_{2,1}$: 17 | | $R_{2,2}$: 23 |
| | | $R_{2,1,1}$: 16 | $R_{2,1,2}$: 17 | |
| $R_1$: 18.5 | | $R_2$: 25.25 | | |
| $R_{1,1}$: 16.25 | $R_{1,2}$: 18.5 | $R_{2,1}$: 18.5 | | $R_{2,2}$: 25.25 |
| | | $R_{2,1,1}$: 16.25 | $R_{2,1,2}$: 18.5 | |

Thus, every rule gains credit after at most three searches. If searching was done with just the basic actions, then $R_{1,1}$ would only gain its first bit of credit after the fifth search, and then it would gain a mere 1/64 of a point of credit.

If we are cunning, we can design the hierarchy so that every path at every level is just two or three steps long. Then at each level, credit is passed back after just a few attempts at search. Suppose each path is just two steps long and there are $D$ levels in the hierarchy, then every rule will have gained credit after $D + 1$ successful searches. By contrast, there are $2^D$ basic operations at the lowest level, so if instead all the basic operations were applied directly, at a single level, the first of them would only gain credit during the $2^D$th successful search. What is more, in the hierarchical algorithm, early rules gain much more credit.

A rule whose action is basic will appear just like the rules above for the straightforward version. A rule with hierarchical action will have a form:

IF *Conditions*
THEN *Subgoal_Conds*
NAME *Name*
WEIGHT *Weight*

where *Subgoal_Conds* is a condition describing the goal of the subtask which this rule sets up.

This hierarchical version can be encoded quite easily. All the essential work is done by a function called *hbba* whose value is the goal, if the algorithm finds one, or else *false*. The other major difference from the straightforward version is that the variable *LR* now stores a list of all rules which helped to set up the current state. The rule at the back of this list will have a basic action. All the others before it will have hierarchical actions, each setting up the subsidiary search which was solved by the following rule. *hbba* accepts as inputs

- the current state, $N$;

- a specification of the current (sub)goal;

and answers

- a pair $(G, LR)$ where $G$ is a goal and $LR$ is a list of rules leading to $G$; or

- (*fail*, [ ]) if there is no such pair $(G, LG)$.


Top level:
   $(N, LR) := hbba\ (start, Goal\_Conds)$
   IF $N \neq fail$
   THEN
      FOR each $R$ in $LR$
        $Weight\ of\ R := Weight\ of\ R + C$
   RETURN $N$


Recursive function $hbba\ (Current\_Node, Goal/Subgoal\_Conds)$:
   $N := Current\_Node$
   $LR := [\quad]$

   WHILE $N$ does not satisfy *Goal/Subgoal_Conds* and $N \neq fail$
      $CS :=$ the conflict set
      IF $CS = \{\quad\}$
      THEN
        $N := fail$
      ELSE
        $R :=$ a rule in $CS$ with highest weight

        FOR each rule $R'$ in $LR$
          $Weight\ of\ R' := Weight\ of\ R' + K \times Weight\ of\ R$
        $Weight\ of\ R := (1 - K) \times Weight\ of\ R$

        IF *Action of R* is basic
        THEN
          $N := Action\ of\ R\ on\ N$
          $LR := [R]$
        ELSE
          $(N, SLR) := hbba\ (N, Subgoal\_Conds\ of\ R)$
          $LR := [R \mid SLR]$

   $hbba := (N, LR)$

## 3.4    BOARD GAMES

One can play board games with rules, rather like the rules used for searching. Such a rule is called a **weighted pattern**. It takes the form

IF a part of the board position matches a pattern $P$
THEN credit this position with weight $W$.

The weights $W$ may be positive or negative.

The key to the analysis of a position is what is called its **static value**. The static value of a position is found by

1.   finding all ways in which all patterns can match the board;

2.   adding the weights. Each pattern's weight $W$ is added once for each occurrence of the pattern $P$ on the board.

The form of a pattern depends on the particular game. In noughts and crosses, a pattern may contain

●   some squares filled with $\bigcirc$s;

●   some squares filled with ×s;

●   some squares which are empty.

A pattern need not cover the whole board; thus in noughts and crosses, a pattern may cover just a row or just a diagonal. In a more complicated game like chess, a square in a pattern may be labelled as just 'filled', or 'filled with a white piece', rather than stipulating precisely which piece. Also, some squares in the pattern may only match a square on the edge of the board.

*Example*: Playing 'go' by pattern matching

There is another board game called go which has been played in China for millennia. The old Chinese army used it to train their generals in strategy. The objective is to win empty space. You have won a bit of space if your opponent cannot play into it without eventually being captured. The rules are beautifully simple, but its branching ratio is 300 or more and it is more subtle than chess.

In this game, opposing groups attack each other by surrounding and touching. The board can be represented in two ways: the low level of board positions and pieces and the higher level of groups (Fig. 3.1). This higher level is a graph. Each group is portrayed by a node, which is labelled with the group's essential characteristics, such as its size and how much free space it has round it. Between any two interacting groups, there is an arc which is labelled with the degree of interaction. In addition, an area of free space may also be represented by a node, with arcs from it to the groups surrounding it.

Since the branching ratio is so great, search at the lower level is a poor playing strategy. An alternative is to choose a 'sensitive region' at the higher level, and

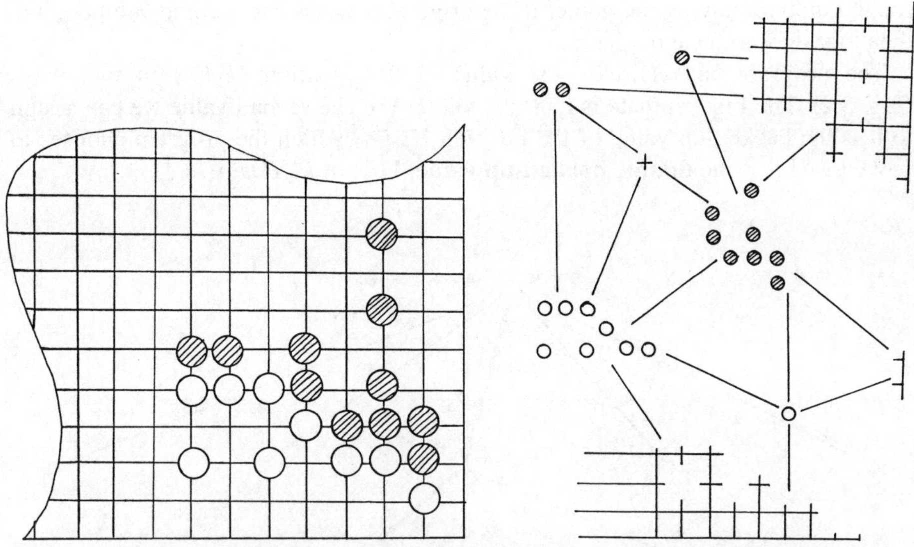

**Fig. 3.1**  Part of a position in the game of go and its high-level graph.

then only explore low-level moves there. A player can find sensitive centres by pattern matching in the high-level graph.

Suppose a program contains such patterns. Assume we are at a board position called *MyPos*, and it is the program's turn to play. We adopt a standard strategy called **two-ply minimax search**. Nilsson's book describes it. The program considers every possible move which the laws of the game allow it to make, and arrives at various positions which we may call *HisPos*. Then, for each such move, it considers each possible move that the opponent might make, leading to positions called *MyNewPos*. (In practice, there are ways of avoiding calculation of all possibilities.)

The program calculates the static value of each position *MyNewPos*. From these, it estimates a backed-up value for each HisPos: the **backed-up value** of *HisPos* is the *minimum* of all static values of positions *MyNewPos* derived from *HisPos*. The move which the program chooses is the one which leads to a position *HisPos* with *maximum* backed-up value.

As an exercise, try writing some such weighted patterns for noughts and crosses, and find out what moves they will lead the minimax algorithm to choose.

### 3.4.1  LEARNING BY DEEPER SEARCH

That is the playing part. Now we come to the learning algorithm. As with the bucket brigade algorithm, we want to learn a better set of weights for the patterns. The principle is this. We assume, to start with, that the weighted patterns provide a rough guide to playing the game. If the original patterns and weights are no good, then this algorithm cannot learn.

The program can estimate the value of the position *MyPos* in two ways (Fig. 3.2). The first estimate is its static value, *Vs*. The second value we can assign to it is the backed-up value of the position *HisPos* which the program chooses to move to. This is the **double backed-up value**, *Vdb*, of *MyPos*.

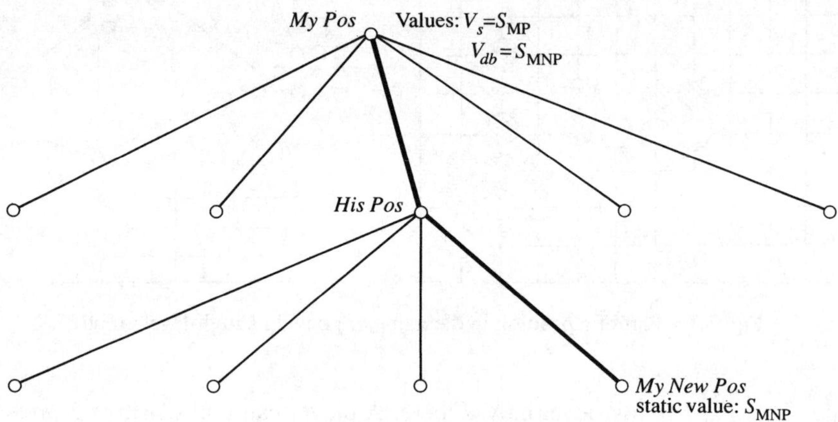


**Fig. 3.2**   Learning by deeper search.

One would expect that *Vdb* is a better estimate of the true worth of *MyPos*. The static value *Vs* depends solely on the patterns, while *Vdb* exploits both the patterns and some search. The search is likely to reveal more about the future course of the game than the patterns alone. The algorithm runs thus:

IF *Vs* > *Vdb*
THEN
    the static value of *MyPos* is probably too high, so decrease the weight of each pattern which matches *MyPos*.

IF *Vs* < *Vdb*
THEN
    increase the weight of each pattern which matches *MyPos*.

This algorithm is applied every time it is the program's turn to move. It is extremely crude. As always, there is a credit assignment problem. At any one position, there may well be a key pattern which captures the critical features of play—but we have no way of finding it. For lack of a better alternative, all patterns matching the position are given equal credit. The amount of the increase or decrease is a matter of preference. It might be proportional to (*Vs* − *Vdb*).

### 3.4.2 LEARNING BY WATCHING THE OPPONENT

Needless to say, there are better learning algorithms for weights of patterns. One such is based on the following idea. If the opponent is a weak player and the program is strong, then the opponent's choice of move will usually be worse than the one leading from *HisPos* to *MyNewPos*. The program's static estimate of *MyNewPos* will be nearly right and the program can ignore the opponent's play. However, if the opponent is strong and the program is weak, the opponent may realize that his best move from *HisPos* is not the one leading to *MyNewPos*. Suppose instead, he chooses to move to position *CorrectNewPos*.

The program will then calculate *Vdb* for *CorrectNewPos* and discover that the opponent's evaluation of it is more accurate than its original static value (Fig. 3.3).

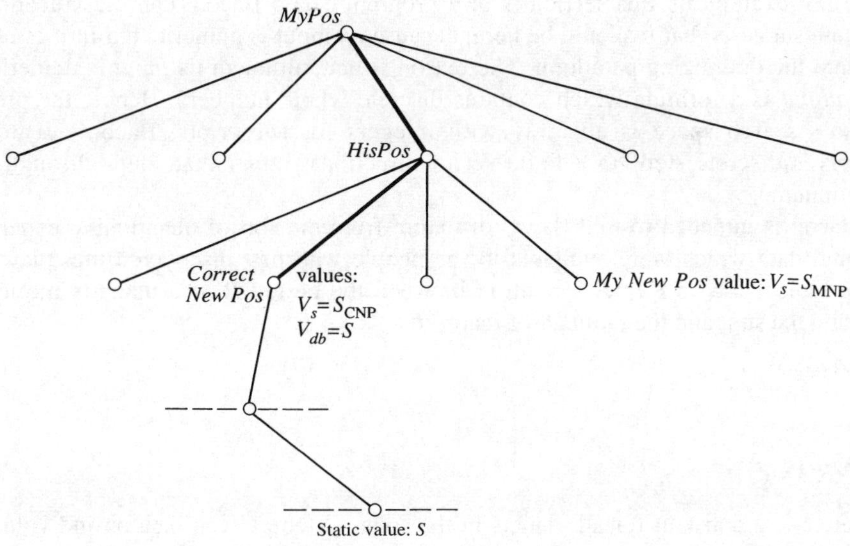


**Fig. 3.3**  Learning by watching the opponent.

If

$$Vdb\ (CorrectNewPos) < Vs(MyNewPos)$$

then the program made a mistake. Both *Vs(MyNewPos)* and *Vs(CorrectNewPos)* should be corrected. The algorithm is:

After each of the opponent's moves,
  IF *CorrectNewPos* is not *MyNewPos*
  AND *Vdb(CorrectNewPos)* < *Vs(MyNewPos)*
  THEN
      increase the weight of each pattern
          which matches *MyNewPos*;
      decrease the weight of each pattern
          which matches *CorrectNewPos*.

Of course, learning weights on their own is not enough. A skilled learner will invent new patterns as well. That is beyond the scope of this section.


## 3.5  DISCOVERING PHYSICAL LAWS: WHERE THE CLASSIFICATION IS FLAWED

The last example in this section is of a program called Bacon. Our classification system suggests that it should be here, because its input is numeric, but it does not fit into the optimizing paradigm. The reason is that, although its *input* is numeric, its *output* is a formula which contains discrete whole numbers. Hence, the program's search space is an array with integers for subscripts. Bacon's search involves discrete steps in a finite set of directions, rather than steps through a continuum.

Bacon is intended to find 'laws' of nature from the sort of quantitative experimental data which were available to the people who first discovered these laws. One typical law is Kepler's relation between the period $P$ of a planet's motion around the sun, and the radius $R$ of its orbit:

$$P^2 = cR^3$$

or

$$P^2R^{-3} = c$$

where $c$ is a constant for all planets in the solar system. Given data on individual planets, Bacon finds the constant $c$ and the powers, 2 and $-3$, of $P$ and $R$.

Bacon searches through the space of such formulae. The more interesting aspect of its search is the part which finds the whole numbers, the powers of $P$ and $R$. During the search, Bacon constructs intermediate approximations to the right powers, all of the form

$$P^iR^j$$

where $i$ and $j$ are whole numbers. In a grossly simplified form, the essential rules are

IF a known quantity $A$ is constant in all cases
THEN
    conjecture a law:
      $A = c$
    and stop.

IF two known quantities $A$ and $B$ are linearly related
THEN
    conjecture a law:
      $A = kB + c$
    and stop.

IF two known quantities $A$ and $B$ increase together
THEN
    invent a new quantity
      $C = A/B$.

IF one known quantity $A$ increases while another $B$ decreases
THEN
    invent a new quantity
      $C = AB$.

If the input is the (fictitious) data for three planets:

| $P$ | $1$ | $8$ | $27$ |
|---|---|---|---|
| $R$ | $0.3$ | $1.2$ | $2.7$ |

then the program's steps will generate

| $A$ | $3.333$ | $6.666$ | $10$ | $(P/R)$ |
|---|---|---|---|---|
| $B$ | $11.11$ | $5.555$ | $3.704$ | $(A/R)$ |
| $C$ | $37.04$ | $37.04$ | $37.04$ | $(AB)$ |

and declare the law:

$$P/R \times (P/R)/R = 37.04$$

in a suitably tidied form.

Even the first version of Bacon is subtler than this summary suggests. It contains sixteen basic rules, which include some for overall control and some which collect data. It avoids re-inventing a quantity. The entire system is goal directed, although the search through the space of formulae is forward. Conflicts are resolved by choosing a rule instance which refers to the most recently created quantity. This leads to a depth-first backtracking search.

In its later versions, Bacon is much cleverer than this. It could:

* invent dependent parameters. If given three or more parameters as input, it can hold all but two constant and discover a law relating these two, then treat the

constants $(c, k)$ within this law as new parameters and relate them to the other inputs. This involves a recursive hierarchical search.

- exploit expectations. In the version with dependent parameters, once Bacon has found a form of law relating two inputs while others have particular settings, it will expect that these two are always related by a law of the same form in which only the values of $c$ and $k$ vary.

- invent intrinsic properties. In the example above, the constant 37.04 depends on the mass of the sun. If we could observe several different solar systems around other stars, the resulting constants would vary. Later versions of Bacon can recognize that each star has an intrinsic property which manifests itself as this 'constant'.

- perform what physicists call dimensional analysis, and what programmers call type checking. Two quantities can only be equal if they contain the same powers of space, time, and mass.

In addition, some versions could find periodicity and other forms of regularity.

Bacon is unusual because its input and output are both from spaces with exceptional structure. The input space is given by real numbers. The output is a cross between real numbers (the constants $c$ and $k$) and integers (powers of $P$ and $R$). There are very few published accounts of learning programs which search in spaces parametrized by integers. Bacon could be discussed here, since its input is numeric, or in the section on correlation and clustering since it searches for correlations. Perhaps it is best viewed as a correlator. It is described here so that you can see as soon as possible that a learning program may have many facets and classification is fraught with subtlety.


## FURTHER READING

Siddall (1982) describes in detail several methods of control theory, with examples and implementations in Fortran.

The experiment with a robot, on learning by remembering, was performed by Bill Clocksin and Andrew Moore, who describe it in their joint paper (1989). Other details appear in Moore's thesis (1990). Learning by remembering is applied in many situations and under different names. The method of *chunking*, described in Chapter 7, is one form of it. Another has been tested by P. T. Breuer (1991), who used it to speed up a game-playing program. Andersen and Jones (1994) formulate a general form, applicable to many imperative programs.

The bucket brigade algorithm appears, among other places, in Booker, Goldberg, and Holland (1989). The hierarchical version is due to Stewart W. Wilson (1987).

Bacon is described by Langley, Simon, Bradshaw, and Zytkow (1987) who give an unusually thorough account of their experiments. Later developments appear in Langley and Zytkow (1989).

# EXERCISES

1.  The code below is the essential part of a Prolog program for forward search without
    backtracking, in which conflicts between rules are resolved with weights. Extend
    this code so that it learns better weights by the elementary bucket brigade
    algorithm.

    ```
    search (Goal, Goal, _) :-
      goal_conds (Goal).

    search (State, Goal, Rules) :-
      find_conflict_set (State, Rules, Conflict_set),
      find_best_rule (Conflict_set, Rule),
      !,
      Rule = rule (_, Action, _, _),
      act (Action, State, Next_state),
      !,
      search (Next_state, Goal, Rules).

    find_best_rule (Conflict_set, Rule) :-
      member (Rule, Conflict_set),
      Rule = rule (_, _, _, Weight),
      not better_rule (Conflict_set, Weight).

    better_rule (Conflict_set, Weight) :-
      member (Better, Conflict_set),
      Better = rule (_, _, _, Better_weight),
      Better_weight > Weight.
    ```

    You can assume that the code for *goal_conds* and *find_conflict_set* and *act* is given.
    In the revised code, *search* will have an extra argument:

    ```
    search (State, Goal, Rules_in, Rules_out)
    ```

2.  A searcher should calculate improved weights with the hierarchical bucket brigade
    algorithm. Suppose that it is presented with a task which can be solved in just five
    basic operations. What is the best way to compose these into hierarchical actions,
    so that learning will occur as quickly as possible?

    Show that any successful sequence of basic operations can be composed into
    hierarchical ones so that the hierarchical bucket brigade algorithm will award credit
    to every action in the sequence after just two successful searches. How will the
    search's branching ratio with these hierarchical operations compare with search
    with just the basic operations?

    When you have studied algorithms for learning in rule-based systems, suggest
    one by which such hierarchical operations might be learned.

3.  Here is a simple game for two players, one with white counters and the other with
    black. The 'board' consists of five columns set side by side. Each column can hold
    five counters. To start, all five columns are empty. The players take turns to drop a
    counter down a column. The winner is the one who first constructs a line (vertical,

horizontal, or diagonal) of four counters of his shade. For instance, the following positions are both won by white:

(a)   Design a set of patterns which are relevant to this game. (For example, one such pattern might consist of two white counters and an empty position in a horizontal line. Another consists of a counter of any shade or the bottom of a column under the empty position.)

(b)   Predict which patterns should have largest weights.

(c)   Assign the weight 2 to each pattern which you think is likely to occur in good positions for white and −2 to all other patterns.

(d)   Play the game a few times. One player should play intuitively. The other should play by the two-ply minimax method, using the static value of each position calculated with the weighted patterns. During play, adjust the weights of patterns by deeper search.

(e)   Check whether the patterns which you predicted in (b) are actually the ones which gained most weight.

Eventually, one hopes, the weights will reach nearly stable values and the algorithm's rate of learning will decrease. If this does not happen, it is probably because there are not enough relevant patterns. This exercise will probably be easier if the algorithmic player is actually encoded as a program, but it can be done with pencil and paper.

4.   Repeat Exercise 3, with the same patterns, but modify the weights by watching the opponent rather than by deeper search.
     Play the game, with *both* players using weighted patterns and the minimax method. One player uses the weights learned by deeper search, and the other with weights learned by watching the opponent. Which wins?

# 4 Association and neural networks

## SUMMARY

This chapter presents a selection of learning algorithms which handle large sets of similar simple attributes. The ones presented are:

- an individual neural node;
- layered perceptrons and the back-propagation algorithm;
- WISARD;
- logical neural nets;
- Boltzmann machines and annealing;
- Kohonen's feature maps;
- reproduction from a single parent, with mutation;
- genetic algorithms.

These algorithms have very different characteristics. We shall see how the behaviour of each depends on:

- the space it searches;
- its degree of heuristic guidance.

We shall also see how they compare with natural minds.

## INTRODUCTION

We come now to the class of learning algorithms which, some proponents claim, most closely depicts the workings of human minds. The input to such an algorithm is a set of values for very many attributes. All the attributes are of the same type. The only distinguishing feature of two different attributes is that they are in different positions or such like, which has no bearing on their possible values. Each value is simple—typically either 1 or 0, or 'yes' or 'no', or a real number.

The prototype for such a situation is the input to the nervous system behind a human eye. This comes from the retina, which consists of very many cells which each 'fire' when illuminated. (Retinal cells are not all the same, so the eye is a bit more complicated than our idea of the algorithm, but the difference is not too great.) Behind the retina is a cascade of other cells, neurones, which are all physiologically of just a few types. Each can receive impulses from other cells and each will fire and so stimulate other neurones if its own inputs trigger it. The subtlety of mental processing is due to the layout of links between neurones.

This chapter will discuss seven kinds of device, representing the major computational models in the field. The first, the **simple perceptron**, is one of the oldest in the class. Its output depends on a weighted sum of its inputs. It is indeed simple, but its basic form is limited. The simple perceptron is the basic ingredient of **multi-layer perceptrons**. The *WISARD* machine and **logical neural nets** have structure and function resembling multi-layer perceptrons, but the individual nodes in them are made from one-bit random access memories. The so-called **Boltzmann machine** is more elaborate and perhaps more versatile, but slow. **Kohohen nets** are designed to invent classifications. The **genetic algorithms** form a separate family which do not involve anything resembling neurones, but they are included here because they take their data in a very low-level form, like the input to a neural net.

## 4.1   PERCEPTRONS

The term **perceptron** has been used by different people to mean different things, but its definition seems to have stabilized by now. A simple perceptron is intended as a model of a single neurone. It is a simple computational unit, sometimes called a 'node' (see Fig. 4.1). One node has many inputs and a single output. There are various kinds of node which differ in the forms of their inputs and output and how the output depends on the inputs. We shall start with the simplest.

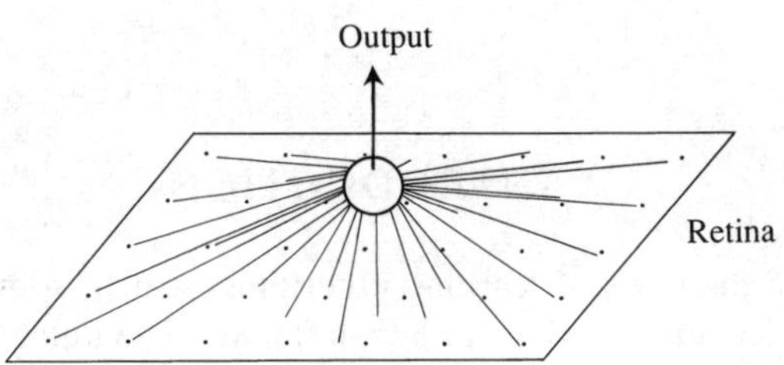


**Fig. 4.1**   A single node.

In this kind, each input is 0 or 1, which you can interpret as 'no' or 'yes'. Its output is similarly 0 or 1. When its output is 1, we say that it **fires**. You can visualize the inputs as coming from an array of photo-electric cells, which we call its retina. Note that in this picture, each photo-cell yields just 0 or 1, not values in between. Thus, it is as if the image on the array of photo-electric cells is a silhouette. (This paradigm is very common, but there is no reason why the inputs should lie in a two-dimensional array. They might form an array of some higher dimension, or they might not have any array-like structure at all. In the brain, the olfactory system which analyses scents appears to be composed of a sequence of layers of nodes with no obvious geometric structure.)

Let us call the node's inputs

$$s_1, s_2, \ldots s_N$$

where $N$ is some number which may be large. Each input has an associated weight

$$w_1, w_2, \ldots w_N.$$

These weights are real numbers. The node also has a **threshold** $t$ which is another real number. The node's output is

1     if $s_1 w_1 + s_2 w_2 + \ldots + s_N w_N > t$;

0     otherwise.

Within the node, knowledge is encoded by the weights and the threshold. In fact, the threshold is redundant. You can always replace it by an extra 'input' $s_0$ whose value is always 1 and a corresponding 'weight' $w_0$ whose value is $-t$. Thus, the modified node's behaviour depends on the sum

$$D = s_0 w_0 + s_1 w_1 + \ldots + s_N w_N$$

and it produces

1     if $D > 0$;

0     otherwise.

Any perceptron **recognizes** a certain class of all its possible inputs. This class, usually called $C$, is

$$\{ \, (s_0, s_1, \ldots, s_N) \mid D > 0 \text{ and } s_j \in \{0, 1\} \text{ for each } j \leq N \, \}$$

It is the class of retinal images which make the perceptron fire.

A node learns by having its weights changed. There is a well-known simple algorithm for doing this. Its input is an infinite sequence of images

$$I_1, I_2, I_3, \ldots$$

on the array of inputs, each supplied with

'yes'     if it is in the class $C$ which the node should learn to recognize;
'no'     if not.

We assume that the node is in the second form, so its threshold is 0. The algorithm runs thus:

> REPEAT, for each input $I_j$ in turn,
>     IF $D(I_j) > 0$ but $I_j$ is not in the desired class $C$
>     THEN
>         replace each $w_n$ by $w_n - s_n(I_j)$;
>     IF $D(I_j) < 0$ but $I_j$ is in the desired class $C$
>     THEN
>         replace each $w_n$ by $w_n + s_n(I_j)$.

The first alternative corrects all weights which are too high and so cause the node to recognize an image not in $C$. The second corrects weights which are too low. Whenever the node classifies an input correctly, its weights do not change. A weight may be modified by both alternatives several times before it approaches a stable value.

This learning algorithm has the following pleasant property:

*Convergence theorem*
   *Suppose that there do exist weights*

$$w'_0, w'_1, w'_2, \ldots, w'_N$$

   *such that the sum*

$$s_0 w'_0 + s_1 w'_1 + \ldots + s_N w'_N$$

   *is > 0 whenever the input $I$ is in the class $C$ and is < 0 whenever $I$ is not in $C$; then, for any initial weights $w_i$ and for any training sequence $I_j$, there is some number $k$ for which the perceptron classifies $I_j$ correctly for all $j > k$.*

That is to say, if *any* node is capable recognizing the concept $C$, then after some limited time, the given node will complete its learning. However, the theorem does not let us recognize when this stage is reached.

This theorem should be used with care. Recall, the node has just a finite (though large) set of inputs $s_i$, and each input can only be 0 or 1. Hence there is only a limited set of possible instances $I$. The sequence $I_j$ may be infinite, but after some point, no new instances can occur. This theorem cannot be relied on in other cases.


## 4.2   MULTI-LAYER PERCEPTRONS

The very simplest kind of perceptron consists of a single node. The simplest form which is commonly considered has two layers of nodes. The lower layer contains many nodes. Their inputs come directly from the retina. A single photo-cell on the retina may be an input to several nodes. The second layer consists of just one node and its inputs are the outputs of all the nodes in the first layer. The perceptron's output is the output of this top node. It is intended to yield 1 if the image on the retina is of a particular kind. For example, it might be 1 if the scene on the array of photo-cells is convex and 0 otherwise.

The word 'perceptron' can mean just one isolated computational node, like a single neurone; but it is often taken to mean an entire device with many layers and many nodes in each layer. A single node will recognize something if at least a certain number of its inputs are on, so it is basically a counting device. A two-layer perceptron (see Fig. 4.2) can recognize convex areas on its retina. A three-layer perceptron can recognize arbitrary polygonal areas, which may be neither convex nor even connected.

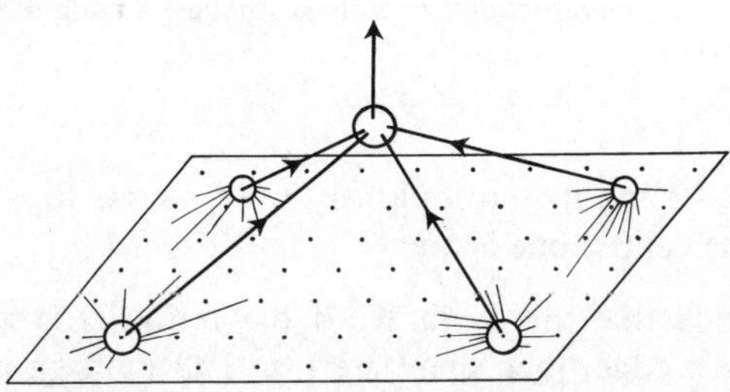

**Fig. 4.2**   A two-layer perceptron with four nodes in the bottom layer.

It is not clear *a priori* what classes can be recognized by any perceptron. One perceptron on its own is limited. For instance, no single node can recognize the class of images consisting of just one dot anywhere on the retina. For any finite class $C$, one can make a two-layer perceptron which recognises $C$. Minsky and Papert studied in some detail how elaborate a perceptron must be in order to recognize a particular class.

There are some simple and interesting restricted forms of layered perceptron. For instance, one can place conditions on each node in the first layer. The device is said to be:

- **order limited** with limit $k$ if each node in the first layer has at most $k$ inputs;

- **diameter limited** with limit $d$ if each node in the first layer can have any number of inputs, but all its input photo-cells lie in some disc of diameter $d$.

*Example: recognizing convex areas*

There is a two-layer order-limited perceptron with limit 3 which can recognize the class of *convex* images (Fig. 4.3). An image is convex if every straight line with both ends in the image is actually entirely in the image.

The perceptron's upper layer contains just one node, with an input from each node in the lower layer. It fires if every node in the lower layer does not fire. Thus, all weights on its inputs might be set to $-2$ and its threshold $t$ may be $-1$.

In the lower layer, there is one node for each set of three distinct collinear points in the retina. Its weight on the middle input is $-2$ and its two outer inputs

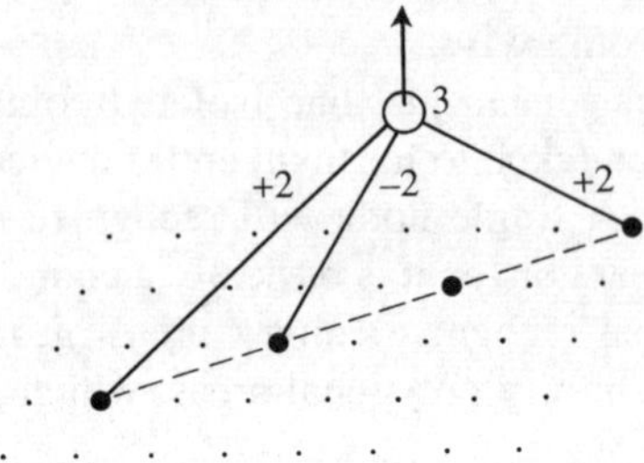

**Fig. 4.3**   The perceptron which recognizes convex images: a node in the lower layer with input weights and threshold.

both have weight +2 and its threshold $t$ is 3. This node fires if the two ends are illuminated but the central one is not.

Even a simple device like this is large. If the retina is a square array with $n$ photocells along each edge, then there are $n(n-1)/2$ pairs in it which can be the ends of a line. On each such line, there are on average some $c \times n$ possible middle points, where $c$ is a number independent of $n$. Hence $K$, the number of nodes in the lower layer, is of order

$$K = O(n^3)$$

*Example: recognizing sets of rectangles*

There is a three-layer diameter-limited perceptron which can recognize the class of sets of non-overlapping solid rectangles whose edges are oriented horizontally and vertically. As before, the top layer consists of a single node which fires if and only if all nodes in layer 2 do not fire.

Layer 2 has a node for each $2 \times 2$ square of four adjacent photocells in the retina. It will fire if the input from this little square could not be inside or outside or part of an edge or corner of any rectangle. Its inputs come from 10 nodes in layer 1, all with inputs from the same $2 \times 2$ square. Among these ten nodes, one fires iff its inputs are all on, one iff its inputs are all off, four recognize the four possible orientations of an edge, and four recognize the four possible corners.

By contrast, it is generally accepted that a diameter-limited perceptron *cannot* recognize the class of connected images. The argument runs thus: consider the four figures shown in Fig. 4.4. The top two are connected but the bottom two are not; but any small portion of either of the bottom two also appears somewhere in one of the top two, so any node whose inputs are restricted to some small region cannot behave differently for connected and disconnected inputs.

It is possible to build layered perceptrons with many nodes in the top layer. Such a device does not 'recognize' any particular concept. Rather, it calculates a

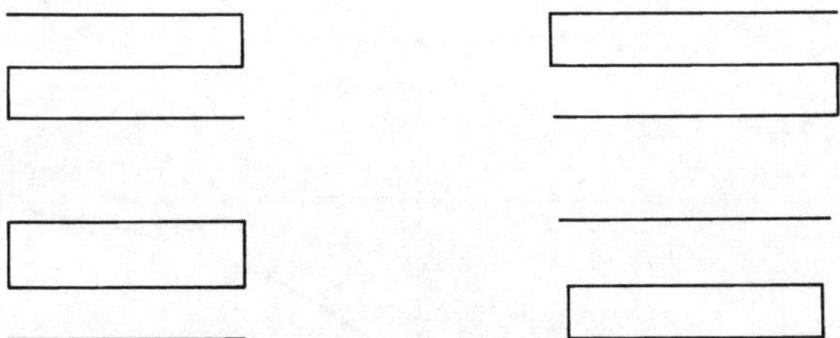

**Fig. 4.4**

function: the function's argument is the set $I$ of inputs to the bottom layer and its value is the resulting set of values of all nodes at the top layer.

### 4.2.1 OTHER FORMS OF PERCEPTRONS

When the inputs to a perceptron

$$s_0, s_1, \ldots s_N$$

can only be 1 or 0, then the possible instances $I$ that it can be shown are the corners of an $N$-dimensional cube. There is no particular reason why input should be so restricted. We can also let any point inside the cube be input too. We just let each $s_j$ be a real number:

$$-0 \le s_j \le 1.$$

Many people design perceptrons thus. Which kind of input is best depends on what the perceptron is for. Plain 0/1 input is better if the perceptron is supposed to classify silhouettes, since then, each point on the retina is black or white. If the image has grey shading then each $s_j$ can take intermediate values.

Recall, the output of our first simple perceptron is a step function:

$$f(D) = \begin{cases} 0 & \text{if } D \le 0 \\ 1 & \text{if } D > 0 \end{cases}$$

where, as before, $D$ is the sum of weighted inputs:

$$D = s_0 w_0 + s_1 w_1 + s_2 w_2 + \ldots + s_N w_N.$$

Another variety of perceptron can have any number in between 0 and 1 for its output too. There is a common model used by theorists, called a **sigmoid** non-linearity. The output of a sigmoid perceptron is

$$f(D) = \frac{e^D}{e^D + 1}.$$

This is always between 0 and 1 and never quite either. The function $f$ has the pleasant property that

$$f(-D) = 1 - f(D).$$

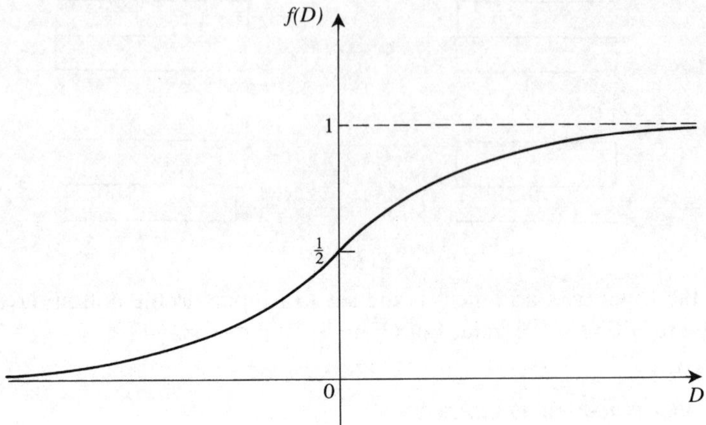

**Fig. 4.5**   The output of a sigmoid node.

We say that this perceptron **recognizes** the class of inputs for which

$f(D) > 1/2.$

One can invent other varieties, with different forms of output. The essential feature is that the output function $f(D)$ should be

- monotonic increasing;

- at or near 0 for large negative values of $D$;

- at or near 1 for large positive values of $D$.

The virtue of the step function perceptrons we discussed first is that they are very simple, so it is possible to derive theoretical results for them, such as the convergence theorem. However, a sigmoid perceptron can embody more information.

### 4.2.2   THE BACK-PROPAGATION ALGORITHM

This learning algorithm is intended for a multi-layer perceptron in which each node is sigmoid. It will work for a recognizer, with just one node in the top layer, but it is phrased as if the top layer contains many nodes. Suppose that there are $M$ layers. The input 'bottom' layer is called layer 1 and the output 'top' layer is numbered $M$. There are $N_k$ nodes in the $k$th layer. Each training instance consists of inputs to the bottom layer:

$s_0, s_1, \ldots s_{N1}$

and the corresponding desired outputs

$d_0, d_1, \ldots d_{NM}$

from the top layer.

REPEAT
   FOR each training instance
      FOR each layer, starting at layer $M$
         and working down to layer 1
        FOR each node in this layer
           Adjust the weights of this node
UNTIL the output is sufficiently close to the desired output
   for every instance.

Weights are adjusted according to the following rule.

$$w_j := w_j + K \times \text{diff} \times x_j.$$

Here,

$w_j$    is the value of the node's weight for its $j$th input.
$K$    is a positive constant, called the **gain**.
$x_j$    is the $j$th input to this node in the current training instance.

When the node is in the first layer, then $x_j$ is $s_j$. In higher layers, $x_j$ is the output from a node in the previous layer.

diff    is a measure of the error in this node's output.

There is a general method for calculating *diff*, outlined in the survey by Hinton (see the *Further reading* section). The following is a simple approach (see Fig. 4.6).

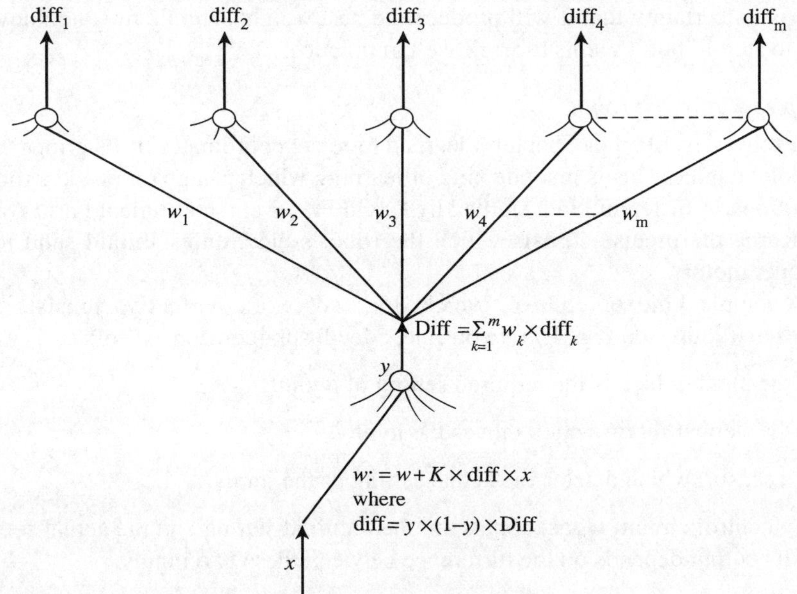


**Fig. 4.6**   A step in the back-propagation algorithm.

Suppose the node's actual output is $y$. If this node is in the top layer and its desired output is $d$, then

$$\text{diff} = y \times (1 - y) \times (d - y).$$

If it is in some lower layer, then

$$\text{diff} = y \times (1 - y) \times \text{Diff}$$

where $Diff$ is a weighted sum of the *diffs* of the nodes in the layer above. Say $w_k$ is the weight on the arc from the current node to the $k$th node in the layer above, and

$diff_k$ is the *diff* for this upper node; then

$$\text{Diff} = \text{diff}_0 \times w_0 + \text{diff}_1 \times w_1 + \ldots \text{diff}_m \times w_m$$

where the sum is over the layer above the current node.

The algorithm calculates *diff* recursively for nodes in successive layers, starting at the top output layer and working back down to the input layer—hence its name, back propagation.

As usual, we have treated the threshold of each node as an extra input from some extra node which is permanently switched on.

This is a hill-climbing algorithm. It is meant to minimize the mean square error of all the output nodes. The gain, $K$, controls its speed of convergence. If $K$ is large, then the algorithm will 'jump to conclusions' quickly, but it may overshoot the optimum weights. Hence, one can expect that it is prone to the canyon problem. There is no certainty that it will produce the best weights and there is no knowing when to stop it, but it seems to work well in practice.

### Example: a different robot

The robot described in Chapter 3 learned to relate coordinates of its gripper with its joint angles. This is just one kind of learning which can go on inside a robot's workings. A different kind, studied by David Fraser and his student Farid Azhar, concerns the precise signals which the robot's electronics should send to its stepper motors.

A simple kind of control, which does not work perfectly, consists of a **feedback loop** (see Fig. 4.7). In outline, a feedback loop consists of:

- the input, which is the required setting of a joint;

- the stepper motor which moves the joint;

- a sensor, which detects the actual setting of the joint;

- a control circuit, whose inputs are the required setting and the actual setting. Its output depends on the difference between these two inputs.

The disadvantage with this design is that, when the robot's joint is loaded, it is never in quite the right position. The circuit only sends the motor an impulse

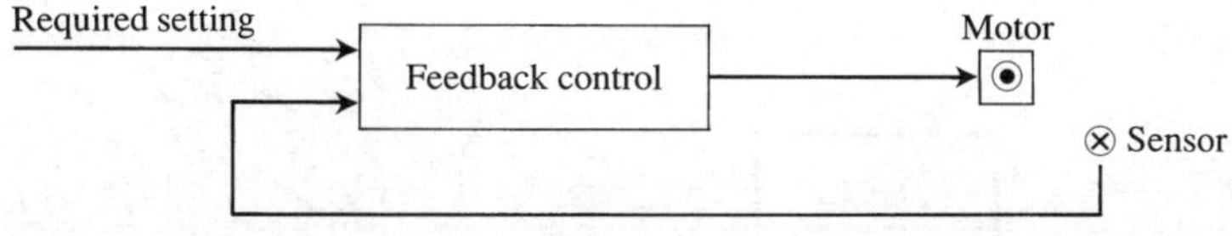

**Fig. 4.7**

when the sensor's output is different from the required setting. The joint always has to sag a bit before the control starts to push it.

A standard improvement on this feedback loop is shown next. It includes an additional **feedforward** control (see Fig. 4.8). This anticipates the inadequacy in the feedback control and provides a supplementary signal. The output from the feedforward control is intended to provide the motor with just the extra push which will bring it to the required setting.

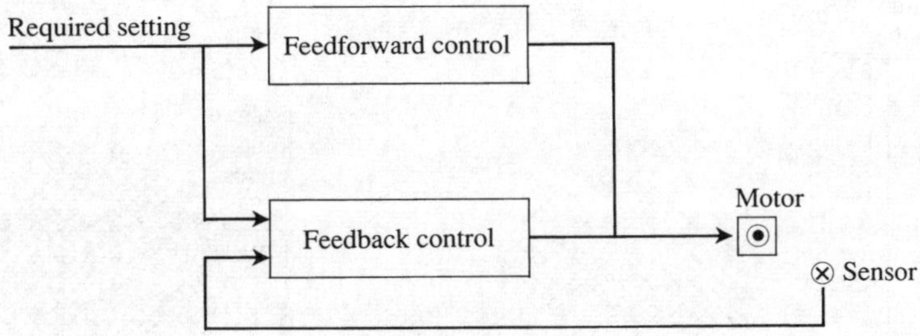

**Fig. 4.8**

The controls are run periodically, three times each second. Both the input required setting and the sensor are sampled once for each cycle. In their implementation, Fraser and Azhar gave the feedforward control a simple memory. The feedforward control calculates its output from the latest value of the required setting and the previous four values too. Thus, the feedforward control can provide one value when the required setting is constant and a different value when the required setting is changing steadily and yet a third value when the required setting is accelerating.

Both control circuits are implemented as layered perceptrons with sigmoid nodes. Their structures are shown in Fig. 4.9. The robot used by Fraser and Azhar actually has three motors for its three joints: waist, shoulder, and elbow. (The gripper was controlled separately.) Hence, each control circuit has three sets of inputs and three output nodes, one for each motor. Internally, each net has nodes with interconnections from all inputs and to all outputs. This allows the output for each motor to depend on the inputs for all three joints.

The feedback net has six inputs. Three are the required settings of the motors and the other three are the observed settings measured by the sensors. These six inputs form the net's retina. Between it and the output layer, there is a single layer of 18 nodes. Each node in this middle layer is connected to all the retinal

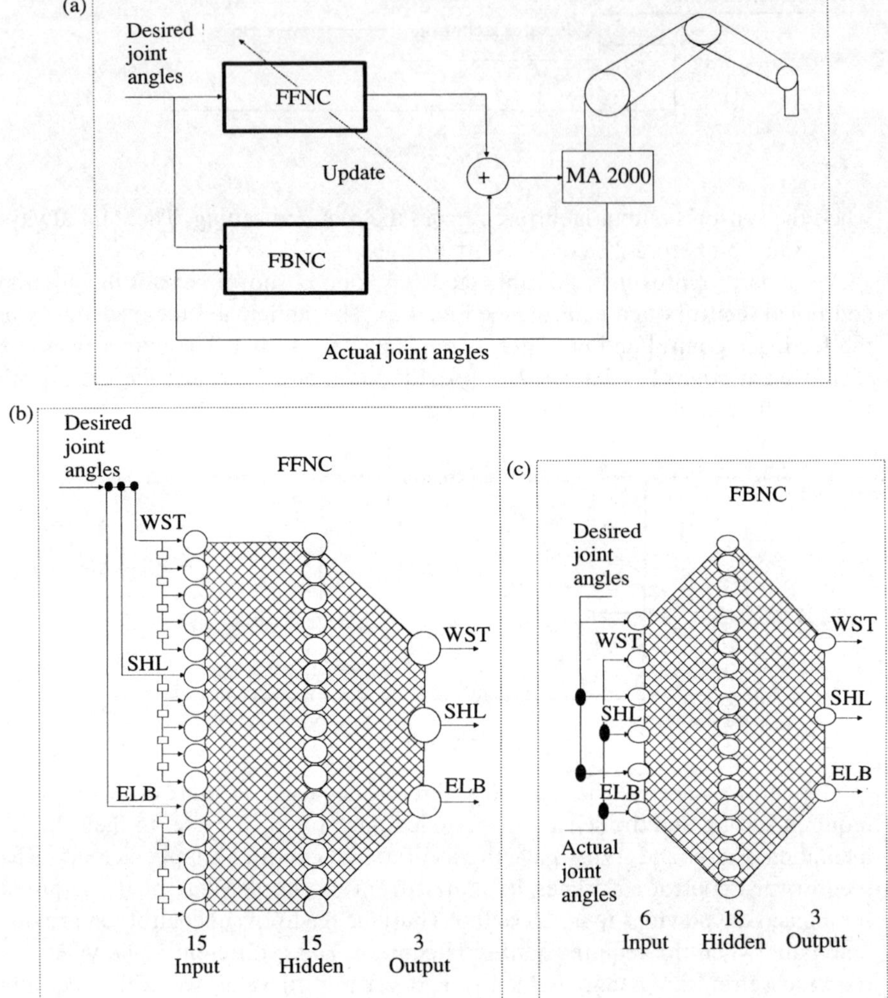



**Fig. 4.9** A robot which learns by back propagation. FFNC: Feedforward neural controller; FBNC: Feedback neural controller.

nodes and all the output nodes. Before this net was connected to the robot, it was trained by the back-propagation algorithm to mimic a standard controller.

The feedforward circuit has a retina consisting of fifteen inputs. At any instant, three of them are the current required settings of the three motors, three are the previous required settings, a third of a second earlier, three are the required settings a third of a second before then, and so on. In the diagram of this net (Fig. 4.9b), the little rectangular boxes on the input lines signify delays. The middle layer also consists of fifteen nodes, each connected to all the inputs and all the outputs.

This net was trained while the robot worked. Its weights were originally set to small random values. The training algorithm is essentially back propagation. It cannot be exactly the algorithm given above because the desired outputs $d$ from the top layer are not known, so it cannot calculate

$$y\,(1-y)\,(d-y).$$

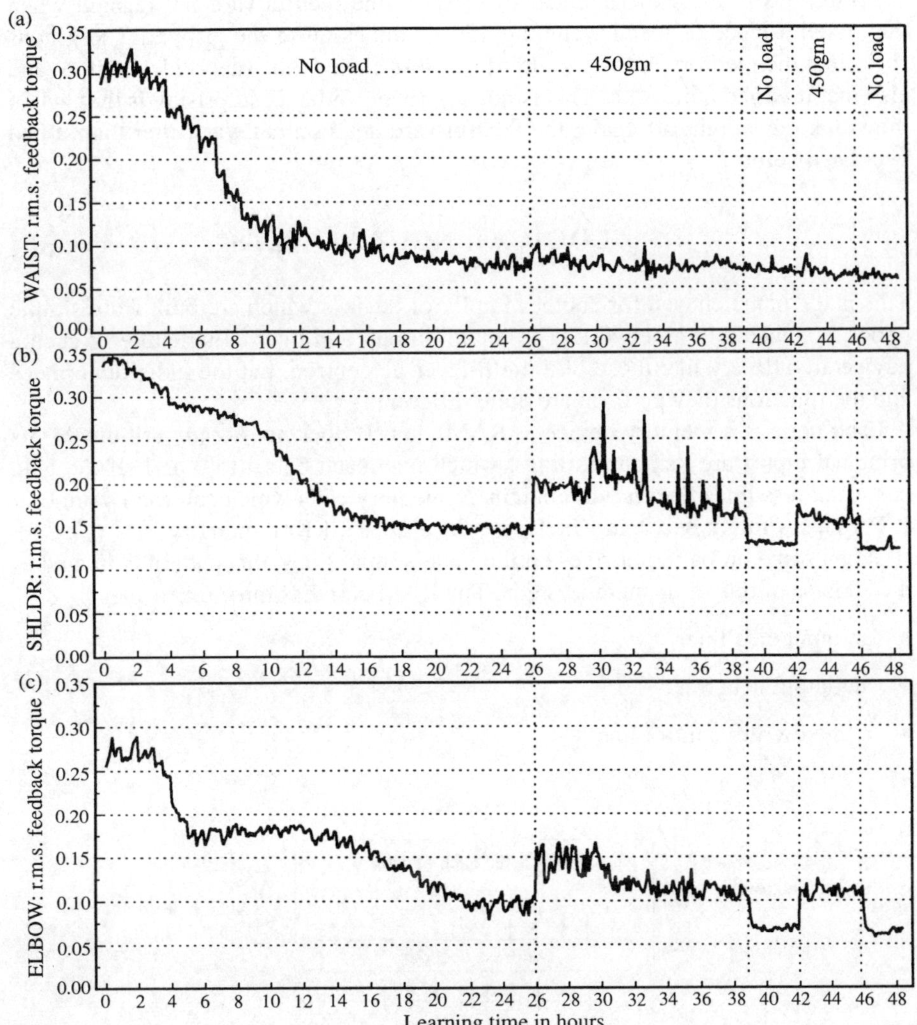


**Fig. 4.10**   Results from the robot's learning.

Instead, *diff* for each node in the top layer is the output from the corresponding node in the top layer of the *feedback* net. Thus, the feedforward net is trained to minimize the output from the feedback net.

The graphs in Fig. 4.10 show how learning proceeded in the feedforward net. Learning was quickest for the motor which moved the waist. The desired torque on this motor does not depend much on the settings of the elbow and the shoulder. The torque at the elbow depends a bit on the setting of the shoulder, so learning to control the elbow took a little longer. The torque at the shoulder depends a lot on the angle at the elbow and so, as one would expect, learning to control the shoulder took longest.

Fraser and Azhar experimented to see how the feedforward net reacted when they fixed a piece of metal weighing 450 grammes on to the gripper. The graphs show that the feedforward net's output was less close to the (new) ideal output and also that it varied quite a lot. This is not surprising. What is surprising is that, when they took the weight off again, the feedforward net's output was better than it had ever been before.

## 4.3   LOGICAL NEURAL NETS

This is the name chosen for another family of devices which are built from simple nodes, connected in a network. The input, output, and internal structure of such a device are all very like those of a multi-layer perceptron, but the individual nodes and the functions they perform are quite different.

Each node is a memory device, a RAM chip. It does not do any arithmetic. Its principal inputs are its address lines, which may each be set to 0 or 1. If the chip has $N$ address lines then it will contain $2^N$ memory cells which can each store 0 or 1. Perhaps $N$ might be 10, in which case the chip is a 1 Kbit memory.

When some bit pattern arrives on the node's input lines, this pattern is treated as the address of one of its memory cells. The RAM chip has three other lines:

- an input data line;

- an output data line;

- a 'read/write' control line.

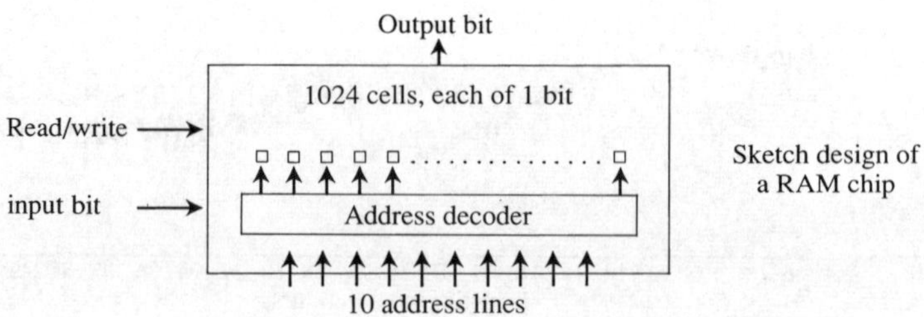


**Fig. 4.11**   Sketch design of a RAM chip.

The control line tells the chip whether to 'read' or to 'write'. When it is told 'read', it copies the contents of the addressed cell to the output data line. If it is set to 'write' then the addressed cell is overwritten with the value supplied on the input data line (see Fig. 4.11).

It might seem that such a node, incapable of any computation, would be a useless building block for a computing device, let alone a learner, but this is not so.

### 4.3.1   WISARD

This is the name of a machine designed and built by a team led by Aleksander. The name WISARD is short for 'Wilkie, Stonham, and Aleksander's Recognition Device'. Some of the nodes in it do a bit of computation, but the learning part is built just of RAM chips.

The inputs to WISARD are taken from a retina, just like the inputs to a perceptron. Each input from the retina is 0 or 1. You can imagine that the inputs to each RAM chip are taken from a random selection of points on the retina. (In a practical implementation, maybe the inputs to the different chips are chosen nonrandomly, so that the different chips survey a carefully chosen range of features.)

WISARD consists of three layers and is capable of three operations: learning to recognize a class of images, recognizing whether a new example is in a learned class, and distinguishing between classes. These three operations are performed in the successive layers. The whole device is connected as a tree. The top node is a 'calculator' which distinguishes between classes. This has one input from each subtree below it. Each of these subtrees is a recognizer for one class, whose top node is a summing device. Below it, the leaves of the tree are the RAM chips, whose principal inputs (address lines) are connected to the retina, as shown in Fig. 4.12.

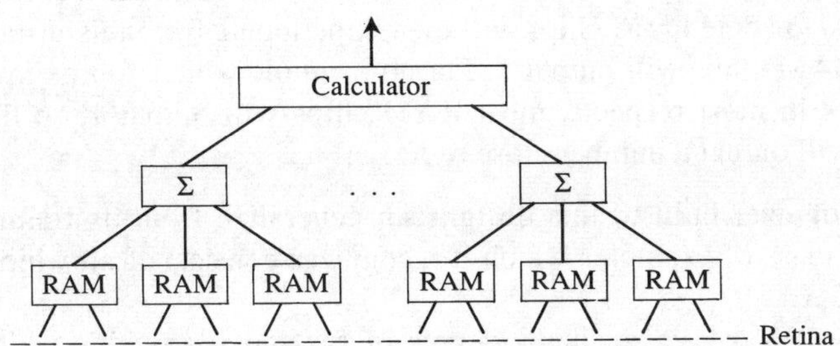


**Fig. 4.12**   The architecture of WISARD.

### *Learning*

Each recognizer learns to recognize just one class. It is taught separately, with its own training set. The training set need only contain positive examples of this recognizer's class, but it should be clean. The learning procedure for one recognizer is:

- initially, all cells in all RAM chips are set to 0;

- an image is projected onto the retina. Each RAM chip treats its inputs from the retina as the address of one of its cells. Each chip is instructed to 'write' the input value 1 to this cell.

## *Recognizing*

- An image is projected onto the retina, as when learning. The chip is instructed to 'read', so it copies the value in its addressed cell to its output line. This value will be

  1  if the current image matches some training instance on this cell's $N$ inputs;
  0  otherwise.

- In each recognizer, all the RAM chips' outputs are fed to the summing device. The output from the summing device is just the number of 1s fed into it.

Suppose that a recognizer contains $K$ RAM chips.

- If the example presented for recognition is exactly the same as a training instance, then all this recognizer's RAM chips will output 1, so the summing device will output the number $K$.

- If the example is quite unlike any training instance, then the values on the $N$ input lines to each RAM chip will be uncorrelated with any of the values it was trained from. The cell addressed inside the chip will be chosen at random and since they were all set to 0 originally, the vast majority of RAM chips will output 0. The output of the summing device will be much less than $K$.

- If the example is similar to a training instance, then there is a good chance that its inputs to some RAM chips will match the inputs from this instance exactly. These RAM chips will output 1. For an example which approximates training instances in most respects, most RAM chips will output 1, so the summing device will output a number close to $K$.

Thus, a recognizer built to this design can generalize from its training set. The generalized class of examples which it recognizes consists of all which produce a sum close to $K$.

The class of a recognizer has a degree of fuzziness: anything which produces exactly the sum $K$ is certainly in the class. An example which produces a smaller sum $k$ is probably in the class if $k$ is nearly $K$, but progressively less likely to be in it as $k$ decreases.

## *Distinguishing between different classes*

WISARD combines several of these recognizers. It can recognize one class with each recognizer. The top layer takes for its inputs all the sums produced by all the recognizers and calculates which ratio $R = k/K$ is greatest. WISARD's output consists of:

- the name of the class corresponding to this greatest ratio;

- the absolute confidence, $R$, that the example belongs to this class;

- the relative confidence that the example is in this class rather than the next most likely class. Say the next largest ratio is $R'$. The relative confidence is taken to be $(R - R') / R$.

When clean training data are available, the WISARD design has substantial advantages over conventional multi-layer perceptrons. It can recognize classes and distinguish between them and generalize from its training set, just like a perceptron. It can also provide measures of confidence in its classification, whereas most perceptrons do not. Its sensitivity is only limited by the number of RAM chips built into it. With enough chips and a large enough training set, it could in principle recognize any class of retinal images with absolute confidence. Its designers have calculated how its reliability increases with more memory. Last but not least, the learning process is fast. During learning, each training instance is presented just once and the learning process consists of writing once to each RAM chip. That takes less than a microsecond!

### 4.3.2 A LOGICAL NET WHICH LEARNS RESPONSES

This device is half-way between WISARD and a multi-layer perceptron. Its nodes are all like WISARD's RAM chips, but the interconnections between them are arranged so that the nodes are in a layered partial order, not necessarily a tree. The arcs between nodes just convey bits of data. They do not have any associated weights. They could be copper wires.

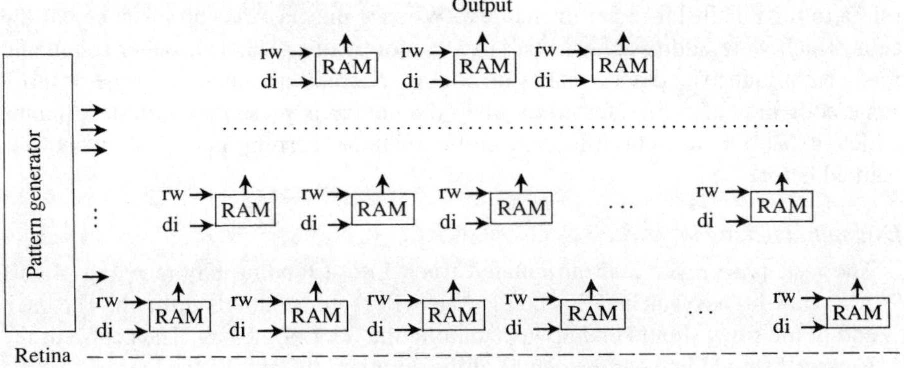


**Fig. 4.13** The architecture of a logical neural net.

As well as the nodes and their interconnections, the net has a separate circuit which can generate random bit patterns. If the net has $K$ nodes, then there will be $K$ bits in each pattern. The $K$ lines from this circuit are linked to the input data lines on the $K$ nodes as shown in Fig. 4.13.

If there is just one node in the top layer, then the device is a recognizer which outputs 1 for examples in its recognized class and 0 for all other inputs. If the top layer contains several nodes, then the device computes a more elaborate function of its input.


### *Problem solving*

An example is projected on to the retina. In each node of the bottom layer, the inputs are treated as the address of a cell. The value stored in this cell is the node's output.

The same procedure is followed in progressively higher layers. The outputs from each layer are taken as inputs to the next layer up, as far as the top layer, whose outputs form the output of the whole device.


### *Learning*

As for WISARD, the training set should be clean. It consists of correctly matched input/output pairs. Thus, if there is just one node in the top layer, so that the device is a recognizer, the training set should contain both positive examples (with expected output 1) and counter-examples (with expected output 0).

Before learning starts, every cell in every node is set to 0 or 1 at random. This is done by writing a random bit from the generator to each cell in turn, in every node. After that, the learning procedure is much like the one above for problem solving. The only difference is that, if the output from the whole device is wrong, then every node in the whole device is instructed to reset its addressed cell to some random value again.

When the device classifies some training instance correctly, then all the cells it refers to for this instance are unchanged. When it misclassifies an instance, all the cells which were addressed are reset to a random value. This is a rather rough and ready technique. The device makes no attempt at credit assignment. Some of those same cells may also be referred to when the device is presented with an instance which it has already classified correctly, so later learning can upset what was learned before.

### *Example: backing up a truck*

The task is to reverse an articulated truck into a loading bay (see Fig. 4.14). Anyone who has watched this being done will realize that it is difficult. The back end of the truck should end up just abutting the door of the bay. The centre of the back end should be near the centre of the door and the axis of the rear half of the truck should be perpendicular to the door.

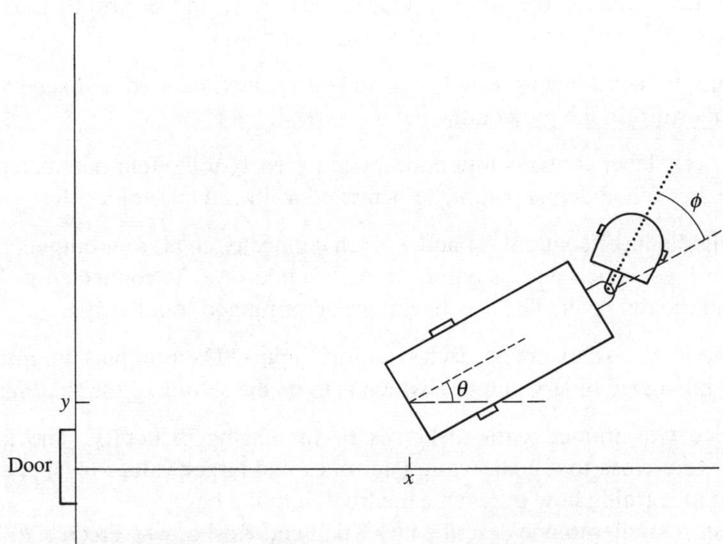



**Fig. 4.14**   Backing up a truck.

A state of the task is described by giving

- the coordinates $(x, y)$ of the centre of the back end, relative to the centre of the loading bay door. In a goal state, $(x, y)$ should be near $(0, 0)$;

- the angle $\theta$ of the axis of the rear half of the truck away from a line perpendicular to the door. In a goal state, $\theta$ should be near 0;

- the angle $\varphi$ between the axis of the driver's cab and the axis of the rear half of the truck. This angle does not matter in a goal state.

Thus, a task state is described by four real numbers. Within the net, these numbers are represented as bit patterns and their numerical values have no significance.

A system which learns to solve this problem consists of two parts:

- a logical neural net, as above. Its input is a current state of the truck. Its output is a setting for the steering wheel in the driver's cab.

- an 'emulator', which is a program whose inputs are a representation of a current state of the truck and the output of the logical neural net. Its output is the state of the truck, after the truck has been reversed a short distance with the steering wheel set as instructed by the net.

The training set consists of a variety of initial positions of the truck. The net and the emulator are run alternately. Each time they are run, the net's output is considered 'good' if the truck's coordinates, $x$, $y$, and $\theta$, are all closer to 0. Otherwise, the net's addressed cells are reset at random.

An example like this was tested as a student project, devised and supervised by Teresa Ludermir. In the project, the net has three layers.

- The lowest layer contains four nodes, one for each of the four parameters $x$, $y$, $\theta$ and $\varphi$. In each node, its parameter is treated as the address of a cell.

- The middle layer contains 24 nodes, each connected to all four outputs from the bottom layer. Each node's output in the middle layer is connected to just one node in the top layer. These connections are arranged 'randomly'.

- The top layer has six nodes, each with four inputs. Their outputs are interpreted as a 6-bit signed binary number, which is to be the setting of the steering wheel.

This device was trained with instances of increasing difficulty. The first few instances were close to a goal, while later ones had larger values of $x$, $y$, and $\theta$. It succeeded in learning how to reverse the truck into the bay.

This task is similar to one described by Kong and Kosko, who credit it to Nguyen and Widrow. Their version of the problem is easier because the truck is not articulated. The steering wheels are mounted directly on the trailer. Kong and Kosko describe an implementation of the task using fuzzy rules and another using a traditional layered perceptron trained by the back-propagation algorithm. The implementation with fuzzy rules worked, but it was provided with substantial prior knowledge in the form of the rules and I am not sure how much of its success was due to learning. It still worked quite well when up to half its original rules were removed.

The neural net controller contained 24 hidden sigmoid neurones. Learning in this net needed more than 100 000 iterations. Kong and Kosko write that the learning algorithm for it did not always converge and the resulting system did not always back the truck up smoothly.

### Design of logical nets

In modern digital computers, it is cheaper to use large RAM chips with many address lines, but in a logical net, small RAM chips may work better. In any learning task, the proportion of cells in each chip which should be set to 1 is likely to be independent of the number of cells in the chip, so the number of learning cycles will be roughly proportional to the size of a chip. If each chip has $N$ input address lines then the chip will contain $2^N$ cells. Hence, the time needed to train the device is likely to rise roughly exponentially with $N$.

Another advantage of small chips with few address lines is that each cell of a chip only stores a very crude fragment of knowledge about the image on the retina. In logical nets, generalization depends on a chance exact match between some portion of the retinal image and a training instance. The more bits surveyed by each chip, the more precise will be the required match and the proportion of matches among all the features surveyed by all nodes will be smaller. Hence, the scope in the net for generalizing will be better if $N$ is small.

## 4.4   BOLTZMANN MACHINES

Within a single node in a perceptron, the flow of information is just one way, from input to output. If perceptrons are layered, then similarly the flow is just one way. This is not always desirable. A computational device of this sort cannot perform any calculation involving steps equivalent to iteration or recursion. For that, one needs some kind of feedback loop. Perceptrons can be connected in loops. Rather than study them, we here describe a kind of network in which information can flow in both directions. This particular sort is called a Boltzmann machine.

A Boltzmann machine (or BM for short) is a finite graph, with various additions.

- Each node $n$ is a *switch* which may be 'on' or 'off', 1 or 0. The setting of this switch will be written $s_n$. Note that a switch is not like a simple perceptron.

- In addition, each node has an attached threshold, $t_n$.

- Each arc is labelled with a *weight*, a real number. The weight on an arc between nodes $i$ and $j$ will be called $w_{ij}$.

- The nodes are partitioned into two sets as shown in Fig. 4.15:
  the *visible* nodes $V$, which behave as input/output links; and the *hidden* nodes $H$ which are involved only in calculation.

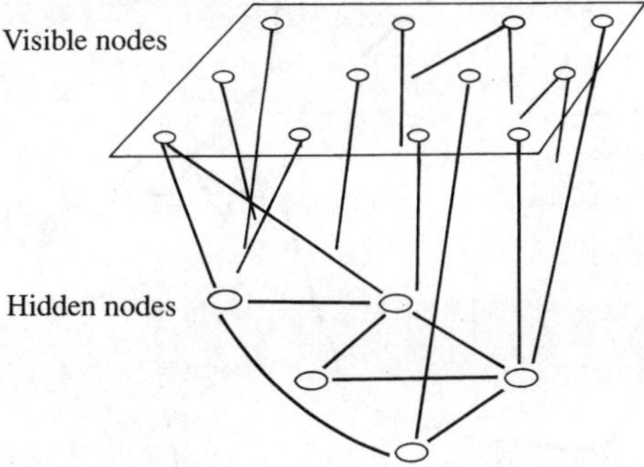


**Fig. 4.15**   Separation of visible and hidden nodes in a Boltzmann machine.

For practical applications, we also assume that the graph is connected and that $V$ is non-empty. Each arc is symmetric in $i$ and $j$. It is not directed and the weight on it may be denoted as either $w_{ij}$ or $w_{ji}$.

The BM is designed to accept as input a setting of some of its visible switches and provide as output settings of the remaining visible switches. Note that there is no fixed division between inputs and outputs. The BM is rather like a Prolog predicate. Each visible switch can be treated as input for one application and output for the next. The

BM tries to find the best settings of all its switches, subject to any restrictions placed on any in *V*. What counts as 'best' depends on the values of its weights.

Each setting of the nodes in *V* can be regarded as a 'fact' or a 'typical situation'. For instance, one can imagine a BM intended to recognize dogs and lamp posts. All but two of its visible nodes will be tied to the output of cells on a retina. Of the remaining two, one should be on if there is a dog on the retinal image. The other should be on if the image includes a lamp post. The machine can be used in different ways. You could show it a scene, while its two special nodes are free, and see if either or both of them come on. Alternatively, you could show the retina an incomplete image of a dog as in Fig. 4.16, with some retinal nodes free, and tie the

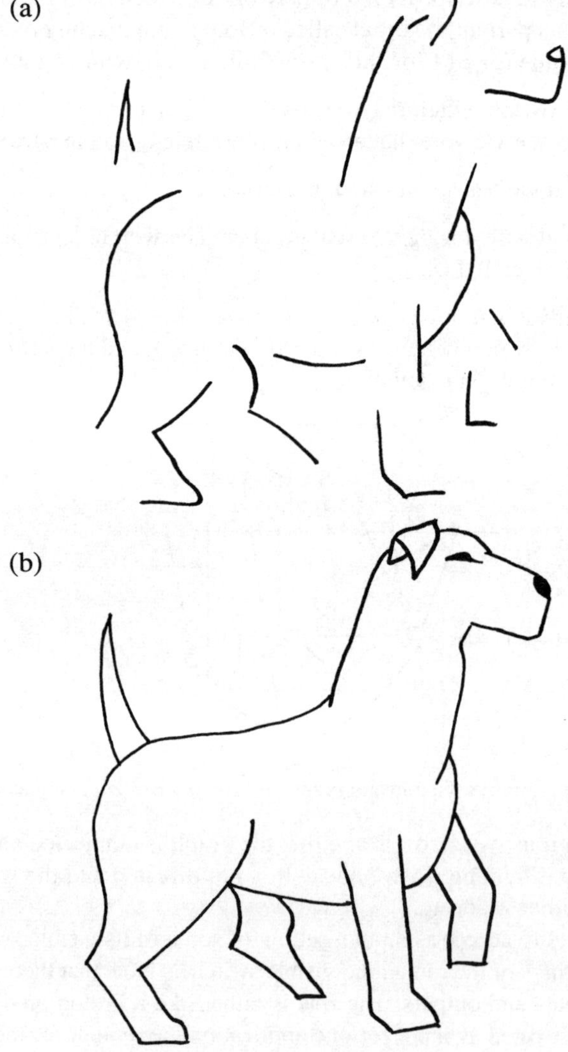


**Fig. 4.16**   (a) Incomplete image of a dog; (b) the completed output image.

special 'doggy' node on. In that case, the machine should complete the image. The free retinal nodes should be switched so that the final image on the retinal nodes is a better image of the dog.

The problem-solving aspect of a BM is a classic case of hill climbing. It actually looks for a lowest point, not a summit. The 'height' which it tries to minimize is called the machine's **energy**, and is defined to be

$$E = \sum_i s_i t_i - \sum_{i<j} s_i s_j w_{ij}.$$

As with perceptrons, the threshold can be replaced by an extra 'switch' called $s_0$, which is permanently on and which is connected to each other node $i$ by an arc with weight

$$w_{0i} = -t_i.$$

Then, the first sum for $E$ can be omitted. We shall do this:

$$E = -\sum_{i<j} s_i s_j w_{ij}.$$

Suppose that some switch $s_i$ is initially 0, at which point the BM has energy $E_0$, and then this switch is turned on and the machine's new energy is $E_1$. The increase in energy is

$$E_1 - E_0 = -\sum_{j \neq i} s_j w_{ij}.$$

This gives a simple rule for deciding how to alter each switch, to lower the energy.

- If $s_i$ is originally 0 and $E_1 < E_0$, then turn this switch on;

- If $s_i$ is originally 1 and $E_1 > E_0$, then turn this switch off;

- Otherwise, leave this switch unchanged.

To begin with, all switches except the inputs are set at random. The BM attempts to minimize $E$ by repeatedly adjusting the settings of all switches except the inputs, according to this rule. When $E$ is at a minimum, the output is the settings of the other visible switches.

The references on BMs leave open the issue of how one chooses which switch to consider when. One might either consider changing each switch in turn, cyclically, until none is changed, or one might consider changing them all simultaneously. The machine's behaviour will be slightly different for different options. The design is well suited for parallelism, with one processor for each node, all switching together (synchronously) or at random (asynchronously).

### 4.4.1 ANNEALING

As with all hill climbing, a BM may end up at a local minimum. To try to avoid this, the above simple rule is changed. The BM introduces some randomness into its switch settings, which is intended to make it sometimes jump out of local minima. The degree of randomness is controlled by a parameter $T$, called the BM's **temperature**. The rule for changing switch settings becomes:

- If $s_i$ is initially 0, then change it to 1 with probability

$$\frac{1}{1 + e^{(E_1 - E_0)/T}}.$$

- If $s_i$ is initially 1, then change it to 0 with probability

$$\frac{1}{1 + e^{(E_0 - E_1)/T}}.$$

When $T$ is very small, this rule is almost the same as the first simple form. It only differs significantly when the change in energy, $E_1 - E_0$, is very small too. When $T$ is progressively larger, so too is the random element. This particular form is symmetric about the case $E_1 = E_0$, at which the switch has an even chance of going either way.

The annealing algorithm runs thus:

Fix the input visible units.
All other switches are initially random.
FOR progressively smaller values of $T$, until $T$ is small
   alter the switches repeatedly according to this probabilistic rule.
The output is now on the free visible nodes.

At any stage, the machine may jump away from any local minimum, but as $T$ gets smaller, it is more and more likely to stay near a particular minimum. The intention is that while $T$ is large, it will jump often between the catchment areas of different minima. As $T$ falls, it will still be able to jump readily out of the catchment of undesirable local minima, but it will be progressively less likely to jump all the way out of the neighbourhood of the global minimum, which is where we want it to finish.


### 4.4.2   LEARNING IN A BOLTZMANN MACHINE

Within a BM, knowledge is stored in the values of its weights. The kind of knowledge it can hold consists of preferred settings of all its visible nodes.

Suppose that there are several such preferred settings,

$$S_1, S_2, \dots S_k, \dots$$

which occur naturally with different frequencies. You can regard each setting $S_k$ as depicting some fact. A BM can be taught many such facts.

Say the natural frequency with which $S_k$ occurs is $F_k$. If the BM is running 'free' at some fixed temperature $T$, with no restrictions on any of its units, then different settings will appear on its visible nodes. Ideally, we would like just these settings $S_k$ to appear, and no others, and each should appear with its natural frequency $F_k$.

The basis of learning in a BM is that we force it so that when it runs free, this is what it does.

The BM's learning algorithm is a cyclic process which never ends. The BM can never be guaranteed to have learned perfectly, so learning continues until the machine's performance appears to be good enough. Each cycle runs as follows:

- Fix the temperature of the machine to some constant value $T$.

- Run the annealing algorithm, lowering its temperature. While it runs, make the input on the visible units jump between the various desired settings. The proportion of time that the input is any particular $S_k$ is $F_k$.

- Also while the machine runs, observe each node. Let $p_{ij}$ be the proportion of time for which $s_i s_j = 1$.

- When the values of the $p_{ij}$s are known fairly accurately, let the machine run free.

- While it runs free, say $p'_{ij}$ is the proportion of time during which $s_i s_j = 1$.

- When the frequencies $p'_{ij}$ are known fairly accurately, stop the machine and adjust its weights.

- If $p_{ij} > p'_{ij}$ then increase $w_{ij}$.

- If $pij < p'_{ij}$ then decrease $w_{ij}$.

The amount by which one changes weights is somewhat arbitrary. One could say that, whatever the values of $p_{ij}$ and $p'_{ij}$,

$$\text{change} \quad w_{ij} \quad \text{to} \quad w_{ij} + (p_{ij} - p'_{ij}).$$

In practice, real BMs sometimes just add or subtract 1.

The logic behind this rule is that if $s_i$ and $s_j$ are switched on together more frequently in the desired state than while the machine is free, then we want them to correlate more, so we increase the weight of the link between them.

In fact, the learning rule can be given a firmer foundation. While the machine runs, it passes through a sequence of states. A state is a set of values for all its switches; so if the BM has $N$ nodes, then it has $2^N$ possible states. We shall denote any state by the letter $G$. Suppose that in the first phase of a learning cycle, while its inputs are the desired settings $S_k$, each state $G$ occurs with frequency $P(G)$. Then in the second phase while the machine is running free, $G$ occurs with frequency $P'(G)$. Ideally, this should be the same as $P(G)$ for each $G$. The discrepancy between them can be quantified as a number, commonly called $C$. This is zero when each $P'(G)$ equals the corresponding $P(G)$. Otherwise, $C$ is greater than zero. The learning algorithm is designed to find a minimum of $C$. It is another example of hill climbing, with height $C$.

Ludwig Boltzmann was an Austrian physicist. The name 'Boltzmann' is chosen for these devices because their algorithms can be justified by a statistical analysis of the sort first devised by Boltzmann. His method shows how one can explain many physical phenomena, including the speed of sound in air and the colour of light from a glowing furnace. However, the analogy should not be taken too far. Boltzmann did not assume that energy is minimized. On the contrary, he assumed that it is constant, as have all physicists before and since. His principle is that another quantity called entropy is maximized. Entropy measures randomness; so Boltzmann's physical principle is that nature maximizes chaos. Intelligence minimizes it.

## 4.5   KOHONEN'S FEATURE MAPS

All the varieties of networks considered so far can learn, but they cannot invent new concepts. If you want a perceptron or a BM to learn to recognize some class of examples, you have to provide it with:

(1)   examples of the class;

(2)   counter-examples;

(3)   for each training instance, the fact that it is an example or a counter-example.

Kohonen's nets are quite different. They invent clusters. A cluster is not the same as a concept, but if a learner can discover natural clusters then this is progress towards finding a concept. The training set does not consist of labelled examples and counter-examples. Rather, it contains a mixture of instances of all sorts, without any distinction between examples and other instances, and the net discovers natural groupings among them. We shall investigate other clustering algorithms later.

A node in a Kohonen net is simpler than a node in a perceptron or a BM. Every node is connected to all the inputs

$$s_0, s_1, \ldots, s_N$$

and for each input, each node has a weight

$$w_0, w_1, \ldots, w_N$$

and nothing else. Each input $s_i$ is a real number. However, the weights are not used to multiply the node's inputs. They are used to measure the 'distance' of the node from an input, as we shall see. When the algorithm begins, nodes are 'scattered' randomly or evenly across a space described by the values of their weights $w$. As it progresses, the nodes' weights are changed a little at a time so that the nodes drift into clusters. The clusters form in areas where there are lots of inputs (see Fig. 4.17). Other regions where inputs are rare become empty of nodes. Thus, a Kohonen net correlates inputs by forming clusters of nodes.

Typically, the net contains many nodes, but each node only has a few inputs: $N$ is quite small. The nodes are laid out in some sort of space. The natural way to do this is an $N$-dimensional rectangular array, but all that matters is that there is some metric $D$ on the set of nodes, so that we can talk about the distance $D(n_1, n_2)$ between any two and about neighbourhoods of a node.

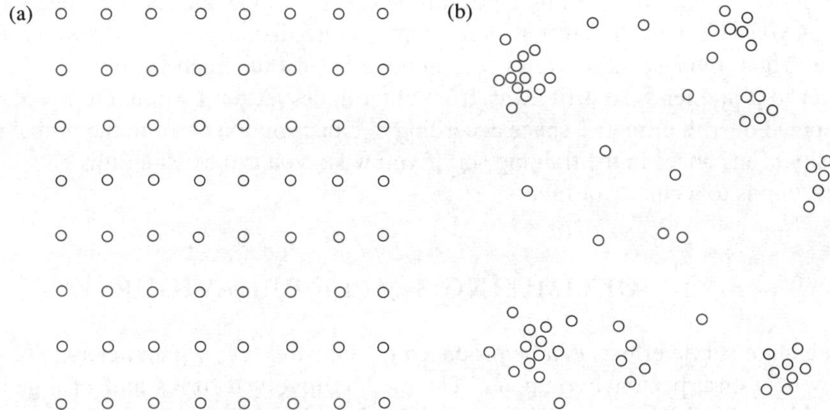

**Fig. 4.17** (a) Initial layout of nodes in a Kohonen net; (b) layout after nodes have drifted into clusters.

Kohonen's clustering process involves two separate metrics. The one we have just discussed is a distance between nodes. It is given before the algorithm begins. The other, $d$, is a distance from a node's weights to an input instance and is calculated at run time. The $D$ metric is a computational convenience. It is the $d$ metric, calculated from weights, which is of most interest.

The algorithm involves two numbers, $K$ and $c$. $K$ is used to decide when two nodes are close. $c$ is a gain term and specifies how much is learned from each input. It proceeds as follows:

Choose random values for all the weights of all nodes. No two nodes should have all their weights the same.

Choose a positive value for $K$
and a value for $c$ in between 0 and 1.

WHILE $K$ and $c$ are not negligible
    FOR each training instance $s_1, \ldots s_N$
      Compute the distance $d$ of each node from this instance:

$$d = \sqrt{\sum_i (s_i - w_i)^2}\,.$$

    Select the node $n$ whose distance $d$ from this input is least.
    FOR each node $n'$ close enough to $n$: $D(n, n') < K$
      Update its weights:

$$w_i' := w_i' + c\,(s_i - w_i').$$

Reduce $K$ and $c$ a bit.

The effect of one cycle of this algorithm is that one node and its near neighbours are moved closer to the current input. Other more distant nodes stay where they were. Thus, over several cycles, the nodes spread out. Each follows the inputs nearer to it and tends to drift away from other nodes. After a while, the nodes will be spread over their metric space according to a distribution close to the probability density of instances in the training set. If you wish, you can consider that each node corresponds to a cluster of inputs.

## 4.6   OPTIMIZING SIMPLE BEHAVIOUR

Some quite subtle effects can be produced by adjusting a few parameters. You can show how on a personal computer. The method involved forms half of a genetic algorithm. The full genetic algorithm will be described below.

The computer screen displays a few white dots, called 'protozoa', and lots of purple dots called 'bacteria'. The program adds bacteria to the screen in a steady stream, scattered at random. Once a bacterium has appeared on the screen, it remains there until a protozoon approaches it and eats it. Each protozoon makes a biased random walk about the screen. It has a certain store of energy. This store is boosted by 40 units each time it finds and eats a bacterium. It loses one unit of energy for each step of its walk. If its energy level falls to zero, it dies and disappears. If its energy ever exceeds 1000, it divides into two offspring.

A protozoon has a head, labelled F. Each step consists of a rotation through some multiple of 60 degrees, followed by movement forward one unit. The six possible rotations are labelled F (for no rotation—just move forward), L (left), HL (hard left), RV (reverse), HR (hard right), and R (right); see Fig. 4.18. Its motion is characterized by an array of six whole numbers, one for each possible angle that it might turn through. On any move, the chance that it might turn through one angle is large if the number for that angle is large in comparison to the numbers for other angles. If the number for direction X is $x$, say, then the probability that it will turn hard left is

$$\frac{2^{hl}}{2^{f} + 2^{r} + 2^{hr} + 2^{rv} + 2^{hl} + 2^{l}}.$$

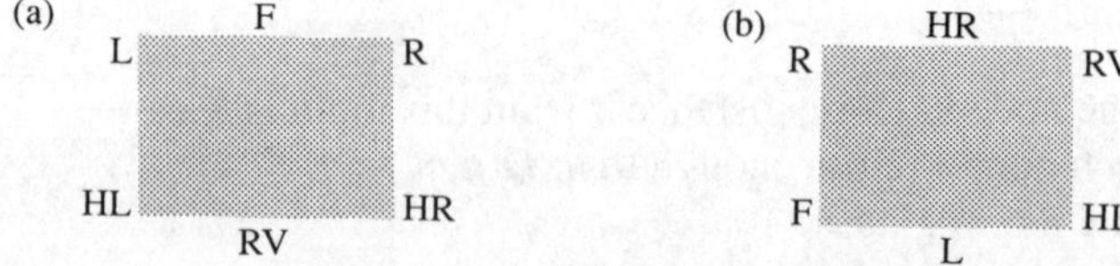


**Fig. 4.18**   (a) A protozoon heading up the page; (b) a protozoon heading towards bottom left.

When a protozoon divides, its two offspring have almost the same numbers as the parent, but one of them has just one number augmented by 1 and the other has just one number decreased by 1.

The system begins with its protozoa jiggling about at random, but after a few generations, they evolve into a race of protozoa which tend to move in straight lines. This is a good foraging tactic when the bacteria, their food, are scattered at random.

There is a variant of the experiment: bacteria drop at random over most of the screen, but in one small region (the *Garden of Eden*) they appear much more frequently. The result is evolution into two distinct populations. The protozoa outside the Garden of Eden tend to travel in straight lines, as before. However, inside it, they evolve into a race of protozoa which spin in tight circles. Thus, they stay in the garden.

## 4.7   GENETIC ALGORITHMS

This is another family of learning algorithms which, its supporters claim, has biological verisimilitude. It works when:

- the learning task can be described as hill-climbing;

- each state in the domain of learning can be described by a (long) string of similar entries, such as real numbers or binary bits, and each such string describes some state;

- there is *lots* of processing power at hand.

It has been used successfully, for example, in the design of jet engines. The idea runs thus.

The algorithm keeps a set of several strings. It is simpler if they all have the same length. Each string portrays one state of a target system (although the state of the learner is the set of all the strings). The initial choice of strings may be random, but the algorithm works better if some of the ones it starts with are varied but fairly sensible.

There is some 'height' function $H$ of states of the target system, which we want to optimize. In each learning cycle, the algorithm calculates $H$ for each of its strings. When it has done this, it replaces the worst few strings with new strings, formed by randomly blending a random selection of the best strings, as illustrated in Fig. 4.19. This cycle repeats, until the best string is good enough.

**Fig. 4.19**   A copy made by blending two strings.

Such an algorithm can provide an alternative learning mechanism in a neural network. Let the arcs of the net be ordered. This can be done in any way you please, but once the order is chosen, it stays fixed. Each arc has a weight, which is a number. Any state of the net is given by specifying its list of weights (in order). If each weight is written in binary form, all to the same precision (say 8 bits), then we can form a bit string by simply concatenating them. Thus, the weights

  15   3   8   −37

correspond to the bit string

  00001111000000011000010001101011011.

Suppose that this is a good string and another good string is

  0001001110111001010001110111000001

which encodes the weights

  19   −71   71   113.

We combine these by choosing just a few random points in them, say two or three, and copying from the first string up to the first point, then copying from the second up to the second point, then copying from the first again, and so on, switching between them at each point. If the chosen points are as shown:

  00001111000000011000010001101011011

                    ×                ×

  0001001110111001010001110111000001

then the resultant string will be

  00001111000010010100011111011011

which encodes

  15   9   71   −37.

If you prefer that the algorithm should not invent new weights, but only select existing weights from the parent strings, then the crossover points marked '×' may only be at the ends of 8-bit sequences.

The idea of genetic algorithms is to mimic natural Darwinian selection of genetic codes. The chance of a given string being chosen to 'reproduce' is related to the worth of its state, $H$. When the two good strings are similar, then any feature which occurs in both will inevitably appear in their 'offspring'. Any feature where they differ is chosen from one of them at random. Thus, features which lead to good states are reproduced and others drop out.

In principle, it would seem that the performance of a genetic algorithm may depend significantly on the order of the basic data in each string. When a particular order is chosen, then any cross between two strings which keeps a weight $w$ from one parent will keep $w$'s neighbours in that parent, unless there is a crossing point ($\times$) just before or just after $w$. Thus, the choice of initial order in the strings imposes correlations between the weights and these may be unnatural. Enthusiasts for genetic algorithms do not seem worried about this. Nature does the same.

Such a method was tested by Whitley on three neural nets. He reports that the algorithm does indeed usually converge. It is often better than a standard learner, such as the Boltzmann algorithm, because the genetic algorithm keeps note of a large set of states, one for each string. The Boltzmann machine may miss the best solution and climb a foothill or be stuck on a knife edge, but the genetic learner has a broader view of many possible states and is less likely to be so caught.

Genetic algorithms have proved their worth commercially. One simple way to design a good fan blade for a turbine is to design a mediocre blade, describe its shape with some numeric parameters, and then test variants of it. The parameters can be put together to form a 'genetic code' for the blade. The variants can be tested by simulations in a large computer and the codes of successive variants can be optimized by genetic recombination and selection. After several cycles, the mediocre blade will be replaced by a better one. Dr Steve Jones cited this example in his 1991 Reith lectures on the BBC—and he is a real biological geneticist.

Genetic algorithms involve a lot of computation. 30 000 recombinations is considered 'a very small number'. The number of strings which the algorithm keeps may range up to a few thousand and each bit string may be a few hundred long. In addition, there is the overhead of calculating the height function, $H$. For a neural net consisting of layered perceptrons, this may not be so heavy, but if it involves annealing, the total process demands a lot of computing time.

The benefit of a broad view over many states only applies if the set of strings is large and varied. The genetic learner may go astray if its strings are mostly similar. As the learner is designed to favour good features, the population will eventually become monoclonal, however big it is and however it starts. The genetic theory postulates two modes of evolution. One works by crossing two parents with slightly different genetic codes and mixing their genes randomly. This is the essence of the algorithm above. The other mode depends on occasional small mutations, like the changes in the parameters of protozoa.

Whitley tested a variant of the above algorithm which incorporated both modes. It only (!) keeps 50 strings. When they seemed to resemble each other rather too closely, he introduced random mutations in the offspring. The number of mutations increased as the parents became more alike. Compared with the original algorithm, this often produced fast, accurate results, but it did sometimes fail to produce any acceptable result at all.

## 4.8   PROPERTIES OF NEURAL NETS AND GENETIC ALGORITHMS

At the end of the last chapter, it was apparent that the classification scheme which we are using is fallible. For neural nets and genetic algorithms, it is not so much fallible as crude. There is an immense difference between WISARD and logical nets at one extreme and Boltzmann machines at another. The reason becomes clear when we consider these devices' search spaces for learning.

Any learning neural net explores a space in which each state is described by a large set of simple parameters. In a logical net, the parameter values are the settings of the bits in the cells of its RAM chips. In a classical perceptron or Boltzmann machine, the parameters are the weights on edges between nodes.

The learning method in WISARD consists of remembering examples. It does no search. Hence, learning takes just as long as it takes to present and record the examples.

A logical net searches while learning, so its learning process is slower than that of WISARD. The learning search of a logical net falls into two levels. The simpler level is a search for the correct output for some particular instance. The higher level is a search for correct outputs for all instances in the training set. Its search has two interesting features:

(1)   Most neural nets search by some process which involves a form of convergence by successive approximations. In a logical net, at the simpler level of search, there is *no* notion of approximation to a correct answer from given inputs. The learner makes no use of anything like a metric on its set of possible states.

(2)   It uses absolutely *no* heuristic guidance. When an instance is misclassified, all bits used in that classification are reset *at random*.

Given these curious aspects, it is a bit surprising that a logical net's learning process ever succeeds. Perhaps it does because its search space contains very many goal states. Also, although the simpler level of search has no metric and involves random jumps, there is a notion of proximity at the higher level: two states are close if the proportion of RAM chip cells with different entries is small. In this sense, the search at the higher level only takes small steps.

All other kinds of neural nets have learning search spaces parametrized by weights, which are real numbers. (In a simple perceptron, even though the output of a node is discrete, the weights on edges are real numbers.) Their learning processes involve convergence of these weights. One simple distinction is between layered nets which learn by back propagation, in which the search optimizes each node's output *locally*, and nets such as the Boltzmann machine which optimize a *global* feature such as the quantity called $C$. Local optimizers usually learn faster than global optimizers. Actually, BMs optimize locally, and their sluggishness is largely due to annealing. Both BMs and the back-propagation algorithm can be described as hill climbers, but back-propagation is not so prone to climbing false foothills. It is designed to optimize weights in one layer of the perceptron at a time, and the procedure for optimizing weights in one layer is controlled by properties of the layer above which it is already partly optimized. Thus, each layer provides heuristic information about how to optimize the next layer down. At each node of a BM, there is no such heuristic guidance.

Kohonen's method minimizes the distance between an input vector and the vectors of nearby nodes in the net. This is a form of local optimization. The degree of locality is controlled by the parameter called $K$, which bounds the set of nodes whose weights change in one learning step.

The search steps taken by a genetic learner are of two kinds. The process of crossing between two genetic codes is random. The selection of relatively successful codes is heuristic. All the onus of learning falls on the heuristic part. The random part is what makes genetic algorithms slow.

### Strengths

- Algorithms of this sort are very widely applicable. The form of input is so basic that almost any application can be encoded as input to a network.

- Except for the ones based on RAM chips (WISARD and logical neural nets), they tolerate noisy training sets. In fact, noise can be very helpful because it gives the network a chance to jump out of local minima.

- Training examples are not saved.

- Perceptrons solve problems quickly, although Boltzmann machines do not.

- Perceptrons require both examples and counter-examples. Boltzmann machines require only positive examples. Kohonen nets do not involve any notion of a concept, so there is no notion of a positive example. Genetic algorithms require some measure $H$ of the quality of each example.

- They are suitable for parallel processing, with one process for each node.

### Limitations

- In these designs, learning is not incremental. (Other designs of learning neural nets may be able to learn incrementally.)

- These algorithms are typically quite good at learning how to respond to one kind of input. In principle, a perceptron should be able to respond correctly to several varieties of input, but in practice, learning later responses may upset what was learned before.

- Although there is a well-developed theory of how these devices learn and solve problems, there is no very useful theory of how they represent their learned knowledge. The knowledge is not in an accessible form. This is particularly true of BMs. As a result, it is very hard to predict the behaviour of a BM in detail. It is an essentially probabilistic 'black box'. A user of a BM has to live with whatever the BM tells him, which will be correct for most of the time but which is never totally reliable. (The one exception is WISARD.)

- As a consequence of the last remark, these devices are incapable of returning 'right' answers.

- For the same reason, learning is never complete. The user must decide when the machine's knowledge is reliable enough, if it ever is.

- Training sets are typically large, for any kind of neural net. (The world champion backgammon playing program, a neural net, learned from about 1,500,000 games.)

- Learning in neural nets is notoriously slow.

## 4.9   COMPARISON WITH BIOLOGICAL SYSTEMS

The resemblance between genetic algorithms and Darwinian evolution seems close, although natural genetic reproduction may be more subtle. There is a suggestion that offspring do not have an even chance of inheriting a trait from either parent. Each natural gene may come with accompanying genetic material which exists solely to block the corresponding gene from the other parent and, furthermore, there may be even more genetic material which sometimes inhibits the other parent's blocking mechanism! The *Economist* (21 March, 1992, pp. 212–22) summarizes the idea. Nothing like this occurs in the artificial form of the algorithm.

The human brain is a network of neurones. A neurone is a cell with a small centre from which stem long strands called **axons** and **dendrites**. There are links, called **synapses**, between axons and dendrites. When a neurone 'fires', it releases chemicals called *neurotransmitters* at each synapse on its axons. These stimulate (or inhibit) the dendrite on the other side of the synapse. Thus, the flow of information across a synapse is one way, as in a perceptron. Quite a lot is known about how neurones are interconnected in different parts of the brain and how they work together as a computing device.

A lot of research in neural nets appears to be driven by researchers' imaginations and the desire to make devices which work, with very little reference to biology. For example, no neurone behaves like a RAM chip and synapses do not appear to be symmetric as in a Boltzmann machine. This may be a wise strategy. Some comparison between the brain and artificial machines can be made purely by scale: each brain contains some 10 000 000 000 neurones; each neurone is linked by some 10 000 synapses; and there are about 100 known kinds of neurotransmitter. In addition, neurones are not linked randomly. There are signs of organized layout among the axons and dendrites. From all this, it will be clear that there is no prospect of building artificial nets with anything like the subtlety of the brain. However, neurophysiology does offer some encouragement.

For a start, the brain of a bee contains only about 300 000 neurones. That sort of number is within the range of current technology.

Many observed functions within the brain resemble what can be done with layers of perceptrons. In the parts which deal with smell and taste, lower layers of neurones show only a little specialization between different stimuli. Later layers show progressively more specialization. The human vision system can recognize a face in about 150 milliseconds. This sort of speed is typical of some twenty layers of neurones.

By contrast, the process of learning is slow and subtle. Learning a manipulative skill, such as typing, takes hours. This is more typical of a Boltzmann machine. There is a region of the brain called the hippocampus which has something to do with recording new memories. Within it, some neurones link back on themselves, as in an autocorrelator.

There are some aspects of psychology which have not yet been fitted into the neural net paradigm. One is storage of memory for events. Nobody knows whereabouts this occurs in the brain. Another involves the process of recall. When you remember a scene, it is as if what you once saw reappears before your eyes. If the memory is particularly vivid, you can actually fail to see what is really in front of you. It seems that the remembered visual image replaces input from the retina.

A third is the question of awareness, emotion, and reasoning. There is a phenomenon called 'blind sight' which sometimes occurs after brain damage. The person involved may look at something and not be aware of it, but if he is asked to *guess* where the object is in his field of vision, he guesses correctly. Evidently, all the intermediate levels of visual processing still work, but there is a broken link between the output of vision and the seat of awareness. That seat of awareness is not understood at all.

## FURTHER READING

Over the past half decade, there has been a flood of designs and theories for different devices of this sort. This short chapter is only intended to offer a taste of the topic. There is no space here to attempt a full survey. Anyone who wants a thorough view should study the *Neural Computation* journal and some of the many new textbooks on neural nets, such as Aleksander and Morton (1990), Judd (1990), Kosko (1992), and Zurada (1992). Hinton (1989) and Lippmann (1987) give good introductory surveys.

For more details of perceptrons, see Minsky and Papert (1969) which is reviewed by Block (1970). The account here is adapted from Bundy *et al.* (1980). Variants of the learning method for a single node, and others, appear on pages 59–73 of Zurada (1992). Ergezinger and Thomsen (1995) report a major improvement on the standard back propagation algorithm. Nilsson (1990) provides two proofs of the convergence theorem.

There have been studies of perceptrons with other styles of output function. One such is based on a node whose output is near zero if the sum of inputs is either large positive or large negative. The graph of the output is a Gaussian hump. Details and further references appear in Hartman, Keeler, and Kowalski (1990), Park and Sandberg (1991), and Poggio and Girosi (1990).

The account of WISARD given here follows Aleksander and Morton (1990). I am grateful to Teresa Ludermir who introduced me to it, and to logical neural nets. She has published various studies of them (1990, 1991). Her version of the truck reversing problem is similar to the one given by Kong and Kosko; see Kosko (1992), although her solution is different.

Boltzmann machines and other similar devices have foundations in information theory. This can justify both the formula for $C$ and the expression for the probability of switching $s_j$ while annealing. For more discussion, see Hinton *et al.* (1984). The theory of neural nets has been developed much further. It now covers stability of the convergence process. This work derives from the theory of dynamical systems which was originally developed by Anosov, Smale, and others in the 1960s. Ruelle (1989) describes some of the underlying mathematics.

For more details of Kohonen's nets, see T. Kohonen (1982, 1984). The experiment with protozoa is described by A. K. Dewdney (1989). Booker, Goldberg, and Holland (1989) describe genetic algorithms as a form of rule-based systems, but this seems an unduly narrow view. Forsyth (1989), Chapter 4, gives an extended discussion with reference to applications. The account here is taken from Whitley (1989), whose observations are given a theoretical basis by Cerf (1994).

Rolls (1987), referred to in Chapter 1, gives an account of real neural processes. The seat of awareness is a subject of philosophical speculation. For some suggestions on what it might do, if we can ever find it, see Johnson-Laird (1983).

## EXERCISES

1. Any point in a plane can be described by its $x$ and $y$ coordinates. Three simple nodes each have these two coordinates as their inputs. Suggest weights and thresholds for them so that:

   * the first recognizes the half-plane above the $x$-axis ($y > 0$);

   * the second recognizes the half-plane $x > 0$;

   * the third recognizes the half-plane

     $\{(x, y) \mid x + y < 1\}$.

   Using these three nodes, design a two-layer perceptron which recognizes the triangle which is formed by intersecting these three half-planes.

2. Write a program which encodes the back-propagation algorithm. Its inputs are a description of a multi-layer perceptron with just one node in the top layer, and a training set of positive and negative instances of some concept. When run, this program should train the perceptron to recognize an approximation to the concept.

   In Pascal, the data types involved might resemble

```
node        =   record
                    threshold:      real;
                    connections:    array [1..size] of Bool;
                    weights:        array [1..size] of real;
                end;
layer       =   record
                    n_of_nodes:     integer;
                    nodes:          array [1..size] of ^node;
                end;
perceptron  =   array [1..depth] of ^layer;
```

Test this program on a two-layer perceptron with one node in the top layer and three in the lower layer. The training data consist of points chosen randomly in the square

$$[-1, +1] \times [-1, +1].$$

The positive instances are the ones in the triangle of Exercise 1.

3.  The XOR function has domain

$$\{0, 1\} \times \{0, 1\}.$$

Its value is 0 if both its two arguments are the same, and 1 if they are different. Show that the XOR function cannot be encoded in a simple perceptron. That is to say, if a node has two inputs, then whatever its threshold, there is no way of assigning weights to the inputs so that the node's output will be 0 if both inputs are the same and 1 if they are different.

Show that the XOR function can be encoded in a two-layer perceptron with two nodes in the bottom layer and one in the top layer. (Hint: in the bottom layer, one node should be on only if both inputs are on and the other should be on only if both inputs are off.)

4.  Suggest a design for a Boltzmann machine which might predict good moves for noughts and crosses (tic-tac-toe) or any other simple board game. The visible units represent the board, with two visible nodes for each board position—one for a $\times$ and the other for a $\bigcirc$. Write down a few training instances for it.

If the game is the one described in Exercise 3 of Chapter 3, then the Boltzmann machine could be designed so that it has an internal node for each occurrence of each pattern.

5.  A Kohonen net has 100 nodes, laid out in a $10 \times 10$ square array and named $N_{ij}$ ($i, j$ = 1, ... 10). The distance between two nodes $N_{ij}$ and $N_{pq}$ is

$$D(N_{ij}, N_{pq}) = |\, i - p \,| + |\, j - q \,|.$$

Each node is connected to two inputs, $S_1$ and $S_2$, by weights $W_{1ij}$ and $W_{2ij}$ which initially have the values

$$W_{1ij} = i \qquad W_{2ij} = j.$$

During the first two cycles of the machine,

$$K = 7/2 \quad c = 1/2.$$

The inputs for the first training instance are

$$S_1 = S_2 = 5.$$

After the net has learned from this instance, how many nodes have changed weights? What are the weights of node $N_{53}$?

If the second instance's inputs are 0 and 5 then, when the net has learned from it too, which nodes have had their weights changed twice?

6.  Consider the logical neural net which learns to reverse an articulated truck. Recall that it has four RAM chips in its bottom layer, 24 in the middle layer, and six in the top layer. Each RAM chip has four address lines, so has 16 cells.

Estimate the number of goal states in its search space for learning.

Suppose that $I$ and $J$ are two distinct training instances with negligible correlation, and $p$ and $q$ are two distinct RAM chips in the top layer. While the net learns with instance $I$, suppose that chip $p$ yields the wrong output, so all cells leading to this output are reset at random. Estimate the chance that this learning step will change:

- at least one cell which is used when calculating the output of $q$ from input $J$;

- the output of $q$ from input $J$.

# 5   Clustering and correlation


SUMMARY

The subject matter of this chapter falls into three broad sections. The first covers the basic principles of clustering:

- top-down and bottom-up clustering techniques;

- bias: the effect of description language on clustering;

- extensional clusters;

- intensional descriptions of clusters;

- names for clusters;

- the contrast between statistical and conceptual clustering;

- simple hierarchies and ontological hierarchies.

The second contains descriptions of five clustering algorithms:

- a method for inferring context-free grammars;

- a top-down decision-tree constructor depending on entropy, which is explained;

- a more sensitive incremental tree constructor;

- a way to decide which properties of a system depend directly on each other;

- an elaborate algorithm which builds a directed graph of concepts incrementally and which can decide where to concentrate its effort to best effect.

The final section is a comparative study of these algorithms and the methods they incorporate. It describes:

- features of the above clusterers and how properties of a clusterer affect its behaviour;

- a general procedure for inventing a clustering algorithm for any particular situation.

There is an appendix on:

- the design of statistical experiments;

- estimating entropy;

- composite properties formed from simple observable ones;

- expected success rates and the relative merit of entropy.

## INTRODUCTION

The paradigm underlying this family of methods is that patterns or clusters appear non-randomly in the training set, for some unknown underlying reason. The learner discovers the clusters among examples by noticing non-random correlations among their features. Usually, he (or it) takes absolutely no account of any possible causes of such features. This sort of learning ignores underlying structure of examples and it does not try to explain why the regularities occur.

The differences between these methods and neural nets are both qualitative and quantitative. All the inputs to a net are simple and similar and there are typically lots of them. A correlator's input consists of examples with just a few features which can be quite elaborate. This means that the correlator can use statistical methods, or approximations to statistics, which would be impractical in a large neural net.

Perhaps the most satisfying kind of learning combines this sort of statistical observation with later analysis. That is what Mendel did. He noticed statistical trends in inherited traits, he postulated a statistical model with nice properties which fitted his data, and then he suggested a simple discrete model which would behave like the statistical model. Many years later, biochemists showed that his discrete model really does occur in the molecular structure of living cells.

## 5.1   HOW CLUSTERING WORKS

A cluster is a set of examples which are close together. The very word 'close' only makes sense when the examples are in a metric space or something like it. If you want to do clustering of any sort, you need a pseudo-metric. A pseudo-metric is a function like a metric, which satisfies all the axioms of a metric except that the distance between two distinct points can be zero. Every metric is a pseudo-metric.

The natural prototype for a metric space is the world we live in, where the metric is Euclidean distance which you can measure with a ruler. This is a bit misleading, for the following reason. Each position in Euclidean space can be described by three coordinates, and coordinates are real numbers which can be added. This means that you can take an *average* of several positions. Hence if, say, you are watching a flock of starlings pecking for worms on a lawn, you can talk about the centre of the flock. This is a point on the lawn where there may not be any starling. In fact, if the starlings are all in a circle (perhaps the lawn has a fairy ring which the worms find congenial), then the centre of the flock will be quite a long way from any of the birds (see Fig. 5.1).

**Fig. 5.1**  A circular flock. The centre is not near any bird.

The Euclidean metric is unduly specialized. If you think instead of a Venn diagram metric, then you will understand that not all metrics permit averages.

*Example: classifying species of British birds*

This is a complicated process, with at least two levels. To start with, imagine you are given just the physical outline of adults of several species. You might look for clusters, using the properties:

- size;

- shape of beak;

- position of eyes;

- shape of wing.

Each of these properties is described by a number or a few numbers. The shape of a beak is described by the ratio of its length to its depth, and its curvature; similarly, the position of eyes may be characterized as the angle which they subtend at the centre of the bird's head.

For each of these properties, there will be a few clusters which can be assigned average characteristics. Thus, a bird's size may be small, medium, or large. These three values are the names of three clusters of birds' sizes. Similarly, beaks may be short, long and straight, or curved. Again, these values are the names of clusters of beaks.

There are many species with short stubby beaks, some with very long beaks, and some with curved beaks (Fig. 5.2). Clustering does not tell us why. We know why: different bill shapes suit different eating habits, but never mind that for the moment. Similarly, there will be a few significant clusters of wing shape. Again, from our vantage point, we know that each is suited to a particular kind of flight and life style, but let us ignore that fact and continue with the primitive analysis.

**Fig. 5.2**   Shapes of beaks.

We can now assign to each species the names of the various clusters for its numeric attributes. Thus, a sparrow

- is small;

- has a short beak;

- has a wide field of vision; and

- has stubby wings.

A kestrel

- is large (as birds go);

- has a curved beak;

- has eyes which look forward; and

- has long wings.

These properties are discrete. In any further analysis, we cannot take averages. Let us try a different pseudo-metric on the space of species. Say the distance between two species is the number of characteristics which they don't have in common. Thus, the distance between a sparrow and a kestrel is 4. (This distance function is a pseudo-metric rather than a metric, because there are different species with identical values for these four properties. Sparrows and blue tits have the same four characteristics, so the distance between them is 0.)

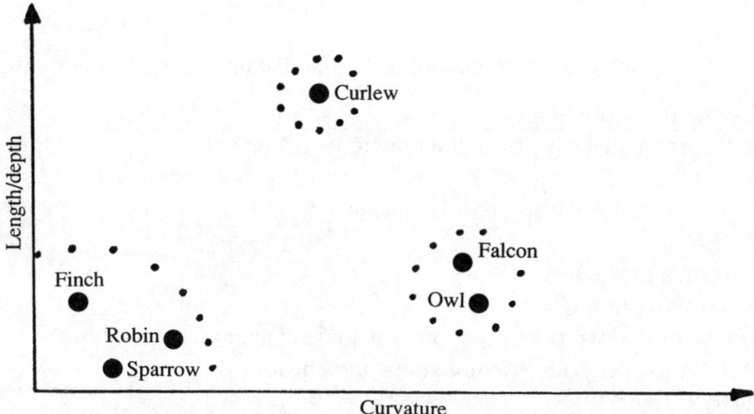



**Fig. 5.3** Clusters of bird species by real-valued traits.

If we look for clusters in this space of species, differentiated by coarse discrete characteristics, then we do indeed find quite striking clusters (Fig. 5.3). There are many small birds with short bills, short wings, and all-round vision. There is a small but distinctive group of larger birds with longer wings, hooked bills, and forward vision. These are the hunting birds, the owls, hawks, and falcons. There are also other possible groups of characteristics which are strikingly absent among real species; for example, there are very few large birds with short wings and short straight beaks. This classification could be refined by adding more discrete properties, such as foot shape and habitat and whether the bird is brightly coloured.

### 5.1.1 A SIMPLE CLUSTERING ALGORITHM

This particular algorithm has no great virtue. It is presented here so that there is no confusion about the clustering process. Its input is a finite set of examples ('points') in a space with a (pseudo-) metric called $d$. Its output is a partition of these points into subsets called clusters and an association of some unique name with each subset in the partition.

Clustering algorithms often *name* their clusters. Later, any other example can be allocated to a cluster by specifying that name. A name is in effect an extension of the language for describing examples.

Associated with any cluster $C$, there is a number called its **diameter**. This is the largest distance $d(y, z)$ between any two points in $C$. We shall say that a point $x$ is **nearly central** in $C$ if for every point $z$ in $C$, the distance $d(x, z)$ is less than $\frac{2}{3}$ of $C$'s diameter.

Form a list of all unordered pairs of distinct input points
    [{x, y} {x, z} {y, z} ...].
Sort this list of pairs, in increasing order of the distance between the points.

WHILE the list is not empty
    take the first pair {x, y} from the remainder of the list
    and
        IF neither $x$ nor $y$ is as yet in a cluster
        THEN
            form a new cluster $C$.
            To start with, $C = \{x, y\}$.
        IF one of the two points, say $x$, is in some cluster $C$
        AND the other point $y$ is not yet in any cluster
        THEN
            add $y$ to $C$.
        IF $x$ and $y$ are in different clusters $C_1$ and $C_2$
        AND $x$ is nearly central in $C_1$
        AND $y$ is nearly central in $C_2$
        AND $d(x, y) < (diameter\ (C_1) + diameter\ (C_2))/2$
        THEN
            amalgamate $C_1$ and $C_2$ into a new combined cluster.
        (Otherwise, do nothing to the clusters.)

Note that the operations performed in this algorithm are not part of mathematics.
The first

    form a new cluster $C$.
    To start with, $C = \{x, y\}$

really means:

    Invent a new name $C$;
    Associate the name $C$ with the set $\{x, y\}$.

Similarly,

    add $y$ to $C$

means

    Associate $C$ with the union of $\{y\}$ and the set previously associated with $C$;
    Forget $C$'s previous association;

and

    amalgamate $C_1$ and $C_2$ into a new combined cluster

means

> Invent a new name $C$;
> Associate $C$ with the union of the sets previously associated with the names $C_1$
> and $C_2$;
> Forget the names $C_1$ and $C_2$.

The names may not be textual. In this sense, a 'name' may be nothing more than an address in a computer's memory, such as a pointer in a Pascal data structure. Although the naming process is basic to clustering, I shall not always say what the names are. They are just the means by which one refers to clusters.

Kohonen's net algorithm does not invent a partition of the training set into clusters. Rather, it invents a sort of simplified version of the training set: when the net has finished learning, the nodes in it form clusters of just a few points which each reflect much larger clusters in the training set. Each node acts as a name for the set of examples near it.

At the end of this little statistical clustering algorithm above, every point is in some cluster of two or more points. That may be undesirable. Some points may be isolated and are best left as such. When $x$ is in a cluster $C$ and $y$ is not in a cluster, then perhaps $y$ should only be added to $C$ if

$$d(x, y) < \text{twice diameter}(C).$$

Also, the choice of the number $\frac{2}{3}$, in the notion of being nearly central, is arbitrary. This number could be anywhere between $\frac{1}{2}$ and 1. It could even be dependent on the cluster: say the number is $\delta$, so a point in $C$ is nearly central if

$$\forall\, z \in C \quad d(x, z) < \delta \times \text{diameter}(C).$$

Then $\delta$ could be quite an elaborate function of $C$, such as

$$\frac{\underset{x \in C}{\text{median}} \left( \underset{z \in C}{\max}\ d(x, z) \right)}{\textit{diameter}\ (C)}.$$

This algorithm depends a bit too much on the sorting stage. If two pairs $\{x, y\}$ and $\{z, w\}$ are the same distance apart, then the sorter could put them in either order. Different choices of order may lead to different clusters.

This algorithm is an instance of **bottom-up** clustering. The algorithm begins with a few small clusters and it enlarges them and creates new clusters as it is presented with more data. There is a contrast between this sort and the so-called **top-down** algorithms. They start with all data in one large set and they subdivide this set into smaller clusters. Below, we shall study a top-down algorithm which presents its output in a form called a decision tree.

This particular algorithm is manifestly not incremental. All the training data have to be collected and paired and then sorted before the main clustering stage can begin. There are incremental approaches. In outline, they proceed thus:

Start with an empty set of clusters.
WHILE there is still some training data
    $x$: = the next instance;
    IF there is a cluster $C$ containing points near $x$
    THEN
        add $x$ to $C$;
        IF $C$ is unbalanced
        THEN
            subdivide $C$ into new clusters $C_1$ and $C_2$
    ELSE
        form a new cluster whose sole point is $x$.

The notion of an **unbalanced** cluster and the operation of **subdividing** a cluster are complicated. We shall come across specific forms of them later. This algorithm is incremental because it creates clusters from the first few data and these early clusters can be used at once, even though they may be improved during later learning.

Clustering algorithms can also be distinguished by the natures of the clusters which they form. An **optimization** technique constructs a partition of the training set. Each component of the partition is a cluster and the algorithm is intended to find clusters which are as far apart as possible. **Clumping** techniques, by contrast, form clusters which may overlap. Any one instance may lie in several clusters. A **hierarchical** technique generates a tree whose leaves are individual instances. Each internal node represents a cluster of the leaves below it.


### 5.1.2 ASPECTS OF THE CLUSTERING PROCESS

The clustering process can consist of as many as three stages. In practice they often overlap and run concurrently:

(1)   adapting the representation of the training set;

(2)   finding extensional descriptions of clusters;

(3)   inventing intensional descriptions of clusters.


***Adapting the representation***

This is a case of adjusting bias. With luck, you won't have to do it. For species of birds, the data are given in a form which will make any natural clusters obvious. A change of representation is necessary, though, in the case of a circular flock of starlings; for if the basic clustering algorithm were applied to the Cartesian coordinates of the birds, then the information that they are in a circle would be lost.

Figure 5.4 shows three different examples of training sets. The first is not suitable for clustering. The points in it are distributed randomly, so it contains no information which could be learned by any process. The second example has three clusters and the basic algorithm will find them easily.

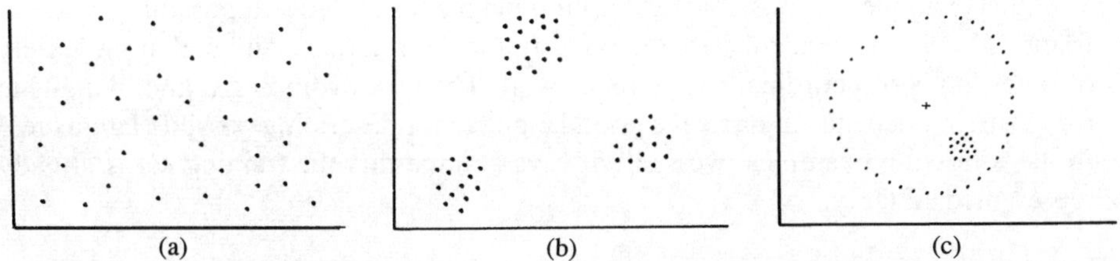

**Fig. 5.4** Three candidate sets for clustering. (a) Not suitable; (b) suitable; (c) requires a change of description language.

The third example is the hard case. The distribution of points in it contains quite a lot of information, but the basic clustering algorithm will lose it because this algorithm will lump all the points into a single cluster. However, the basic algorithm will find the right clusters if the example is transformed into polar coordinates centred at the place marked '+', for then, all the points on the circle are transformed onto a line which projects to a small region on the $r$-axis (Fig. 5.5). This is a case of a more general problem: *the ease of solution of any task depends on the language in which it is expressed.*

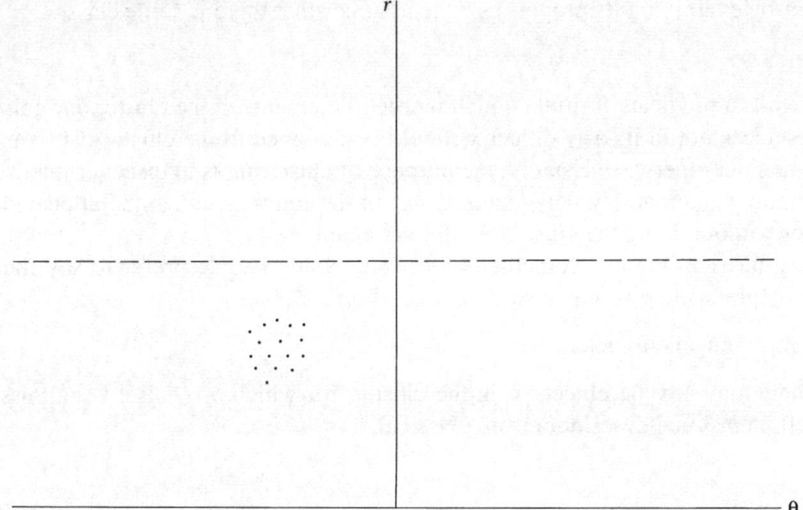

**Fig. 5.5** The same data as in Fig. 5.3(c) but in polar coordinates.

The difficulty is to find a good *description language* in which to express the learning situation. I do not know of any recipe. The human eye is exceptionally good at recognizing patterns such as circles. Computers are much less versatile. Even if an algorithm embodied knowledge that perhaps the training set should be transformed to polar coordinates, still the algorithm would have to search a two-dimensional continuum of possible centres.

In this example, the possible description languages include all possible Cartesian frames of reference and all possible polar frames of reference and perhaps a variety of other kinds of coordinate systems as well. The task of finding a good language for clustering depends in turn on a good language for describing possible languages! In the following examples, we shall always assume that the training set is already represented suitably.


### Finding extensional descriptions of clusters

This amounts to finding clusters in the training set by an algorithm such as the one above. When the data are presented suitably, it is easy. Otherwise it is virtually impossible. This stage produces a family of subsets of the training set, the clusters, but it does not identify each cluster except by stating which example is in which cluster.


### Inventing intensional descriptions of clusters

An extensional description of a cluster just states what its elements are. An intensional description provides a way to decide whether any particular thing is in the cluster by examining its properties. One assumes that there is some class

$$\{x \mid p(x)\}$$

where $p$ is a predicate formula which is true of elements in the cluster and false for all instances not in it. Any object $x$ should be assigned to the cluster if it is in this class and not otherwise. Usually, the purpose of clustering is to discover intensional descriptions, so that any new example not in the training set can be allocated to a cluster without doing the clustering all over again.

This classical view is sometimes too rigid. Some people prefer to say that the defining predicate $p$ is only a sufficient condition:

If $p(x)$ then $x$ is in the cluster

but there may also be objects $x$, in the cluster, for which $p$ is false. I shall assume that all intensional descriptions are classical.


### Prototypes

Intensional descriptions are usually invented by some sort of generalizing method. However, there is a simple algorithm which invents intensional descriptions from raw clusters:

Given a finite cluster $C$, described extensionally,
$\quad c := $ a point in $C$ for which
$\qquad \max \{ d(x, c) \mid x \in C \}$    is least
$\quad D := \max \{ d(x, c) \mid x \in C \}$.

The intensional description of the class corresponding to $C$ is then

$$\{x \mid d(x, c) \leq D\}.$$

The selected element $c$ is called a **prototype** for $C$. The intensional descriptions which we shall come across do not involve prototypes. Perhaps this is a weakness in our presentation. Prototypes can be very natural ways to describe clusters.


### 5.1.3 STATISTICAL AND CONCEPTUAL CLUSTERING

Consider again the classification of birds. It falls into two parts. The first clusters according to numeric properties and the second according to the names of the clusters invented during the first part.

These two levels of classification exhibit two quite different sorts of clustering. The first level, which discovers general trends in the forms of wings and beaks, is a case of statistical clustering. The other, which depends on discrete features, is often called conceptual clustering. These are two extreme points on a spectrum of algorithms. In practice, almost any useful algorithm incorporates aspects of both approaches, but for the moment, we shall explain them by emphasizing their differences.

**Statistical clustering** is intuitively simple. The learner observes extensional clusters of traits in examples and may perhaps name them. The learned information just makes these traits explicit, so that they can be used for prediction. The traits themselves are basic observed numeric features, maybe numbers with decimal points in them. The metric used for clustering is derived directly from these traits.

The term '**conceptual clustering**' has been adopted because the class

$$\{x \mid p(x)\}$$

of an intensional description is often referred to as a 'concept'. There is a continuing discussion as to what a concept really is. The concepts which we humans are aware of are certainly much more subtle, but let that pass.

Conceptual clustering involves more detailed analysis of relationships between traits. Each cluster has an associated intensional definition. For instance, if there are two functions $f$ and $g$ defined on the set of all possible examples and $f$ can take the value $a$ and $g$ can take the value $b$, then one such cluster might be defined as the set of all examples $x$ for which

$$f(x) = a \wedge g(x) = b.$$

In the case of birds, $f$ and $g$ might be *length_of_beak* and *shape_of_beak*. The cluster defined by the property

$$length\_of\_beak(x) = short \wedge shape\_of\_beak(x) = straight$$

includes sparrows and finches. The learner's search space is a space of sets of definitions, in some suitable description language, and the goal is a set of succinct definitions which partition the training set suitably.

The properties of examples are not always described with functions. For instance, one can either describe a species of bird with functions called *length_of_beak* and *shape_of_beak*, so

> *length_of_beak*(*sparrow*)  =  *short*
> *shape_of_beak*(*sparrow*)  =  *straight*

or alternatively one can use binary relations called *beak_length* and *beak_shape*:

> *beak_length*(*sparrow*, *short*)
> *beak_shape*(*sparrow*, *straight*).

The second style is common in the literature.

Since definitions are tidier when they only involve discrete alternatives, the languages used for these definitions almost always contain only functions $f$ with finite ranges. One does not meet so-called conceptual clusterers which generate definitions of the form

$$a < f(x) < b.$$

This feature of the language makes the clustering algorithms simpler.

The ease with which one can learn a concept depends critically on the structure of the predicate $p$, which in turn depends on the description language. Let us assume that we have fixed on a particular language. The concept is called **conjunctive** if $p$ is a conjunct of atomic formulas, perhaps preceded by quantifiers:

$$p \equiv \forall \ldots \exists \ldots (A \wedge B \wedge \ldots C).$$

If $p$ contains the connective $\vee$ as well as $\wedge$ then it is called **disjunctive**. Many of the methods presented in later chapters only work with conjunctive concepts. One virtue of clustering is that it can often invent disjunctive ones. It is possible to imagine concepts in which $A$, $B$, ... could be negated atomic formulas, but in practice, learning algorithms do not invent such concepts.

Note that the input to a statistical clusterer is a set of instances which are not classified in advance. Perhaps the instances are all examples of some large class which we suspect may have meaningful subclasses and the statistical clusterer should invent the subclasses, or perhaps there is no given initial class at all. There is not usually any notion of a counter-example when the clustering is statistical. By contrast, the input to a conceptual clusterer consists of examples which have already been labelled with the names of the clusters they lie in. The conceptual clusterer's input contains both examples and counter-examples of each concept.

Another essential difference between conceptual and statistical clustering lies in the form of the metric. In conceptual clustering, the metric is not always made explicit. Its clustering process is usually equivalent to straightforward clustering using a pseudo-metric $d$, where for any two examples $x$ and $y$,

> $d(x, y) =$ the number of functions $f$ in the description language for which
> $$f(x) \neq f(y).$$

This $d$ resembles the Hamming distance.

When people speak of conceptual clustering, they often mean more than just a clustering process. The possible properties of examples form a partially ordered set: property $P$ is less than property $P'$ if $\forall\, x\, (P(x) \Rightarrow P'(x))$. Conceptual clustering often generates a tree of properties. The top node is the property which is *true* of all possible examples. Each node defines a cluster. The cluster defined by a node is a subset of the cluster defined by any other node above it. Thus, the property

$$f(x) = a \wedge g(x) = b$$

might define a leaf below an internal node

$$f(x) = a$$

and this in turn might lie below the nodes

$$f(x) \in \{a,c,d\}$$

and *true* (see Fig. 5.6).

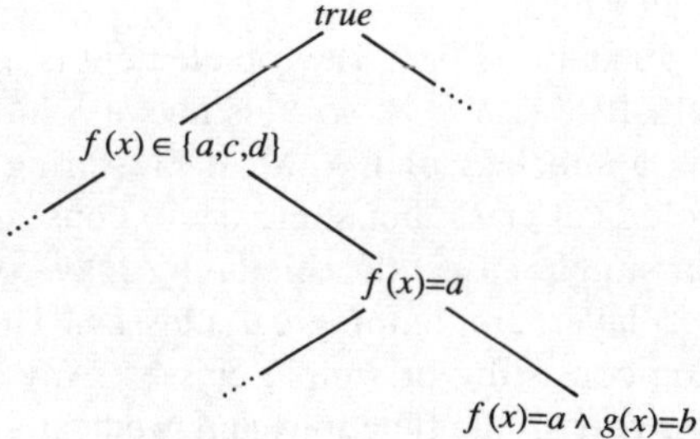


**Fig. 5.6**   A hierarchy of concepts.

Given any such tree of concepts, it is easy to construct a pseudo-metric whose clusters are the nodes of this tree. If two definitions $p_1$ and $p_2$ describe nodes $n_1$ and $n_2$ in the tree, then the distance $d(p_1, p_2)$ is the length of the path in the tree from $n_1$ to $n_2$. However, such metrics do not always appear to be natural. Usually, conceptual clusterers employ the natural hierarchy of predicates directly, and make no mention of a metric.

Some conceptual clusterers invent more general partial orders, not just trees. The term 'conceptual' arises because each node is a 'concept' which is defined by some predicate. A good conceptual clusterer is one which finds a succinct meaningful hierarchy of succinct definitions of meaningful concepts. Both bottom-up and top-down methods have been used.

### 5.1.4   ONTOLOGICAL CLUSTERS

Clustering is easier when every function and predicate which we might consider applies to every conceivable training instance. In the cases considered so far, this is so. Every bird has a beak and wings, so it is meaningful to ask what are the lengths of beaks and wings for all birds. If the training set also includes cats, dogs and cows, then we could not always ask for these values.

Suppose that there is a set called $A$ of all functions and predicates which we might consider using in tests on instances. Each subset $S$ of $A$ has an associated class of all conceivable things $x$ which all the functions and predicates in $S$ can be applied to:

$$C_S = \{x \mid \forall f \in S \text{ it is meaningful to ask, what is } f(x)?\}.$$

If $S$ is empty then $C_S$ contains absolutely everything. If $S$ contains 'length of wing' then $C_S$ includes all birds, but not cats and dogs. $C_S$ is the (extensional) **ontological** cluster associated with $S$.

Within a single ontological cluster $C_S$ one can use all the elements of $S$ to distinguish subclasses of $C_S$. When a training set $TS$ is a subset of $C_S$ and all the tests made on instances use only functions and predicates from $S$, then the invented clusters are called **simple**. All the clusters discussed above are simple. Two simple clusters $C_1$ and $C_2$ are distinguished by some function (or predicate) $f$ which takes a value $v_1$ on all points $x$ in $C_1$ and a different value $v_2$ on all $x$ in $C_2$. Two ontological clusters $C$ and $C'$ are distinguished by some $f$ which has values on all $x$ in $C$ but which is meaningless on all $y$ in $C'$.

The subsets of $A$ form a natural lattice. The top element is the empty set and the bottom element is $A$ itself. If $S' \subseteq S \subseteq A$, so $S'$ is above $S$ in this lattice, then the ontological cluster $C_S$ is a subclass of $C_{S'}$. Most clustering algorithms restrict attention to a single ontological class, but some try to construct a more elaborate hierarchy of concepts. In simple cases, this hierarchy may consist of two layers. The nodes of the upper layer are ontological classes. The lower layer is a conventional classification consisting of simple classes. Any simple class $C_1$ lies below an ontological class $C$ if all the functions and predicates used to describe $C_1$ are in the set $S$ for $C$.

*Example: classification by form of coat*

Suppose that we are classifying things according to parameters in a set $A$ which contains the functions

*form-of-fur*
*length-of-feathers.*

These two functions are only partial. Birds do not have fur and cats and dogs do not have feathers. Therefore, the classification must have an ontological layer, consisting of the concepts above the dotted line in Fig. 5.7. The concepts below the line are simple.

There are classification tasks for which even this view is too elementary. Suppose that we are classifying living creatures. The top level of the classification is by phylum. 'Phylum' is a function of animal species, whose values include

*mammal*
*bird*
*insect*
*arachnid*
*fish*

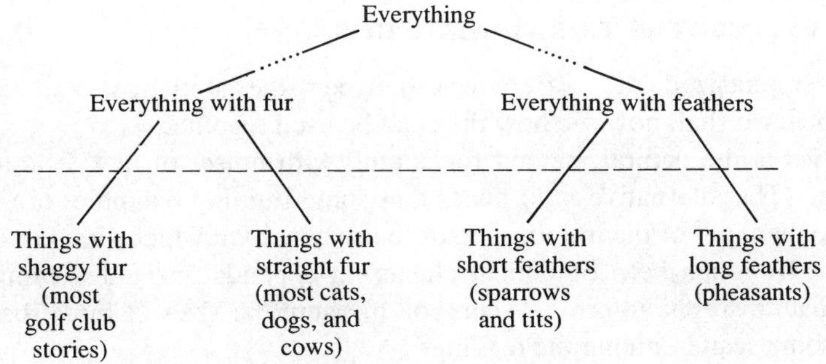

**Fig. 5.7**   The separation between an ontological layer and a simple layer.

The top few levels of such a classification are shown in Fig. 5.8.

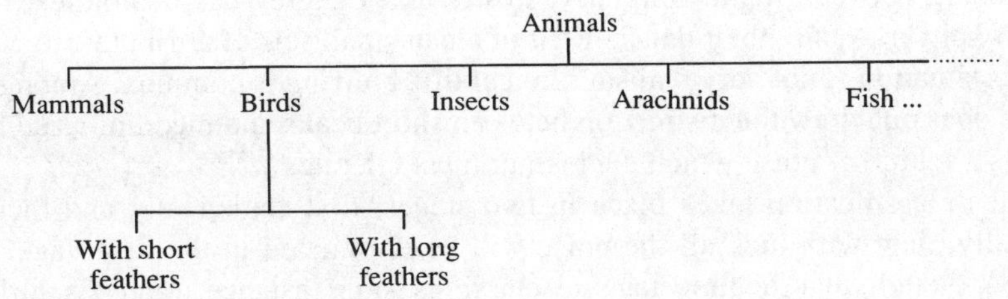

**Fig. 5.8**   A more complex hierarchy.

The complication in this example is that the concept *birds* appears to be simple, because it is the subclass of *animals* on which the function *phylum* has the value *bird*, but at the same time, the concept of birds is ontological because the function *length-of-feathers* is defined on it. In general, it seems that an ontological concept can occur below a simple one. The lattice of concepts does not always fall neatly into two layers.

Whether we regard a concept as ontological or simple depends on our choice of description language. This example is based on the assumption that $A$ includes the functions *form-of-fur* and *length-of-beak*, but no function such as *kind-of-coat*. If we choose a different set $A$, containing *kind-of-coat* but omitting *form-of-fur* and *length-of-beak*, and if the possible values for *kind-of-coat* are

*shaggy fur*
*straight fur*
*short feathers*
*long feathers*

then the whole classification can be completed with just simple concepts because the function *kind-of-coat* is total.

### 5.1.5 COMBINING CONCEPTS AND STATISTICS

Having emphasized the differences between the statistical and conceptual approaches, we shall now see how they can be used together.

Statistics is the natural method for coping with noise. In fact, it is almost the only way. (The alternative is to guess that some training examples are erroneous and leave them out of the training set, or to use a method which simulates statistics, such as a Kohonen net.) Statistical clustering depends on values which may be approximate or prone to error. When you measure the sizes of birds' beaks, there may be some scatter among the readings.

Conceptual clustering cannot handle noise so easily. In the conceptual case, each example's properties are supposed to be precise and discrete. The prime example discussed by Stepp and Michalski involved classifying simple pictures of goods trains. Their aim was to partition the set of pictures by the shapes of the engines and wagons. The pictures show just a few shapes of wagon. There is no possibility of blurring between them. Similarly, Fisher and Langley use anatomical data to classify phyla. Again, their data consist of clean small sets of attributes.

This situation is not very realistic. In real life, blurring is common. Among birds, where does one draw the distinction between short beaks and medium-sized beaks? This is a source of noise which the learner must tolerate.

When classification takes place in two stages, first statistically and then conceptually, then with luck all the noise will be eliminated in the first stage. It can happen, though, that the first stage misclassifies some instance. Analysis during the second stage may suggest revision. This actually happened while chemists were classifying the elements. Dobereiner, Newlands, and Mendeleev grouped the then-known elements by chemical properties and found that their masses formed a steady sequence, with very few exceptions. Among the exceptions were chromium, indium, platinum, and gold. Mendeleev suggested that the given masses might be in error. The experimenters repeated their tests and found that their first results were indeed wrong and Mendeleev was right.

Clustering is rarely used in isolation. A learner is usually searching for some property which fulfils criteria which have nothing to do directly with the observed data. This is particularly true of conceptual clustering. In the case of birds, we know that meaningful clusters will be associated with birds' life styles. In Chapter 7, on learning in rule-based systems, we shall come across a technique called explanation-based generalization (EBG). This relates an example's features to its function and purpose. Using EBG, one might observe just one species, a sparrow, and deduce that any bird with short wings and a short beak could survive comfortably on a diet of seeds and insects. EBG would then suggest that there may be many bird species with these characteristics, and that they all have such a diet. Such deductive methods complement observation and statistics. Each approach on its own is somewhat unreliable. The statistics may have low confidence levels, and the assumptions for deduction may be ill founded, but if they both lead to the same classification, then a learner can trust it.

What Mendel did was rather different. His analysis was more like the algorithm embodied in one of the later versions of Bacon. Given his statistical results, he conjectured the existence of some otherwise completely unknown feature—the genetic code. Pat Langley calls such a conjectured feature an 'intrinsic property'. For many years, until biochemists performed their wonders, there was no other way to justify Mendel's theory. It had two foundations. The first was its close fit with observed inheritance. The other was and is its elegance.

In the rest of this chapter, we shall study five algorithms. They are chosen because they are practical and valuable and they are relatively clearly described in the literature. All five show some aspects of both statistical and conceptual clustering. They all classify according to discrete properties, but they also have statistical features and so they can all handle noisy data.

1.  The first, due to Gerry Wolff, synthesizes *context-free grammars*. Such grammars are used to define languages, which themselves are necessary for describing other tasks.

2.  The second builds *decision trees*. Variants of it have been used in many situations.

3.  The third, following work of Al-Mathami, builds somewhat more elaborate trees which can be used to predict properties of examples which are not completely described.

4.  The fourth constructs a graph of properties in which there is an arc between two properties if they are essentially correlated. It was invented by Fung and Crawford.

5.  The fifth algorithm, due to Scott and Markovitch, invents an efficient hierarchical classification of objects according to their behaviours under actions.

All these algorithms are examples of a class of clusterers studied by Buntine. He has sketched a procedure for inventing classifiers in many learning situations, and he suggests a Bayesian approach which helps find good heuristic components for them.

Any clusterer should be founded on some kind of theory based on probabilities. Among these five, the one with least justification is the grammar learner. It is a hill-climbing process which attempts to find a compact encoding of a language which is not too dissimilar from its training set. Al-Mathami's method attempts to maximize the chance that its predictions will be correct. The Fung–Crawford graph constructor chooses constructions with the greatest experimental backing. The decision tree learner can be justified by a notion called *entropy*, which is also used in the theory of Boltzmann machines and in the Scott–Markovitch learner. Entropy measures the expected surprise of a situation, and all these algorithms are designed to search for explanations which make the world appear more coherent and less surprising. This notion will be explained further.

## 5.2   INFERENCE OF GRAMMARS

This case of clustering was investigated by Gerry Wolff. His method invents a grammar for a written language. In principle, it might be applied to English. The grammar is used to describe the structure of a passage of text. It consists of **context-free productions**, which are rules of the form

$$A \rightarrow X \quad Y \quad \ldots \quad Z.$$

The symbols $X$, $Y$, …, on the right may be symbols which occur in the text (**terminal** symbols) or they may be other symbols which are invented for use just within the grammar (**variables**). The symbol on the left of the arrow, $A$, is always a variable.

*Example*

Say the terminal symbols are

  *Jane   John   likes   chases   dog   ball   the*

and the variables are

  NP   Proper   Common   Verb   Art   S.

The productions might be

| | | |
|---|---|---|
| Proper | $\rightarrow$ | *Jane* |
| Proper | $\rightarrow$ | *John* |
| Common | $\rightarrow$ | *dog* |
| Common | $\rightarrow$ | *ball* |
| Verb | $\rightarrow$ | *likes* |
| Verb | $\rightarrow$ | *chases* |
| Art | $\rightarrow$ | *the* |
| NP | $\rightarrow$ | Proper |
| NP | $\rightarrow$ | Art   Common |
| S | $\rightarrow$ | NP   Verb   NP |

Given any symbol, $A$, we can speak of all the **sentences generated** from $A$ by the grammar. Here, the term sentence does not mean what you were taught in school. In this sense, a *sentence* can be any finite sequence of terminal symbols.

If $A$ is a terminal symbol, then this set of sentences consists of just $A$ itself. Otherwise, $A$ may give rise to lots of sentences.

- Choose any production for $A$

$$A \rightarrow X \quad Y \quad \ldots \quad Z;$$

- For each symbol on the right side, choose some sentence generated from this symbol by the grammar ($S_X$ for $X$, $S_Y$ for $Y$, and so on);

- Concatenate all these sentences, in the order of their symbols in the production:

$$S_X \quad S_Y \quad \ldots \quad S_Z.$$

This combined sentence is now a sentence generated from $A$. We sometimes write

$$A \xrightarrow{*} S_X \quad S_Y \quad \ldots \quad S_Z.$$

*Example*

With the grammar above,

$\quad$ NP $\xrightarrow{*}$ *John*

$\quad$ NP $\xrightarrow{*}$ *the ball*

$\quad$ S $\xrightarrow{*}$ *John likes the ball*.

(Note: Such a grammar is almost the same as a Chomsky context-free grammar (CFG). The only difference is that in a CFG, one variable symbol is singled out and called the *start* symbol. The only sentences generated by a CFG are from its start symbol. Wolff originally described a way of learning CFGs. The algorithm discussed here omits the start symbol.)

Each way of generating a sentence gives rise to a rooted tree, called its **parse tree**. The root is labelled by the variable, $A$, you generate from. All the internal nodes are labelled with variables and the leaves are labelled with terminals. At each internal node, labelled $X$ say, there is one child for each symbol on the right side of the chosen production for $X$. These children are ordered in the same way as their labels are ordered in $X$'s production. Thus, a parse tree for

$\quad$ S $\xrightarrow{*}$ *John likes the ball*

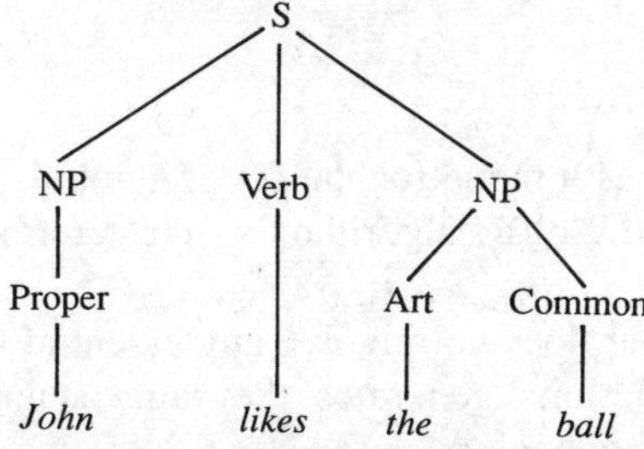


**Fig. 5.9** A parse tree.

is shown in Fig. 5.9. In this case, the grammar permits only one parse for this sentence.

The whole process can be turned round. Rather than start from a symbol and find what sentences can be generated from it, one can begin with a sentence and a grammar and find a way of generating that sentence. This process is called **parsing** the sentence. A **parser** is a program which takes as input a sentence and a grammar and which constructs such a parse tree.

Grammars and parsers were invented for the study of human languages. They are now also ubiquitous throughout computing. They are basic to the growing subject of computational linguistics, which aims to let computers talk and make sense of human language, but before that arose, they proved invaluable for computer languages like Basic, Pascal, and Prolog. They have other applications too.


### 5.2.1  WOLFF'S ALGORITHM

The input to the algorithm consists of a text and a grammar, like the ones described above. Its output is another grammar extending the input grammar. While processing, it also produces a sequence of parse trees for abutting segments of the input text.

The input grammar may be completely empty to start with. By repeating the algorithm cyclically, it can form ever larger, more comprehensive grammars.

The grammar found by Wolff's method is a bit special. It has two distinct kinds of variables: *AND symbols* and *OR symbols*. An AND symbol $A$ has just one production, but otherwise it is unrestricted.

$$A \rightarrow B \quad C \quad D \quad \ldots$$

An OR symbol $X$ may have lots of productions, but they are all **unit** productions. That is, each has only one symbol on its right side:

$$X \rightarrow P$$
$$X \rightarrow Q$$
$$X \rightarrow R$$

All these different productions are sometimes summarized by writing them thus on a single line:

$$X \rightarrow P \mid Q \mid R \mid \ldots$$

Each OR symbol acts as a name for the set of symbols $\{P, Q, R, \ldots\}$ on the right sides of its productions. Wolff's algorithm is a clusterer because it forms and names these sets.

This kind of grammar does not involve any essential loss of generality. Any set of context-free productions generates the same sentences as one of Wolff's grammars. The algorithm begins:

Find a sequence of abutting parses of the input text, using the input grammar:
    WHILE some text remains unparsed,
        IF the grammar is inadequate for parsing any initial
          segment of the remaining text
        THEN
          skip one terminal symbol of the text
        ELSE
          find the longest possible initial segment of the remaining text which can
          be parsed with this grammar, and a parse tree for it.
    Record properties of the resulting sequence of parse trees.

There are two sorts of information recorded:

(1)  the sequence of top symbols of parse trees. (For this purpose, a skipped terminal counts as such a top symbol. If the algorithm begins with an empty grammar, then this recorded sequence of top symbols is simply the original input text.)

(2)  within all parse trees, all the contexts in which each symbol occurs.

Suppose some symbol $B$ occurs in a parse tree whose top symbol is $A$. Within this tree, $B$ is the top node of a smaller parse tree. Let us say $S_A$ is the sentence covered by the whole tree and $S_B$ is the shorter sentence generated within it from $B$. $S_B$ is a substring of $S_A$. Say $L$ is the part of $S_A$ to the left of $S_B$ and $R$ is the part to the right (see Fig. 5.10). There are various notions of context for $B$ in this parse tree. Different versions of the algorithm employ different notions of context. There is room for experiment here. The simplest is the pair $(L, R)$.

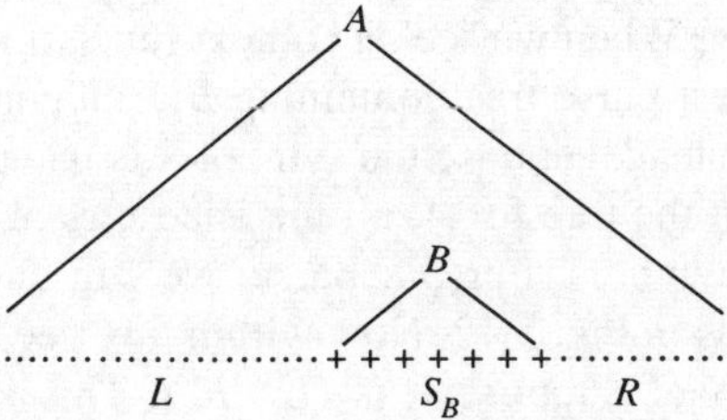

**Fig. 5.10**  A simple context of terminals.

For some purposes, this is too crude. $L$ and $R$ are strings of terminals. If the algorithm is to do all we want, contexts should be allowed to include some variables too. Suppose that $X$ is an OR symbol generating $S_X$ in the tree, and $S_X$ is disjoint from $S_B$ as shown in Fig. 5.11. Say $L'$ is formed from $L$ by replacing each such string $S_X$ in $L$ by the variable symbol $X$ which spans it. Similarly, $R'$ is got by substituting OR symbols for their spans in $R$. The second kind of context is the pair $(L', R')$.

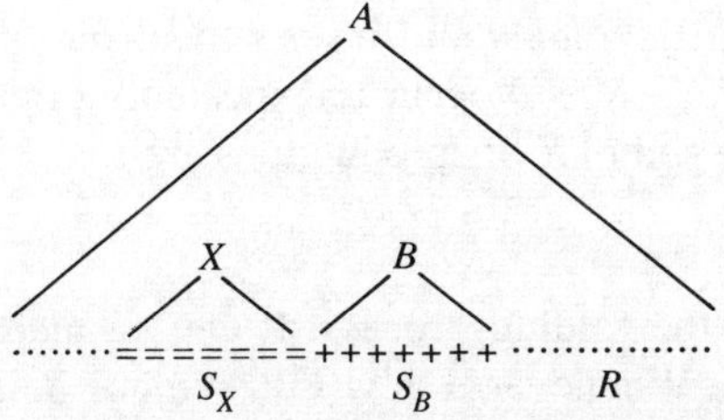

**Fig. 5.11**  A context including OR symbols.

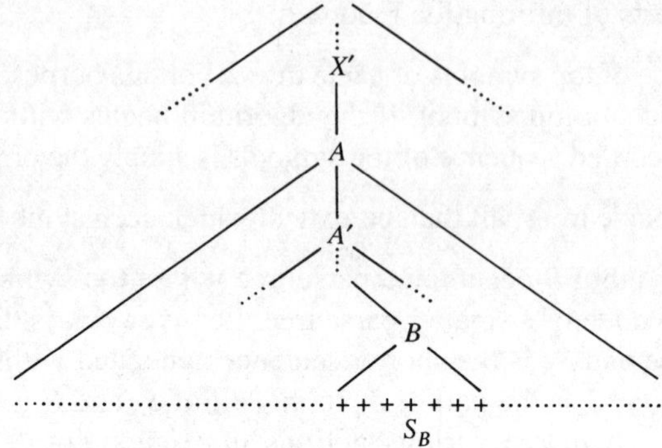


**Fig. 5.12**   An AND-tree context.

As the grammar grows larger and larger, with repeated cycles of the algorithm, the parse trees also grow. When we seek a context for $B$, it may not be wise always to work within the largest parse tree containing $B$. The contexts found from large trees are too specific. Instead, suppose the symbol $A$ is taken to be an AND symbol and, furthermore, within the tree for $A$, all the ancestors of $B$ (symbols between $B$ and $A$) are also AND symbols. For our parse, we can choose the largest which satisfies this condition, as in Fig. 5.12. Now within this tree, we can find the second kind of context, as above. Suppose it is $(L'', R'')$. This is called the AND-tree context of $B$.

Once it has recorded features of the sequence of parses, the algorithm continues:

Revise the grammar
  *Build* new AND symbols;
  *Fold* new OR symbols;
  *Rebuild* over-general OR symbols.

**Building** is simple. The algorithm scans the sequence of top symbols of parses and looks for pairs $M$, $P$ which crop up together unexpectedly often. Whenever it finds such a pair, it invents a new AND symbol, say $A$, and a production for it:

$$A \to M \quad P.$$

**Folding** invents new OR symbols and inserts them into previous productions. If several symbols (say $P$, $Q$, $R$, …) occur unexpectedly often in a context, then this step invents a new OR symbol $X$. Its productions are

$$X \to P \mid Q \mid R \mid \ldots$$

(There are variants on the folding process. It can be made to depend on several contexts, not just one.) Whenever a new OR symbol $X$ is folded, the algorithm modifies every AND production which contains an alternative for $X$. In the right-hand side of each such production, it replaces every occurrence of $P$ or $Q$ or $R$ or … by $X$. Thus, the production

$$A \rightarrow M \quad P$$

would become

$$A \rightarrow M \quad X.$$

Wolff calls this process **generalizing**.

**Rebuilding** is necessary because folding sometimes groups too many symbols together as alternatives for a single OR symbol. Maybe $P$ and $Q$ and $R$ behave alike, but there may be times when the algorithm folds a set $\{P, Q, R\}$ where, say, $Q$ sometimes occurs in a context which never contains $P$ or $R$. The algorithm looks for any context which contains some of $X$'s children but not the others. If it finds just one such context, then it forms a new OR symbol $Y$ with productions

$$Y \rightarrow P \mid R$$

and it revises the productions for $X$:

$$X \rightarrow Y \mid Q \mid ...$$

The first stage, parsing the text, is a preliminary to learning. It compiles information from which one can calculate a metric (although no explicit metric is ever actually calculated). The main clustering activity is folding. Rebuilding is a case of subdividing a cluster which has become unbalanced.

Wolff's algorithm works well with artificial data. It has also been tried with natural text, with some success. It builds sensible AND symbols, but finding meaningful OR symbols is somewhat harder. One notable feature of it is that it has succeeded in discovering *recursive* productions such as

$$A \rightarrow X \quad l$$

$$X \rightarrow A \mid a.$$

Within this grammar, $A$ can generate a whole family of strings

*al all alll allll alllll ....*

Building extends the grammar, by correlation; but it can also be looked on as a way of extending the vocabulary of the learner. When the algorithm starts with an empty grammar, the only symbols it has are those in the text. These are not enough. It also needs to study short strings within the text. Building gives it a way to refer to these strings.

In this algorithm, the clustering process is folding. It depends on contexts in parse trees and the algorithm can only calculate parse trees if it has enough AND symbols, so clustering depends on the set of available AND symbols. Any initial restriction on the form of a problem is a kind of bias. In Wolff's algorithm, the building stage adjusts the bias so that it is more suitable for folding.

There are some rigorous theorems which prove that context-free grammars cannot be learned, in a certain sense. However long a text you take, you can never be certain that a grammar you find by *any* algorithm is the one the text is generated by. If we only allow ourselves plain text as input, the output is always something of a guess. These theorems, due to Gold, will be discussed further in Chapter 8. Besides that, some features of natural human languages cannot be captured by any context-free grammar. They require a more subtle description.

In practice, people do succeed in learning to talk and understand each other. There is an old argument between so-called *rationalists* and *empiricists*. The empiricists believe that language should be learnable by experience, without any prior knowledge. The only prior knowledge a baby has is what is in its chromosomes, and chromosomes do not encode that sort of knowledge. Rationalists argue that this is impossible. Language is just too subtle. Never mind what we think is in chromosomes—if babies can learn to talk and listen, then some sort of knowledge must be in there somewhere. Wolff's approach tends to the empiricist view. It incorporates some knowledge, to wit the algorithm itself, but not more.

In real life, the situation is complicated by other factors. For a start, nobody is quite sure what is English or whatever language it is we are trying to learn. Native speakers of English all think they know, but almost any two will differ somewhat in their usage and idiom. One person's usage may be another's notion of dialect and a third person's idea of a mistake. The best definition that linguists can come up with is: your native language is what you speak. If you say it or accept it, then it is grammatically correct. (This definition upsets some people and it has caused arguments among teachers and politicians.)

When children learn, they have much more information at hand than what is provided to Wolff's algorithm. Children hear speech, which is a very subtle medium. Volume, pitch, speed, and rhythm all convey more than mere printed words. Besides that, a child learns from his mother's behaviour as they converse. The two of them look in the same direction and are interested in the same things. The mother sometimes repeats what her child has just said, asks questions, and extends a conversation. All these processes aid the learner; see the discussion in Chapter 1 on natural learning.


## 5.3   DECISION TREES

A decision tree is suitable for classifying things when each example is characterized by values for a few distinct attributes.

*Example: will it rain?*

On any particular day in Britain

- it may be sunny;

- it may have been sunny for the past week;

- it may be raining already;

- the sky last evening may have been red or yellow or some other shade;

- the visibility may be good or it may be hazy;

- there may be no wind at all;

- if there is some wind,

  - the wind may be rising or falling;

  - the wind may be predominantly from north, south, east or west;

  - the direction of the wind may be steady or it may be veering (moving towards the right) or it may be backing (moving towards the left);

and so on.

From all such evidence, one can make a reasonable guess as to whether it will soon rain. An algorithm for predicting rain might go as follows.

Is it raining now?
*Yes*: It will rain.
*No*: Is it sunny now?
    *Yes*: Has it been sunny for the past week?
      *Yes*: It won't rain.
      *No*: Is the visibility good?
        *Yes*: It will rain.
        *No*: It won't rain.
    *No*: Is there any wind?
      *Yes*: Is the wind rising?
        *Yes*: It will rain.
        *No*: Is the wind backing?
          *Yes*: It will rain.
          *No*: It won't rain.
      *No*: Was the sky red last night?
        *Yes*: It won't rain
        *No*: Is the visibility good?
          *Yes*: It will rain.
          *No*: It won't rain.

A decision tree is an algorithm in this form. It is a directed tree whose internal nodes are labelled with questions and whose arcs are labelled with the possible answers. You start at the root and follow a path through it to a leaf. At each internal node, you answer the question, and take the appropriate arc to the next question. All the leaves of the tree are labelled with assertions of the form: *It will* or *won't rain*. Notice that a question can occur more than once at different nodes in the tree, but, on any one path from the root to a leaf, each question can only occur once.

Such trees may be a little more detailed than this one. Some questions can have more subtle answers than just *Yes* or *No*, in which case the node with this question may have more than two children. Also, you might add a confidence level to each leaf:

*It will* or *won't rain, with probability ...*

The tree shown here predicts with absolute certainty that once it has started to rain, it will never stop. Not everyone believes this.

### 5.3.1   INVENTING A DECISION TREE

The learning algorithm presented here is exceptionally simple.

- It uses the Hamming metric, whereas other more elaborate versions cite information theory.

- Its questions are all supposed to have *Yes/No* answers.

- This version will not find confidence levels for its conclusions.

- It assumes that every property can be measured for every example. This is not always true. For instance, on a calm day, it makes no sense to ask whether the wind is backing, so this learning process cannot even invent the little tree above.

However, you can extend this learner easily to suit your own needs.

The algorithm is recursive. It starts with the whole training set and finds the most significant question. This is the question at the root of the tree. It splits the training set into two subsets, according to this question's answer: *Yes* or *No*. Each subset is used to build one of the two child subtrees, in the same way. The process stops when all sensible questions have been asked.

*To start:*
Set the current set of examples, *TS*, to be the whole training set.
Set the current node to be the root.

*Recursive algorithm:*
    IF all instances in *TS* have the characteristic property
    THEN
       This node is a leaf. Label it:
         the example has the property
    ELSE
    IF all instances in *TS* don't have the property
    THEN
       This node is a leaf. Label it:
         the example doesn't have the property
    ELSE
       Find the best question, if there is one (see below).
       IF there is no best question
       THEN
         the current node is a leaf.
         IF the majority of instances in *TS* have the interesting property

```
THEN
    label this leaf:
        the example has the property
ELSE
IF most don't have it
THEN
    label the leaf:
        the example doesn't have it
ELSE
    the best question labels the current node.
    Split the set TS of instances into
        TSY: those which have answer Yes
    and
        TSN: those which have answer No
    to the best question.
```

Calculate the child down the *Yes* arc recursively, with *TSY* for its *TS*.

Calculate the child down the *No* arc recursively, with *TSN* for its *TS*.

### Finding the best question: a simple approach

Suppose that there are $K$ instances in the set *TS*. We construct a Hamming pseudo-metric on the set of questions.

Put *TS* into any order.

```
FOR each question Q
    Construct a K-tuple of 0s and 1s.
        The jth entry in this tuple is
            1  if the jth instance in TS has answer Yes to Q
            0  if it has answer No.
```

The distance between any two questions is the Hamming distance between their tuples. (Two distinct questions may have the same answers for all instances, so this is only a pseudo-metric.)

Discard all questions which give just a single answer for all instances in *TS*. That is, discard it if its $K$-tuple consists all of 0s or all of 1s.

```
IF each question gives a single answer
THEN
    there is no best question
ELSE
    Construct a K-tuple for the property P which the tree is supposed to
    characterize. Its jth entry is
        1  if the jth instance has the property
        0  otherwise
```

Say the *connection* between $P$ and a question $Q$ is

   min { Hd$(P, Q)$,   $K -$ Hd$(P, Q)$ }.

This is the smaller of the Hamming distances from $Q$ to $P$ and from $Q$ to $(\neg P)$.

Among all remaining questions
say a question $Q$ is *good* if its connection with $P$ is least.

Make an arbitrary choice from among the good questions.
The best question is any one of them.

The learner's objective is always to choose the question which extracts most relevant information. Many decision trees will all lead to the same classification. You can generate lots of trees by rearranging the order in which questions are asked. A good decision tree is one which is shallow and so always classifies quickly by asking as few questions as possible.

*Example: building a decision tree — the London stock market*
   Let us assume that the major factors affecting the stock market are:

- what it did yesterday;
- what the New York market is doing today;
- bank interest rate;
- unemployment rate;
- England's prospects at cricket.

The training set is represented by Table 5.1, with one column for each instance. Attributes of instances are listed down the left margin. A Y in the table says that this instance has that attribute.The column labelled *Hd* is the Hamming distance between an attribute and the property *It rises today* which the tree is meant to characterize. Thus

   Hd (*It rises today, Bank rate high*) = 4

because there are four instances where these two have different values.

**Table 5.1**

| Instance no.: | 1 | 2 | 3 | 4 | 5 | 6 | Hd | pm |
|---|---|---|---|---|---|---|---|---|
| It rises today | Y | Y | Y | N | N | N | | |
| It rose yesterday | Y | Y | N | Y | N | N | 2 | 2 |
| NY rises today | Y | N | N | N | N | N | 2 | 2 |
| Bank rate high | N | Y | N | Y | N | Y | 4 | 2 |
| Unemployment high | N | Y | Y | N | N | N | 1 | 1 |
| England is losing | Y | Y | Y | Y | Y | Y | 3 | 3 |

The column headed *pm* is the pseudo-metric distance between an attribute and *It rises today*. Thus,

pm (*It rises today, Bank rate high*) = 2

because the training set contains 6 instances and 2 is the smaller of

Hd (*It rises today, Bank rate high*)

and

6 – Hd (*It rises today, Bank rate high*).

The question which labels the root of the tree is the one with the smallest entry in the *pm* column:

Is unemployment high?

This splits the training set into

*TSY*: instances 2, 3

*TSN*: instances 1, 4, 5, 6.

*TSY* is a leaf, because the stock market rose for all instances in this set. For *TSN*, we repeat the calculation in Table 5.2.

**Table 5.2**

| Instance no.: | 1 | 4 | 5 | 6 | | |
|---|---|---|---|---|---|---|
| It rises today | Y | N | N | N | | |
| | | | | | Hd | pm |
| It rose yesterday | Y | Y | N | N | 1 | 1 |
| NY rises today | Y | N | N | N | 0 | 0 |
| Bank rate high | N | Y | N | Y | 3 | 1 |
| Unemployment high | N | N | N | N | 1 | 1 |
| England is losing | Y | Y | Y | Y | 3 | 1 |

From these data, it appears that when unemployment is low, then the London market follows the New York market slavishly. The complete decision tree is:

Is unemployment high?

*Yes*: The London market will rise today

*No*: Is the NY market rising today?

    *Yes*: The London market will rise today

    *No*: The London market will not rise today

If you are skilled at visualizing *N*-cubes, for large values of *N*, then the following description may help. Think of each question as a dimension of such a cube, so the cube has dimension *N* if there are *N* questions. Each training instance falls on one corner of the cube. Several instances may all fall on the same corner. Positive instances, which have the property (*It will rain* or *The London market will rise*), are pink. Counter-examples are blue. Any one question cleaves the cube in half.

Finding the best question is equivalent to finding how best to cleave this cube, so that almost all pink instances fall in one half and almost all blue ones fall in the other.

The various forms of algorithms like this one tend to differ in the last bit. There are many ways of choosing the best question. Is it better to have 231 pink instances in one half and no blues, and 18 blues mixed with four pinks on the other, or is it better to have 234 pinks with three blues on the left and just 15 blues with one pink on the right? The learner described here says that the two cases are equally good. A thorough statistical analysis would distinguish between them. If ever England wins, the stock market might shoot up fantastically, but the crude algorithm given here will ignore such rare events.

### 5.3.2  ONTOLOGICAL NODES IN DECISION TREES

These arise often. There may well be questions that could occur in a classification scheme which do not apply to all possible instances.

*Example*

Imagine that you are committed to entertaining a guest for supper. You have to plan the meal. One question which may be very relevant to your cooking schedule is

Does the guest like meat rare rather than well done?

If the guest is vegetarian then this question will not apply at all, so it must be preceded by another question

Does the guest eat meat?

Vegetarians are not in the ontological class of the first question.

It often happens that one cannot choose the order of questions in a tree arbitrarily. Certain questions must be answered, and answered in a certain way, before other questions become meaningful.

This phenomenon is not unique to decision trees. It occurs in all sorts of conceptual clusterers, but it will not be discussed again. All the published accounts of such algorithms that I know of appear to be written on the assumption that all properties apply to all instances. This assumption is sometimes false, but the algorithms can easily be extended to cover cases where ontological nodes are necessary.

### 5.3.3  ENTROPY: WHY THE HAMMING DISTANCE GIVES A GOOD CHOICE OF QUESTION

The algorithm above for choosing $Q$ is usually effective but a bit crude, and you may be wondering if it is reliable. It has a sound foundation, based on the notion of entropy, which we present here, but be warned that it is not entirely satisfactory. We shall see why later.

The idea of entropy is pretty as well as practical. It is often invoked by designers of learning experiments. It was invented for information theory and in that role it provides a basis for the algorithms of Boltzmann machines. You will probably notice other situations where it can be used.

There are various criteria for deciding which question $Q$ is best at predicting $P$. One such criterion asserts that, when the answer to $Q$ is known, the answer to $P$ should surprise us as little or as rarely as possible. Entropy is a measure of surprise. A question $Q$ is good if, when we know the answer to $Q$, the answer to $P$ is not likely to surprise us.

Say $S(p)$ is the surprise evoked by the occurrence of an event which has probability $p$. Whatever 'surprise' may be, we would expect that

$$S(1) = 0$$

because then the event is bound to occur. Also

$$S(p) \text{ increases as } p \text{ decreases;}$$

and the surprise of two independent events is the sum of their individual surprises, so

$$S(pq) = S(p) + S(q).$$

It is a fact that any such function must be

$$S(p) = -\log_b p$$

for some base $b$. We opt for the particular case $b = 2$, so

$$S(p) = -\log_2 p.$$

Suppose that $X$ is a random variable (an experiment) whose possible outcomes may be

$$x_n \text{ with probability } p_n \qquad (n = 1, 2, 3, \ldots).$$

The **entropy** of $X$ is the expected value of the surprise which we shall enjoy when we perform the experiment and observe the outcome, one of the values $x_n$. This expected surprise is

$$
\begin{aligned}
H(X) &= \sum_n p_n S(p_n) \\
&= -\sum_n p_n \log_2 p_n.
\end{aligned}
$$

How can entropy be used to choose a best question? Suppose that the available questions are $Q_1, Q_2, Q_3, \ldots$ and each question's answer is just yes or no. For each $Q_j$, we can make the prediction

$$X_j: \text{ If } x \text{ has } Q_j \text{ then } x \text{ will have } P; \text{ and}$$
$$\text{if } x \text{ doesn't have } Q_j \text{ then } x \text{ won't have } P.$$

This $X_j$ is a random variable whose possible values are also yes and no. The probability that its value will be yes is, say, $\pi_j$. Then $X_j$ has entropy

$$H(X_j) = -(\pi_j \log_2 \pi_j + (1-\pi_j) \log_2 (1-\pi_j)).$$

The best question $Q_j$ is the one for which $H(X_j)$ is least. Incidentally, observe that the entropy of $X_j$ is the same as the entropy of its negation, so as one would expect, $Q_j$ is equally good at predicting whether property $P$ is true or false.

Any experimental evidence, such as was tabulated in the example above for the London stock market, will let us guess the most likely values of these probabilities. In most discussions of decision trees, authors estimate the most likely values of these probabilities and then calculate $H(X_j)$ from these estimates. Thus if there are two questions, and the sizes of the various subsets of instances are as shown in Table 5.3,

**Table 5.3**

| $P$: | yes | | no | |
|------|-----|-----|-----|-----|
| $Q$: | yes | no | yes | no |
| $Q_1$: | 231 | 4 | 0 | 18 |
| $Q_2$: | 234 | 1 | 3 | 15 |

then the estimates usually adopted would be

$$\pi_1 = \frac{\text{the number of instances which have both } P \text{ and } Q_1 + \text{ the number of instances which have neither}}{\text{the size of the training set}}$$

$$= \frac{231+18}{253}$$

$$= \frac{249}{253}$$

and

$$\pi_2 = \frac{249}{253},$$

similarly. Hence the two experiments $X_1$ and $X_2$ will have the same estimated entropy and these data suggest that $Q_1$ and $Q_2$ are equally good.

Note that the formula for the probability $\pi_j$ is just the connection between $P$ and $Q_j$ divided by the size of the training set. The graph of entropy against probability, $H$ against $\pi$, is shaped like a symmetric upside-down bulge and is illustrated in Fig. 5.13.

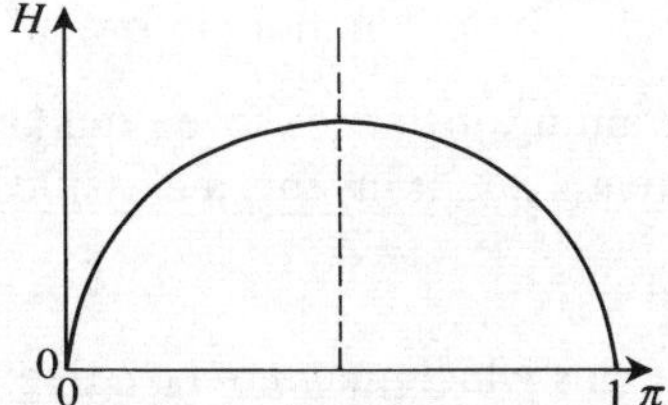

**Fig. 5.13**   The graph of entropy for an experiment with just two possible outcomes.

$H$ is zero when $\pi$ is 0 or 1 and is positive in between. The best question $Q$ is one whose $X$ has a probability $\pi$ which is nearly 0 or nearly 1. That is to say, the Hamming distance between $Q$ and $P$ should be very small or very large — just as we said before.

A word of warning: if you calculate probabilities and entropies by formula, as I did on my home computer, your answers will contain *rounding errors*. For the figures above, my machine worked out that

$$H(X_1) \approx 0.1172208$$
$$H(X_2) \approx 0.1172212.$$

It appears that $Q_1$ is a better question than $Q_2$. This is *wrong*.

Another word of warning: the estimates of entropy calculated here are not the best. The reason is that entropy does not depend linearly on $\pi$. Besides that, the entropy technique may not even be the best criterion to choose between alternative experiments. More details can be found in the Appendix, Section 5.8.

## 5.4   BUILDING HIERARCHIES: WINNER TAKES ALL

We shall next study a simplified form of Al-Mathami's method. The input to his algorithm is a sequence of training instances and its output is a hierarchy (a partial order) which can be used to predict properties of further instances.

Of all the algorithms which we shall discuss, this is perhaps the closest to a pure conceptual clusterer. Like the one which builds decision trees, it extends its hierarchy by a recursive process beginning at a root node. However, there the similarity ends. Unlike the decision-tree builder, this algorithm

- is incremental. The algorithm allocates instances to nodes in the hierarchy one at a time. Any such allocation may cause the hierarchy to be extended and the hierarchy can be used for classification at any stage during learning.

- allocates instances to nodes on the basis of more than just one feature. At each node in a decision tree, there is just one question which distinguishes between the node's children. In the hierarchy produced by Al-Mathami's algorithm, the choice between children is based on more information.

- does not require its training instances to be marked as positive and negative examples. Like Kohonen's net, it invents a classification to suit its data.

Al-Mathami's full version

- forms a hierarchy of nodes which may not be a tree;

- can inherit much information from parent nodes to children;

- is a chunking process—Sibling clusters may overlap,

but the simple version presented here just constructs a tree with no inheritance and disjoint siblings.

Each instance is described by a finite set of values for attributes, so an instance is described by a set of the form

$$\{a{:}v,\ a'{:}v',\ \dots\ a''{:}v''\}.$$

All the attribute types $a$ form a set, called $A$. Each attribute type $a$ in $A$ has just a finite set $V_a$ of possible values:

$$V_a = \{v_1^a,\ v_2^a,\ v_3^a,\ \dots\ v_{na}^a\}.$$

One novel feature of this method is that the description of any instance may be *incomplete*. The method correlates the known attribute values with others, not known, and can guess what the unknown ones should be.

Each node of the hierarchy represents a cluster. The top node corresponds to all instances, the cluster containing everything, and lower nodes represent progressively smaller refinements. Particular instances are not recorded in the hierarchy, so the node does not contain an extensional description of its cluster. In fact, it does not contain a full intensional description either. Assignment of an instance to a node's cluster depends on that node and on its ancestors and siblings.

Each node consists of

- pointers to its parent and its own children;

- a set of attribute types

$$a_1, a_2, \dots a_k.$$

If the attribute type $a$ is in the set for a node $C$, then every instance in $C$'s cluster must be able to have a value for $a$ (although this value may not be known);

- the numbers

$$e_{v_1,\ v_2,\ \dots\ v_k}$$

of instances in $C$'s cluster which are known to have attribute value $v_i$ for type $a_i$, for each $i \leq k$.

The quantities of prime interest in $C$ are the numbers $e_{v_1}, \ldots{}_{v_k}$. These numbers are used to describe $C$'s cluster and also to make predictions about unknown attribute values. Thus, the node has an array with $k$ dimensions, one for each of its attributes. The coordinate values along the dimension $a$ are the possible attribute values $v$ for $a$, and the entry indexed by values $(v_1, v_2, \ldots v_k)$ is $e_{v_1, v_2, \ldots v_k}$.

Al-Mathami chose sometimes not to store all of this array. Instead, he let his program estimate the entries in it. Say, $X_v^a$ is the number of instances in $C$'s cluster with known value $v$ for $a$; thus, for each attribute $a$ in $C$,

$$\sum_{v \in V_a} X_v^a = N_a$$

where $N_a$ is the number of instances in the cluster with a known value for attribute $a$. The node $C$ contains pointers to its parents and its children, and these numbers $X_v^a$, and it may also contain the array of counts $e_{v_1, v_2, \ldots v_k}$.

*Example*

Let $A$ be the set of discrete attributes of birds:

$A = \{\text{size, vision, wing length, beak shape}\}$

where

$$
\begin{aligned}
V_{\text{size}} &= \{\text{small, medium, large}\} \\
V_{\text{vision}} &= \{\text{wide, forward}\} \\
V_{\text{wing length}} &= \{\text{short, long}\} \\
V_{\text{beak shape}} &= \{\text{short, long, curved}\}.
\end{aligned}
$$

A certain node $C$ might have attributes *size* and *vision* and *wing length*. For these three attributes, its cluster contains

$$
\begin{aligned}
&\text{size:} \\
&\quad \text{small} \qquad X_{\text{small}}^{\text{size}} = 15 \text{ birds} \\
&\quad \text{medium} \qquad X_{\text{medium}}^{\text{size}} = 6 \text{ birds} \\
&\quad \text{large} \qquad X_{\text{large}}^{\text{size}} = 0 \text{ birds} \\
&\text{vision:} \\
&\quad \text{wide} \qquad X_{\text{wide}}^{\text{vision}} = 21 \text{ birds} \\
&\quad \text{forward} \qquad X_{\text{forward}}^{\text{vision}} = 0 \text{ birds} \\
&\text{wing length:} \\
&\quad \text{short} \qquad X_{\text{short}}^{\text{wing}} = 18 \text{ birds} \\
&\quad \text{long} \qquad X_{\text{long}}^{\text{wing}} = 3 \text{ birds}
\end{aligned}
$$

In this case, $N_{\text{size}} = N_{\text{vision}} = N_{\text{wing}} = 21$. This is the number of birds in the cluster. One of the birds in its cluster might have

size: small
vision: wide
wing length: long.

It might also have a short beak, but that detail will not be recorded in $C$.

The node $C$ might also record precisely how many known instances have each combination of attribute values. In the above example, this would involve noting within $C$ all the numbers
How many birds
    are small, have wide vision, and short wings;
    are small, have wide vision, and long wings;
    are small, have forward vision, and short wings;
    are small, have forward vision, and long wings;
    are medium,...

### 5.4.1 CLASSIFICATION AND PREDICTION

Suppose the algorithm has found that some instance $x$ should be assigned to a node $C'$ with children and it has to allocate $x$ further to some child of $C'$. It will opt for the child $C$ which, according to estimates, already has the most instances matching $x$. This is the **winner-takes-all** strategy.

When $C$ does not include the numbers $e_{v_1,\ v_2,\ ...,\ v_k}$, the estimates used for them are the ones with greatest experimental backing. This notion of backing was explained in Chapter 2. In this case there are $N$ birds in the cluster for $C$. We also know that their numbers are constrained by the counts $X_j^i$. In this situation, the most likely value is about

$$e_{v_1,\ v_2,\ ...\ v_k} \quad \approx \quad N \times \frac{X_{v_1}^{a_1}}{N} \times \frac{X_{v_2}^{a_2}}{N} \times ... \frac{X_{v_k}^{a_k}}{N}$$

and this is the value used. (The proof that this is the most likely value uses a technique called Lagrange's method of undetermined multipliers.)

*Example*

Say $C$ is the node in the example above, and $C$ has a parent called $C'$. Also suppose that $C'$ has another child node called $D$ whose attribute types are as in $C$ but whose numbers are

size:
    small               4 birds
    medium            4 birds
    large              7 birds
vision:
    wide               3 birds
    forward        12 birds
wing length:
    short            6 birds
    long             9 birds

The new instance $x$ could be a bird with attributes

   size: small
   vision: wide
   wing length: long

For $C$, the estimate of $e_{\text{small, wide, long}}$ is

$$21 \times \frac{15}{21} \times \frac{21}{21} \times \frac{3}{21}$$

which is about 2. For $D$, the equivalent estimate is

$$15 \times \frac{4}{15} \times \frac{3}{15} \times \frac{9}{15}$$

which is about $\frac{1}{2}$. Hence, the winner-takes-all strategy decrees that $x$ is allocated to the child $C$ of $C'$, not to $D$.

An instance can still be assigned to a child when not all its attributes are known. Suppose that $y$ is an instance with the attributes

   size: medium
   vision: forward

but unknown wing length. In that case, the system can estimate the numbers $e_{\text{medium,forward}}$ for the classes $C$ and $D$. They are

$$21 \times \frac{6}{21} \times \frac{0}{21}$$

which is 0 for $C$, and

$$15 \times \frac{4}{15} \times \frac{12}{15}$$

which is about 3 for $D$. Hence $y$ will be assigned to $D$. The system can then guess $y$'s wing length. The value of this attribute, $v$, is the one which maximizes $e_{\text{medium, forward, } v}$ in $D$. Since

$$e_{\text{medium, forward, short}} \simeq 15 \times \frac{4}{15} \times \frac{12}{15} \times \frac{6}{15}$$

$$e_{\text{medium, forward, long}} \simeq 15 \times \frac{4}{15} \times \frac{12}{15} \times \frac{9}{15}$$

the system will conjecture, if asked, that $y$ has long wings.

Al-Mathami mentions an alternative approach to instances with unknown attribute values. It is to include an extra attribute value called *unknown* for each attribute type. This may produce better results than the one above. Long wings are more conspicuous than short ones, so the very fact that the wing length was not observed is evidence that $y$'s wings are actually short.

To summarize: the performance aspect of the algorithm takes as input a hierarchy and an instance. It produces a node of the hierarchy which best classifies the instance. In the process, it adds to its counts of known instances in this output node and all its ancestors. It runs

$C := $ the root node.
$x := $ the instance.
FOR each relevant attribute $a_i$
    $v_i := $ the value of $a_i$ in $x$.
IF $x$ has values for all attributes $a$ in $C$
THEN
    FOR each such attribute $a_i$

$$X_{vi}^{ai} := X_{vi}^{ai} + 1$$

IF $C$ includes the count $e_{v_1, v_2, \ldots}$
THEN

$$e_{v_1, v_2, \ldots} := e_{v_1, v_2, \ldots} + 1$$

WHILE $C$ has children
    $D := $ the child of $C$ for which $e_{v_1, v_2, \ldots v_k}$ is greatest
            where $a_1, a_2, \ldots a_k$ are the attribute types common
                to $x$ and $D$, and $v_1, v_2, \ldots v_k$ are the values
                for $x$ of these attributes;
    $C := D$;
    IF $x$ has values for all attributes $a$ in $C$
    THEN
        FOR each such attribute $a_i$

$$X_{vi}^{ai} := X_{vi}^{ai} + 1$$

IF $C$ includes the count $e_{v_1, v_2, \ldots}$
THEN

$$e_{v_1, v_2, \ldots} := e_{v_1, v_2, \ldots} + 1$$

Return $C$

This version of the algorithm on its own incorporates a little learning, as it augments the counts $X$ and $e$ which are used in classification. However, the main learning procedure is more elaborate.

### 5.4.2   EXTENDING THE HIERARCHY

At each state of the above loop, the algorithm can leave the structure of the hierarchy unchanged. Alternatively, it can modify the hierarchy so that the classification process is improved. There are three basic operators which do this:

- *create* a new child;

- *fuse* two (or perhaps more) existing children;

- *divide* a child.

The first two are shown in diagrammatic form in Fig. 5.14.

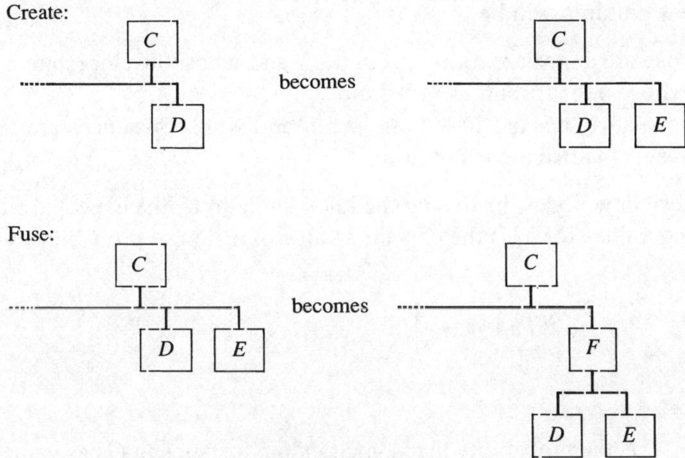


**Fig. 5.14**   Operations in Al-Mathami's clusterer.

*Divide* is the reverse of *fuse*. Al-Mathami notes a few similarities and differences between these operators and analogous ones in programs called CLUSTER/2 (Michalski) and UNIMEM (Lebowitz), but he does not state how the operators set up attributes and counts in new nodes, and he does not give precise criteria for deciding when to apply an operator. Hence, what follows may not be an accurate description of any particular program.

The program has two functions: it *classifies* instances and it *predicts* unknown attribute values. Prediction matters more than classification. The program can be said to work well if its predictions have a high success rate.

The best possible predictor would be the one with a 'hierarchy' consisting of a single node, using all possible attributes and recording all the counts $e_{v_1, \ldots v_k}$ exactly. The reason people invent other kinds of predictor is that this ideal one requires too much computation and storage, and it does not classify. One can often get very good results without testing all attributes or storing all counts. Also, often, there is no need to be able to predict values for all attributes. One can almost always see whether a bird is large or small, so this attribute can be regarded as known. Each classifier is designed so that it is good at predicting values of some particular attribute. In future, we shall concentrate on predicting the value of $a_1$.

The art of designing a good hierarchy lies in deciding what will be the best way to introduce new children and new attributes. Let us investigate a process for constructing hierarchies, using just a single operation. Say $C$ is a node in some hierarchy, presumably a leaf, and we want to invent two children of $C$. The operation does this by

(1)   selecting some attribute $a_i$ where $2 \leq i \leq k$; and

(2)   selecting some non-empty subset $W$ of the set $V_i$ of possible values of $a_i$; and

(3)   selecting a new attribute $a_{k+1}$ which is not tested by $C$.

The two new children will be

> $D$     whose attributes are $a_1, a_2, \ldots a_k, a_{k+1}$ and whose instances are all those of $C$ whose $a_i$th attribute is in $W$; and
>
> $E$     whose attributes are just $a_1, a_2, \ldots a_k$ and whose instances are all the rest, whose $a_i$th attribute is not in $W$.

Without these new nodes, just using the knowledge in $C$, the expected success rate of predicting values for $a_1$ is the expected value of a sum of probabilities

$$E\left( \sum_{(v_2, \ldots v_k)} p_{v_2, \ldots v_k} \times p_{a_1 | v_2, \ldots v_k} \right).$$

Here,

$p_{v_2, \ldots v_k}$     is the probability that some random instance in $C$ has values $v_2, \ldots v_k$ for attributes $a_2, \ldots a_k$;

$p_{a_1 | v_2, \ldots v_k}$     is the probability that the system will correctly predict the value of attribute $a_1$ for any random instance with known values $v_2, \ldots v_k$ for attributes $a_2, \ldots a_k$.

These probabilities are not known. All we know is the total number $N$ of observed instances for $C$ and the counts $e_{v_1, v_2, \ldots v_k}$. Say

$$e_{v_2 \ldots v_k} = \sum_{v_1 \in V_1} e_{v_1, v_2, \ldots v_k}.$$

If $N$ were fixed in advance and all these counts were independent, and we assume in advance for lack of more information that $p_{v_2, \ldots v_k}$ could be anywhere between 0 and 1 with even probability density, then the expected value of $p_{v_2, \ldots v_k}$ would be

$$\frac{e_{v_2, \ldots v_k} + 1}{N + 2}.$$

(This is shown in the Appendix, section 5.8.4.) Similarly, if $e_{v_2, \ldots v_k}$ were known in advance and $e_{v_1, v_2, \ldots v_k}$ were independent of it, then $p_{a_1 | v_2, \ldots v_k}$ would be

$$\max_{v \in V_1} \left\{ \frac{e_{v, v_2, \ldots v_k} + 1}{e_{v_2, \ldots v_k} + 2} \right\}.$$

The necessary assumptions do not all hold, but, when there is at least one observed instance with attribute values $v_2, \ldots v_k$, let us take these as good working estimates. When there is no such known instance, then $p_{a_1 \mid v_2, \ldots v_k}$ is perhaps best forgotten. Thus, our estimate of the success rate of the original predictor is

$$\sum_{(v_2, \ldots v_k)} n_{v_2, \ldots v_k}$$

where the sum is only over the subscripts $(v_2, \ldots v_k)$ for which

$$e_{v_2, \ldots v_k} \geq 1$$

(so there is an instance with these attribute values) and

$$n_{v_2, \ldots v_k} = \frac{e_{v_2, \ldots v_k} + 1}{N + 2} \times \max_{v \in V_1} \left\{ \frac{e_{v, v_2, \ldots v_k} + 1}{e_{v_2, \ldots v_k} + 2} \right\}.$$

Any improvement in the predictive power of the new classifier can only arise because the child $D$ has an extra attribute. If it didn't, then it would make exactly the same predictions as the original one. The contribution from the node $E$ will be just the part of the sum over values $(v_2, \ldots v_k)$ when $v_i \notin W$. The contribution from $D$ will be a sum over values $(v_2, \ldots v_k)$ when $v_i \notin W$, but the summands are different. Say $n_{v_2, \ldots v_k, v_{k+1}}$ is

$$\frac{e_{v_2, \ldots v_k, v_{k+1}} + 1}{N + 2} \times \max_{v \in V_1} \left\{ \frac{e_{v, v_2, \ldots v_k, v_{k+1}} + 1}{e_{v_2, \ldots v_k, v_{k+1}} + 2} \right\}.$$

Our estimate of the success rate for predictions by the new classifier is

$$\sum_{\substack{(v_2, \ldots v_k): \\ v_i \notin W \text{ and there is} \\ \text{such an instance}}} n_{v_2, \ldots v_k} \quad + \quad \sum_{\substack{(v_2, \ldots v_k, v_{k+1}): \\ v_i \in W \text{ and there is} \\ \text{such an instance}}} n_{v_2, \ldots v_k, v_{k+1}}.$$

The reasoning behind this estimate depends on some approximations which should be quite good in most cases.

*Example*

Say the attributes in $C$ are *size* and *vision*, and the array of counts is

|  |  |  | vision: |
|---|---|---|---|
| 6 | 2 | 0 | all round |
| 0 | 3 | 1 | forward |
| size: small | medium | large | |

If *vision* is $a_1$, the attribute to be predicted, and so *size* is $a_2$, then

$$n_{\text{small}} = \frac{7}{14} \times \max\left\{\frac{7}{8}, \frac{1}{8}\right\} = \frac{7}{16} \simeq 0.439$$

$$n_{\text{medium}} = \frac{6}{14} \times \max\left\{\frac{3}{7}, \frac{4}{7}\right\} = \frac{12}{49} \simeq 0.245$$

$$n_{\text{large}} = \frac{2}{14} \times \max\left\{\frac{1}{3}, \frac{2}{3}\right\} = \frac{2}{21} \simeq 0.095$$

so our estimate of the expected success rate is 78 per cent.

   Now suppose that the operator is applied, with

$$i = 2 \qquad W = \{medium\}.$$

Also suppose that the observed instances actually all have known wing length as
well. The numbers of instances with various wing lengths are shown in Table 5.4,
which is obtained from the one above for $C$ by splitting each count into two parts.

**Table 5.4** Number with short wings/number with long wings.

|  |  |  | vision: |
|---|---|---|---|
| 5/1 | 2/0 | 0/0 | all round |
| 0/0 | 1/2 | 0/1 | forward |
| size: small | medium | large | |

Thus, the equivalent table for $D$ is

|  |  |  |  | vision: |
|---|---|---|---|---|
| 0/0 | 2/0 | 0/0 | | all round |
| 0/0 | 1/2 | 0/0 | | forward |
| size: | small | medium | large | |

and the one for $E$ is

|  |  |  |  | vision: |
|---|---|---|---|---|
| 6 | 0 | 0 | | all round |
| 0 | 0 | 1 | | forward |
| size: | small | medium | large | |

The various summands $n$ which contribute are:
from $D$,

$$n_{\text{medium, short}} = \frac{4}{14} \times \max \left\{ \frac{3}{5}, \frac{2}{5} \right\} = \frac{6}{35} \simeq 0.171$$

$$n_{\text{medium, long}} = \frac{3}{14} \times \max \left\{ \frac{3}{4}, \frac{1}{4} \right\} = \frac{9}{56} \simeq 0.161$$

and from $E$,

$$n_{\text{small}} = \frac{7}{14} \times \max \left\{ \frac{7}{8}, \frac{1}{8} \right\} = \frac{7}{16} \simeq 0.439$$

$$n_{\text{large}} = \frac{2}{14} \times \max \left\{ \frac{1}{2}, \frac{2}{3} \right\} = \frac{2}{21} \simeq 0.095$$

Hence, the extended hierarchy will allow predictions of birds' vision with an expected success rate of about 87 per cent.

The operation can be applied in several different ways to this node $C$. The set $W$ could be any one of the seven non-empty subsets of {*small, medium, large*}.

The value chosen for $W$ in this example, {*medium*}, is quite a good choice. It produces a new hierarchy with substantially better predictive power. Note that the choice {*small, medium, large*} for $W$ would make $E$ an empty node, which can be ignored. This is tantamount to adding an extra attribute to $C$.

The full version of the algorithm, with this operator for extending the hierarchy, is

The root node    : = a new node with no counts and no attributes;
The hierarchy    : = {the root node};
WHILE there remains a training instance
    $C$ : = the root node;
    $x$ : = the next instance;
    FOR each relevant attribute $a_i$
        $v_i$ : = the value of $a_i$ in $x$;
    IF $x$ has values for all attributes $a$ in $C$
    THEN
        FOR each such attribute $a$

        $$X_v^a := X_v^a + 1$$

        where $v$ is $x$'s value for $a$;
        IF $C$ includes the count $e_{v_1, v_2, \ldots}$
        THEN

        $$e_{v_1, v_2, \ldots} := e_{v_1, v_2, \ldots} + 1$$

    WHILE $C$ has children
        $D$ : =  the child of $C$ for which $e_{v_1, \ldots v_k}$ is greatest
            where $a_1, a_2, \ldots a_k$ are the attribute types common to $x$ and $D$,
            and $v_1, v_2, \ldots v_k$ are the values for $x$ of these attributes;
        $C := D$;
        IF $x$ has values for all attributes $a$ in $C$
        THEN
            FOR each such attribute $a$

            $$X_v^a := X_v^a + 1$$

            where $v$ is $x$'s value for $a$;
            IF $C$ includes the count $e_{v_1, v_2, \ldots}$
            THEN

            $$e_{v_1, v_2, \ldots} := e_{v_1, v_2, \ldots} + 1;$$
    (At this point, $C$ is a leaf.)
    $E_C$   : =   the expected success rate of predictions, using counts in $C$;
    FOR each pair, $(a, W)$
        $E_{(a, W)}$  : = the expected success rate of predictions if $C$ is extended
                by the operator with parameters $a$ and $W$;

    IF there is any pair $(a, W)$ for which $\dfrac{1 - E_{(a,W)}}{1 - E_C} < 0.8$

    THEN
        extend the node $C$ with the best such pair;
    Return $C$.

In the last test, the constant 0.8 is arbitrary. It restricts extensions of the hierarchy to cases where it will produce a 20 per cent improvement in the expected failure rate of predictions. If this test is not made and the training set is noisy, then any bit of noise might cause an erroneous extension of the tree.

Most conceptual clusterers have a form similar to this one. The general structure is

    Set up an empty hierarchy;
    WHILE there is another training instance
        classify it, at a certain leaf node $C$;
        IF there is an operator which can usefully modify the hierarchy at $C$
        THEN
            apply the best such operator;
    Return $C$.

Some may also modify their hierarchies internally, at ancestors of the node $C$.

This discussion has emphasized the test to decide whether an operator will be useful. The particular test used here is designed to optimize prediction. It may not lead to a particularly neat classification. Most published descriptions of clusterers emphasize the operators themselves, and the operators are designed to classify rather than to predict. Some accounts only give sketchy summaries of the tests. For details of another clusterer with many more operators, capable of generating a more elaborate hierarchy, see Section 5.6 below.

## 5.5  GRAPHS OF PROPERTIES

Imagine that your employer makes motor cars and you want an expert system which helps decide which potential customer will buy which sort of car. It is a fact that rich men tend to buy big cars. Hence, it is tempting to put the rule

$$\text{male } (X) \wedge \text{rich } (X) \wedge \text{big\_car } (C) \Rightarrow \text{may\_buy } (X, C)$$

into the expert system. That would be a mistake. The point is, although this rule is correct, it does not embody the full logic of the customer's choice. A more subtle sensitive system would contain the rules

$$
\begin{aligned}
\text{male } (X) \wedge \text{rich } (X) &\Rightarrow \text{large\_family } (X) \\
\text{large\_family } (X) \wedge \text{big\_car } (C) &\Rightarrow \text{may\_buy } (X, C) \\
\text{male\_} (X) \wedge \text{rich } (X) &\Rightarrow \text{vain } (X) \\
\text{vain } (X) \wedge \text{big\_car } (C) &\Rightarrow \text{may\_buy } (X, C).
\end{aligned}
$$

These imply the first rule.

A learner will be of little use if it clutters up the expert system with redundant derived rules. The learner should detect which relationships are fundamental and which others can be derived. The next algorithm is designed to find the fundamental relationships. It was described by Fung and Crawford. Its inputs consist of

- a finite set $F$ of possible features;

- for each feature $n$, its set $V_n$ of possible values, which is finite too;

- a training set $E$ of examples. Each example is described by stating the value that each feature has for it.

Its output is

- a graph whose nodes are the features. There is an arc between two features if they correlate and if the correlation between them is not conditional on the value of any other feature.

Such a graph is called a **Markov network** or a **causal model**.

*Example*

A keypad has ten keys, for the digits 0, 1, 2, ... 9. It is connected to an LED display consisting of eight components which should show the digit pressed (see Fig. 5.15).

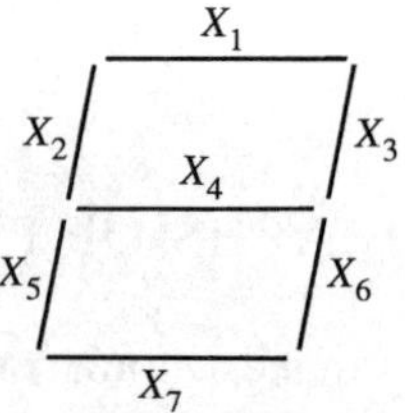


**Fig. 5.15**   An LED display.

For instance, if the digit 5 is pressed, then the segments marked $X_1$, $X_2$, $X_4$, $X_6$, and $X_7$ should go on and display a 5 as shown in Fig. 5.16.

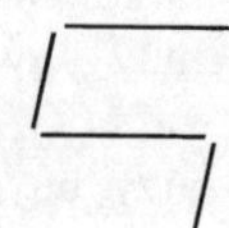

**Fig. 5.16**   An LED display switched to '5'.

In this case, there are eight features: the seven components which can each be *on* or *off* and the key which can be any one of the digits 0, 1, ... 9.

The training set consists of several depressions, roughly evenly distributed over the keys, with, for each depression, a note of which components came on. Thus, one training instance might be

(Key: 5, $X_1$: *on*, $X_2$: *on*, $X_3$: *off*, $X_4$: *on*, $X_5$: *off*, $X_6$: *on*, $X_7$: *on*)

The seven components are correlated. For instance, whenever $X_5$ is on then $X_7$ should be too (in digits 0, 2, 6, and 8). However, there will not be an arc in the output graph between $X_5$ and $X_7$ because the correlation between them is conditional on the value of the pressed key. If we know what key is pressed, then the observed values of $X_5$ and $X_7$ can be explained without assuming any correlations between them. The Markov network for this system is shown in Fig. 5.17. The algorithm can recover this network, even when the display is a bit faulty and some components do not come on when they should or do when they shouldn't, as long as the faults between components are not correlated.

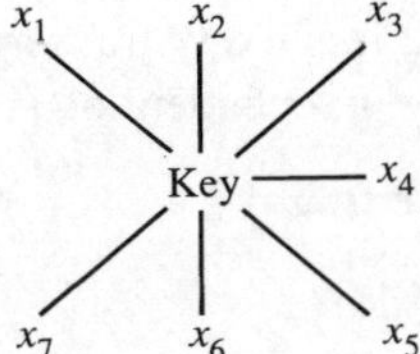


**Fig. 5.17**   The Markov network for an LED display.

The arcs in this graph are not directed. That means that the graph only contains information about correlations and not about causal relationships. If the events recorded by the various features occur in a known time sequence, then of course it is reasonable to assume that an early event causes a later one, if they are correlated at all. In this example, the components of the LED display come on at the same instant that the key is pressed, so this algorithm does not tell us anything about causes.

The key notion in this method is **conditional independence**. Suppose than $n$ is a node in the graph (a feature), and $S$ is a set of features not including $n$, and $T$ is another set of features disjoint from $S$. We say that

$n$ is *conditionally independent* of $T$ given $S$, and write this as $I(n, S, T)$

if, when the values of features in $S$ are known, knowledge of the values of features in $T$ does not help to predict the value of $n$.

In the example, say $n$ is $X_5$, and $S$ is $\{Key\}$ and $T$ is $\{X_7\}$. Then $n$ does correlate with $T$; but if the value of $Key$ is known then the value of $n$ is known too, so knowing the value of $X_7$, will not tell us any more about $n$. Hence, this $n$ is conditionally independent of this $T$, given $S$. We write

$I(X_5, Key, X_7)$.

There is not much point in the curly brackets when the sets $S$ and $T$ are singletons.

The idea of the algorithm is that each feature $n$ has an associated *smallest* set $S$, say $B_n$, which shields it from all other features.

$$I(n, B_n, F \setminus (B_n \cup \{n\}) ).$$

This smallest $B_n$ is called the **Markov boundary** of $n$ in the graph. When the graph's arcs are not directed, so the graph does not embody any causal information, this Markov boundary contains just those features which $n$ is linked to. The core of the algorithm finds the Markov boundary of each node.

The algorithm is statistical. Its outcome depends on the size of the training set and on the numbers of times that each feature has each value. In the example of the LED display, it will only work reliably if every digit is pressed at least a few times. If we want to be reasonably confident that we have found the real Markov boundary of each node, then we must choose some explicit confidence level, say $\alpha$. For any $n$, $S$, and $T$, we shall take it that $I(n, S, T)$ is true if the backing for the hypothesis that it is true is at least $\alpha$.

If, on the assumption that $n$ is conditionally independent of $T$,

the probability of observing a less likely training set is at least $\alpha$

then infer that $n$ is conditionally independent of $T$.

In symbols, $I(n, S, T)$ is taken to be *true* if

$$\sum_{E':P_I(E')<P_I(E)} P_I(E') \geq \alpha.$$

Here as always, $E$ is the actual observed training set and the sum is over alternative conceivable training sets $E'$. $P_I(X)$ is the probability of a conceivable outcome $X$ estimated on the assumption that $n$ is conditionally independent of $T$, given $S$. In their implementation, Fung and Crawford estimated these sums by an approximation called the $\chi^2$ test.

*Example: how to estimate $P_I(E)$*

There are three features of the weather:

   $p$    the atmospheric pressure is low;
   $w$   there is a lot of wind;
   $r$    there is rain.

Each may take the value $Y$ or $N$, so there are altogether eight possible outcomes of a single experiment. Imagine that a meteorologist makes 20 observations on different days and the numbers of these outcomes are as shown in Table 5.5.

**Table 5.5**

|       |   |   |   |   |   |   |   |   | *Total Y* | *Total N* |
|-------|---|---|---|---|---|---|---|---|-----------|-----------|
| $p$   | Y | Y | Y | Y | N | N | N | N | 11        | 9         |
| $w$   | Y | Y | N | N | Y | Y | N | N | 9         | 11        |
| $r$   | Y | N | Y | N | Y | N | Y | N | 10        | 10        |
| *no.* | 5 | 2 | 4 | 0 | 0 | 2 | 1 | 6 |           |           |

From these figures, he estimates the probabilities of the various possible events. For instance, the probability that it will be windy on any random day is taken as

$$p_w \simeq 9/20$$

and the probability that it will be dry and not windy is

$$p_{\neg w \wedge \neg r} \simeq 6/20.$$

These estimates of probabilities are a bit haphazard, but they will do for the moment. (The study in the appendix (p.221) suggests that the best estimate of $p_w$ is 10/22 and the best estimate of $p_{\neg w \wedge \neg r}$ is 7/22.)

Ideally, we would like to have estimates of these probabilities which are independent of the training set used to learn the Markov network. In practice, we do not. The same sequence of observations is also the training set which we call $E$. Each observation in $E$, which we can call $O$, is described by the three values:

$v_r$   which is $Y$ if it rained and otherwise $N$;
$v_p$   which is $Y$ if pressure was low and otherwise $N$;
$v_w$   which is $Y$ if it was windy and otherwise $N$.

The probability $P(O)$ of $O$ is

$$P(v_p)\, P(v_w \,|\, v_p)\, P(v_r \,|\, v_p \wedge v_w).$$

Under the hypothesis $I(r, p, w)$, $P(v_r|v_p \wedge v_w)$ is just $P(v_r|v_p)$ so the probability $P_I(O)$ is

$$P(v_p)\, P(v_w|v_p)\, P(v_r|v_p).$$

Note that the conditional probability $P(v_r \,|\, v_p)$ is

$$\frac{P(v_r \wedge v_p)}{P(v_p)}$$

and similarly for $P(v_w|v_p)$. Hence,

$$P_I(O) = \frac{P(v_w \wedge v_p)P(v_r \wedge v_p)}{P(v_p)}.$$

The quantities $P(v_p)$ and $P(v_r \wedge v_p)$ are just the numbers $p$ which were estimated from the first observation sequence. For instance, if $v_p$ is $Y$, $v_w$ is $N$, and $v_r$ is $Y$ then

$$P_I(O) \;=\; \frac{P_{p \wedge \neg w} P_{p \wedge r}}{P_p} \;\simeq\; \frac{(4/20)\times(9/20)}{11/20} \;=\; \frac{36}{220}$$

$$\simeq 0.164$$

If all observations in the sequence $E$ are independent (for instance, if there was at least a week between any two), the probability $P_I(E)$ of the entire sequence $E$ is the product of the probabilities $P_I(O)$ of the individual observations in $E$.

Now we revert back to the general case. Different values of $\alpha$ will yield different networks. Fung and Crawford use an ingenious means for choosing $\alpha$. They split the training set $E$ arbitrarily into two parts, $E_1$ and $E_2$. They calculate several versions of the graph, using $E_1$ and several different values of $\alpha$. Then, of all these candidate graphs, they choose the one which maximises $P(E_2)$.

That is an outline of the algorithm. In practice, estimating the independence relation $I$ involves a lot of heavy calculation. To simplify matters and avoid considering all subsets of $F \setminus \{n\}$ as candidate sets $T$ on which $n$ might depend, Fung and Crawford assume that $n$ is conditionally independent of $T$, given $S$, whenever $n$ is conditionally independent of each element $t$ in $T$:

$$I(n, S, T) \quad \Leftrightarrow \quad \forall t \in T \quad I(n, S, t)$$

for every $n$ and $S$ and $T$. An independence relation $I$ which satisfies this condition is said to be **composable**. When $I$ is composable, the set $F$ of all features can be partitioned into subsets $F_1, F_2, \ldots$ which are mutually independent:

$$\forall n \in F_i \quad \forall n' \in F_j \quad I(n, \{\ \}, n') \qquad \text{if } i \neq j.$$

Then the algorithm can be applied to each subset $F_i$ separately.


### 5.5.1 FINDING MARKOV BOUNDARIES

For each node $n$, the algorithm starts from the assumption that its $B_n$ is as small as possible. It keeps a set of candidates for $B_n$; say the set of candidates is called $SC$. Initially, there is just one candidate, $C_0$, so $SC = \{C_0\}$. When $n$ is the first node considered, $C_0$ will be empty. When $n$ is a later node, $B_n$ will have to contain all previous nodes whose boundaries include $n$, since

$$n' \in B_n \Leftrightarrow n \in B_{n'}$$

so $C_0$ may be larger. In general,

$$C_0 = \{\ n' \mid B_{n'} \text{ is known and } n \in B_{n'}\ \}.$$

If $SC$ contains some candidate $C$ such that every other node is conditionally independent of $n$, given $C$, then $B_n = C$. Otherwise, the algorithm chooses some smallest $C$ in $SC$ and extends it in all possible ways with one new node $n'$. The conditions on $n'$ are that

$$n' \neq n, \qquad n' \notin C, \qquad \neg I(n, C, n').$$

Since the chosen $C$ is smallest in $SC$ and it is replaced with larger sets, the search is breadth first. The full algorithm involves two other sets:

$U$ is the set of nodes in $F$ whose Markov boundaries are not yet known.
$K$ is the set of pairs $(n, B_n)$ of nodes $n$ with known boundary $B_n$.

It goes:

$$U := F$$
$$K := \{\ \}$$

$$\text{WHILE} \quad U \neq \{\ \}$$
$$n \quad := \text{any feature in } U$$
$$U \quad := U \setminus \{n\}$$
$$C_0 := \{n' \in F \mid (n', B_{n'}) \in K \wedge n \in B_{n'}\}$$
$$SC := \{C_0\}$$

$$\text{IF} \quad I(n, C_0, F \setminus \{n\} \setminus C_0)$$
$$\text{THEN } B_n := C_0$$
$$\text{ELSE } B_n \text{ is not yet known}$$

$$\text{WHILE } B_n \text{ is not known}$$
$$C \quad := \text{any smallest set in } SC$$
$$SC := SC \setminus \{C\}$$

$$\text{IF } C \cup \{n\} = F$$
$$\text{THEN } B_n := C \quad (\text{so } B_n \text{ becomes known})$$
$$\text{FOR ALL } n' \text{ in } F \setminus C \setminus \{n\} \text{ WHILE } B_n \text{ is not known}$$
$$\text{IF } \neg I(n, C, n')$$
$$\text{THEN}$$
$$C' := C \cup \{n'\}$$
$$SC := SC \cup \{C'\}$$
$$\text{IF } I(n, C', F \setminus \{n\} \setminus C')$$
$$\text{THEN}$$
$$B_n := C'$$
$$K := K \cup \{(n, B_n)\}.$$

When this stops, $K$ holds one pair $(n, B_n)$ for each feature $n$ in $F$.
The test

$$\neg I(n, C, n')$$

is made as described above.

The efficiency of the algorithm depends on the choices of $n$ from $U$ and of $C$ from $SC$ and of $n'$ from $F \setminus C \setminus \{n\}$. Fung and Crawford suggest two heuristic rules to keep it efficient.

The first states that the best $n$ is likely to be one which is dependent on very few other nodes:

Choose $n$ so that $\{\, n' \mid \neg I(n, \{\ \}, n')\}$ is smallest.

This $n$ will probably have a small boundary, so the inner WHILE loop will end quickly. Of course, every $n$ must be chosen at some stage; but by the time the algorithm selects other features $n$ with lots of dependencies, many of the nodes $n'$ that $n$ is dependent on will have known boundaries $B_{n'}$, so $n'$ will be included in $C_0$

and hence in every $C$ in $SC$. Hence, although $B_n$ may be large, the search for it will probably be shallow.

The second rule states that the best $n'$ is likely to be one which is highly dependent on $n$:

Choose $n'$ so that the correlation between $n$ and $n'$ is largest.

Since the search is breadth first, this rule will only shorten it during the last cycle of the FOR loop. It is probably still worthwhile.


## 5.6    CLASSIFYING BY THE EFFECTS OF ACTIONS

As a last example, here is a method due to Scott and Markovitch. It is statistical, in that it estimates probabilities, depends on entropy, and can cope with noisy data. At the same time, it resembles conceptual clustering in that it invents predicates which describe concepts and it arranges these concepts in a partial order. However, unlike most conceptual clusterers, it does not depend on any notion of counter examples. It can work in a domain of data for which there is no initial classification scheme. It invents its own classification from unclassified instances.

Another major feature of the method is that it can choose its training instances. One way to describe this is to say that the algorithm performs experiments and, while it runs, it can choose which variety of experiment it will do next. Alternatively, you can imagine that it is provided with several training sets and at any stage of learning, it can decide which set its next training instance will come from. There is no other algorithm I know of with this feature. The effect is that the learner can concentrate attention on the aspect of its domain where learning is likely to be most profitable. As a result, learning is much faster.

The Scott–Markovitch learner is called Dido. It investigates entities which it can examine and which it can alter. It has two kinds of operations: *sensory* operations (**properties**) and *motor* operations (**actions**).

1.   A **property** is a predicate such as

   Shoes($X$, Slippers).

   This means that the thing called $X$ is wearing slippers for shoes. Each property describes a function of its first argument. The second argument is the function's value.

2.   An **action** such as

   Wave-Wand($X$)

   has side effects. This particular action is taken from the following example. It means that the player waves a wand at the entity $X$. It will kill $X$ if $X$ was previously alive and wearing boots. Otherwise, it will turn $X$ into a magician if $X$ was previously a dwarf.

Dido knows nothing about entities except their names and what it can observe with its sensory operations. Sensory operations have no side effects. By contrast, when Dido operates on an entity with an action, it changes the entity, so that in future the sensory operations on this entity may yield different values.

*Example* (Scott and Markovitch)

In a computer game, there are five properties and three possible actions listed in Table 5.6.

**Table 5.6**

| *Property* | *Possible values* | | | | | |
| --- | --- | --- | --- | --- | --- | --- |
| Type: | Dwarf | Magician | Pirate | | | |
| Size: | Small | Medium | Large | | | |
| Vitality: | Alive | Dead | | | | |
| Coat Colour: | Red | Blue | Green | Yellow | Black | White | Gold |
| Shoes: | Boots | Slippers | Sandals | None | | |

| *Action* | *Effect on its argument* |
| --- | --- |
| Hit-With-Axe | Kills a living thing wearing slippers. |
| Wave-Wand | Kills a living thing wearing boots; otherwise turns a magician into a dwarf. |
| Rub-Lamp | Turns a dwarf into a pirate; Turns a pirate into a magician; Turns a magician into a dwarf. |

In general, the value of a property on an entity can be either a constant or another entity, but in this example, the values of properties are all constants, never entities. Each action changes at most one property of the entity it acts on. These features make the resulting classification simple and I am not sure how Dido might behave in a more complex case.

### 5.6.1 DIDO'S REPRESENTATION HIERARCHY

When the learner starts, it knows nothing about its operations except their names (and perhaps the possible values of each property). Its objective, in outline, is to discover the precise effects of any action on any entity, given the entity's properties. Dido constructs a hierarchical classification of all entities in its domain. The hierarchy is designed to reflect the effects of its actions and to be as simple as possible.

The hierarchy can be a general partial order. It need not be a tree. In it, each node has three essential parts:

1. a predicate called an **intension**. This is a definition of a class of entities for the node. It is a conjunction of basic properties, such as

Type ($X$, Dwarf) $\wedge$ Vitality ($X$, Alive) $\wedge$ Shoes ($X$, Slippers).

This particular predicate describes the class of all live dwarfs wearing slippers.

2.   a set of **practical conditionals**. A practical conditional consists of:

   - an action;

   - the set of possible effects of this action on anything fulfilling this node's intension;

   - for each possible effect, an estimate of the probability that the action will have this effect;

   - the *uncertainty* of this action. The action has an associated experiment: perform it on a randomly chosen entity which satisfies the intension and observe any resulting effect. The given probabilities let one estimate the expected surprise of this experiment, its entropy. This entropy is the uncertainty.

3.   the class of known entities which satisfy the intension. This class is sometimes called the node's **extension**.

In the diagrams published by Scott and Markovitch about Dido, extensions do not appear in the pictures of nodes. Thus, a node might appear as its intension, followed by all its practical conditionals, as shown in Table 5.7.

**Table 5.7**

---

*Intension*
Type ($X$, Dwarf)   $\wedge$   Vitality ($X$, Alive)   $\wedge$   Shoes ($X$, Slippers)

---

Wave wand:                                                                 Uncertainty: 0.722
$P$ (Null)        = 0.8
$P$ (*dummy*)   = 0.2

---

Hit with axe:                                                              Uncertainty: 0.817
$P$ (Null)        = 0.11
$P$ (*dummy*)   = 0.06
$P$ (Vitality ($X$, Alive) $\rightarrow$ Vitality ($X$, Dead)) =  0.83

---

Rub lamp:                                                                  Uncertainty: 1.617
$P$ (Null)        = 0.13
$P$ (*dummy*)   = 0.08
$P$ (Vitality ($X$, Alive) $\rightarrow$ Vitality ($X$, Dead)) =  0.22
$P$ (Type ($X$, Dwarf) $\rightarrow$ Type ($X$, Pirate)) = 0.57

---

Null means that this action produces no change.
*dummy* is a special value representing unknown outcomes.
Effects with zero estimated probability are omitted.

If a practical conditional has any effect other than Null and *dummy*, then the one with highest estimated probability is called its expected outcome. Note that Null is never an expected outcome.

The problem solver in Dido predicts the effects of actions on entities. Any entity $X$ satisfies the intension of some lowest node in the hierarchy. If this node has a practical conditional for an action $A$ then it makes a prediction about the effect of $A$ on $X$. If the node has no such practical conditional, then one is inherited from some ancestor node. Scott and Markovitch write that inheritance conflicts do not arise. The hierarchy can be used to predict the effects of actions while it is incomplete, so Dido is an incremental learner. Note that there may be more than one lowest node whose class contains $X$. In that case, the obvious course of action is to use the one whose practical conditional for $A$ is least uncertain.

Dido begins with a hierarchy consisting of a single root node shown in Table 5.8.

**Table 5.8**

| *Intension* |  |
| --- | --- |
| TRUE |  |

| Wave wand: | Uncertainty: 1 |
| --- | --- |
| $P$ (Null)     =   0.5 |  |
| $P$ (*dummy*) =   0.5 |  |

| Hit with axe | Uncertainty: 1 |
| --- | --- |
| $P$ (Null)     =   0.5 |  |
| $P$ (*dummy*) =   0.5 |  |

| Rub lamp: | Uncertainty: 1 |
| --- | --- |
| $P$ (Null)     =   0.5 |  |
| $P$ (*dummy*) =   0.5 |  |

A practical conditional is said to be **fixated** if it has just one action with non-zero probability. Of course, this action's probability will be 1 and the practical conditional's uncertainty will be 0.

Dido performs 'experiments' repeatedly until all practical conditionals in all nodes are fixated. During and after each experiment, it adjusts probabilities and maybe creates new nodes and removes old ones. The experiments and the manner of adjustments will be described below.

In principle, Dido does not have to remember anything except what is in this hierarchy. It does not keep records of outcomes of past actions. In practice, for efficiency, it keeps records of certain features of the hierarchy. It may also keep a note at each node of the names of entities known to be in the node's class.

### 5.6.2   INHERITANCE IN THE HIERARCHY

Since there are three aspects to any node, its intension, extension, and practical conditionals, there are three kinds of feature which might be inherited.

1.   Intensions are inherited downwards, from parent node to child.

2.   Classes are inherited upwards, from child to parent.

3.   Practical conditionals are inherited downwards for problem solving, but not while learning.

The intension of a node includes all the conditions of the node's ancestors. Hence, not all conjuncts are actually recorded at the node. If all the conjuncts in the intension can be inherited, then the node could contain no conditions at all. When this happens, the condition shown for the node is TRUE.

Since each child inherits the intensions of its parents, any entity in the child's class is also in the class of any of its parents.

If a node $N$ has a practical conditional $P$ for some action $A$, then, as explained above, $P$ can be used to predict the effect of $A$ on any entity $X$ in $N$'s class. When $N$ does not have such a $P$ for $A$, but $N$ has an ancestor with an practical conditional $P'$ with action $A$, then $P'$ can be used for such predictions.

However, while learning, the situation is different. A node $N$ and a child $C$ of $N$ can both have practical conditionals with the same action, but the probability estimates in these two practical conditionals may be different. Hence, for the purpose of learning, practical conditionals are *not* inherited.


### 5.6.3   DIDO'S EXPERIMENTS

The algorithm in Dido performs a sequence of *rounds*. In a single round, it chooses a particular node and a particular action and then it performs this action on a few randomly chosen entities from among the node's class. It stops the round when either it has enough results or it has performed seven experiments. The entities which are chosen are residual for this node and action: $X$ is **residual** for node $N$ and action $A$ if

- $X$ satisfies $N$'s intension, and

- $N$ has a practical conditional whose action is $A$, and

- $X$ is not residual for any descendent of $N$ and $A$.

The chosen node and action are those with highest uncertainty. Thus, Dido always explores the part of its network about which it is least confident. Its experiments are chosen to minimize all uncertainties as quickly as possible. The top level of its algorithms is:

Set up the root node.
In it, the possible outcomes of each action are
*Null*     with probability 0.5
*dummy*  with probability 0.5
 so the uncertainty of every action is 1.

WHILE there is a practical conditional whose uncertainty > 0
  $P$ : = a practical conditional with maximum uncertainty
  $A$ : = $P$'s action
  $C$ : = the class of the node containing $P$

  WHILE  the number of entities tested < $b$
    AND  $P$ has not been removed
    AND  $P$'s uncertainty > 0
    AND  there is some untested residual entity in $C$ for $A$

      $X$ : = a random untested residual entity for $A$ in $C$
      $E$ : = the observed effect of the action $A$ on $X$

$$P(E) := \pi\left(1 - \frac{1}{\tau}\right) + \frac{1}{\tau}$$

      FOR ALL other possible effects $E'$ in $P$

$$P(E') := \pi'\left(1 - \frac{1}{\tau}\right)$$

        where
            $n$ is the number of possible effects in $P$
            $\pi$ is the previous probability of $E$
            $\pi'$ is the previous probability of $E'$

        IF $P(E') < \delta$
        THEN remove $E'$ from $P$

      IF any $E'$ was removed from $P$
      THEN
          normalize all probability estimates of remaining effects in $P$

      Remove any useless practical conditionals
      Retract any useless nodes

      IF at least two experiments produced $P$'s expected outcome
      AND one produced some other outcome
      THEN
          perform conjectural specialization

The algorithm depends on the parameters in Table 5.9.

**Table 5.9**

| Symbol | Value in Dido | Interpretation |
|---|---|---|
| $b$ | 7 | The largest possible number of experiments in a round. |
| $\tau$ | 10 | The 'time constant' for adjusting probabilities. If $\tau$ is large then the algorithm depends heavily on past results and each new result has only a small effect on probability estimates. |
| $\delta$ | 0.05 | The probability threshold for deleting an effect from a practical conditional. |

Within this algorithm, there are four major subroutines.

1. *Normalize* all probability estimates of remaining effects in $P$. This adjusts all probability estimates so that they sum to 1. Scott and Markovitch do not give precise details.

2. *Remove* any useless practical conditionals. A practical conditional $P$ is *useless* if its node has a parent with a practical conditional $P'$ with the same action as $P$ and the expected outcome of $P'$ is the same as that of $P$ and the probability estimate of this expected outcome is at least as great in $P'$ as in $P$.

3. *Retract* any useless nodes. A node is *useless* if it has no practical conditionals and no children.

4. Perform *conjectural specialization*. This is the principal means for extending the hierarchy of nodes. It will be described below.

Dido has several other subroutines, not shown in the algorithm above. It will work as shown, but the others improve its performance. These others will be explained below and it will become clear where they fit into the basic form of Dido.

### 5.6.4 DEVELOPING THE HIERARCHY

The full implementation of Dido has seven procedures for modifying its hierarchy. *Conjectural specialization* forms a new child of the chosen node $N$. *Retraction* removes useless nodes. These are the two major processes. Dido is designed to generate new nodes and practical conditionals on thin evidence and then to retract them again if they turn out to be unjustified. This makes for quick evolution of the hierarchy. The alternative would be to extend the hierarchy only after accumulating lots of evidence. That would be a much slower process.

In what follows, in any round, the standard notation will be

$P$ for the chosen practical conditional;

$N$ for the node containing it;

$A$ for the action of $P'$

$O$ for its expected outcome.

### Conjectural specialization

In each round, say that an experiment is a 'success' if it produces the expected outcome $O$. Any other experiment is a 'failure'. Conjectural specialization is designed to construct a child of $N$ with a higher proportion of successes.

For each property and for each value of that property, Dido calculates

$S$   the number of times this value occurred among successes;

and

$F$   the number of times this value occurred among failures.

Say $p$ is the property and $v$ the value for which $(S - F)$ is largest. Conjectural specialization finds a node $C$ whose intention is

$$I(X) \wedge p(X, v)$$

where $I(X)$ is the intension of $N$. This $C$ will be a child of $N$. If it does not already exist, it is created.

If $N$ is the only parent of $C$, then $C$ gains a copy of $P$ from $N$. (This will be $C$'s only practical conditional). However, if $C$ already exists, then it may have several parents which all have practical conditionals $P'$ with the same action $A$. In that case, $C$ gains a copy of the $P'$ in which $O$ has highest probability.

When it is formed, $C$ does not immediately help Dido to make better predictions. However, with luck, entities in $C$'s class will have expected outcome $O$ more often then other entities in $N$'s class; so in a later round, the inherited practical conditional in $C$ is likely to become less uncertain than $P$.

### Retraction

This was described above. It removes useless nodes.

### Conjunctive specialization

This may occur whenever Dido chooses a residual entity, $X$, in the inner WHILE loop of its algorithm. Dido then checks to see if there is another node $N'$ in which $X$ is residual and which has a practical conditional $P'$ and expected outcome $O'$ and the action of $P'$ is $A$ but $O' \neq O$.

When this happens, Dido finds a node $C$ whose intension is the conjunction of the intensions of $N$ and $N'$. If $C$ does not already exist, it is created. It is a child of both $N$ and $N'$. It inherits one practical conditional, either $P$ or $P'$. (Scott and Markovitch do not say which.)

This procedure can convert the hierarchy into something other than a tree. After $C$ has been created, the entity $X$ is no longer residual in $N$ or $N'$.

### Intensional generalization

Whenever a practical conditional $P$ in $N$ fixates, the algorithm looks through its hierarchy for another node $N'$ with a practical conditional $P'$ for which:

- $P$ and $P'$ have the same action, $A$; and

- the intensions $I$ and $I'$ or $N$ and $N'$ test with the same properties and are identical except that, for just one property $p$, these properties have the same values, so for some $J$,

$$I \equiv J \wedge p\,(X, v) \qquad \text{and} \qquad I' \equiv J \wedge p\,(X, v') \qquad \text{where } v \neq v'.$$

If there is such a node $N'$, then Dido finds a node $C$ whose intension is $J$, and it gives $C$ a copy of the practical conditional $P'$. Thus, $C$ is a parent of both $N$ and $N'$. If $C$ did not already exist, then Dido creates it.

This step removes a form of redundancy from the representation. If it is not carried out, the final form of the hierarchy may contain several nodes all predicting the same results of an action. For example, there might be three nodes whose classes are:

(1)   all live magicians wearing slippers;

(2)   all live dwarves wearing slippers;

(3)   all live pirates wearing slippers;

and all predicting that they will die if hit by an axe. Intensional generalization, followed by retraction, will summarize these three results into a single node.

### Useless partition removal

Suppose there is some property $p$ whose possible values are

$$v, \quad v', \quad \ldots \quad v''$$

and there are nodes $N, N', \ldots N''$ whose intensions are

$$J \wedge p\,(X, v), \qquad J \wedge p\,(X, v'), \qquad \ldots \qquad J \wedge p\,(X, v''),$$

respectively. These nodes are said to form a **partition** for the property $p$. In order to save later searching, Dido keeps a note of every partition it creates.

If all the expected outcomes in all these nodes are the same, for all actions, then the partition is *useless*. A useless partition can only be formed when either a practical conditional is added to a class or a practical conditional changes its expected value. Whenever either of these events happens, Dido examines its hierarchy for useless partitions and removes any it finds.

If ever a useless partition occurs, all its nodes $N, N', \ldots N''$ have practical conditionals with the same actions. For each such action $A$, say $P_A$ is the one with least uncertainty. The algorithm

- finds a node $C$ with intension $J$, and

- for each such action $A$, copies $P_A$ into $C$, and

- for each node in the useless partition,
  removes its practical conditional with action $A$.

$C$ is a parent of all nodes in the useless partition.

This procedure saves Dido from fixating all the practical conditionals for $A$ in all these nodes separately. It may also cause some of the nodes to become useless, in which case they can be retracted.

### Fixated partition procedure

Suppose that $N, N', \ldots N''$ is a useful partition of some node $C$. Also suppose that, for some action $A$, all nodes in it have practical conditionals, all fixated, but not all with the same expected outcome. In that case, any practical conditional for $A$ in $C$ is useless. Furthermore, if $C$ has any other descendant with a practical conditional $P$ for $A$, but not in the partition, then $P$ is also useless.

Dido tests for this state of affairs whenever any practical conditional fixates. All useless practical conditionals are removed.

### Residual specialization

This procedure actually makes the hierarchy larger. Its virtue is that it also makes it symmetric, when there is a property $p$ which appears to partition some node $M$ symmetrically for some action $A$.

Suppose that the possible values of $p$ are $v, \ldots v', v''$ and that $M$ has children $N, \ldots N'$ with intensions

$$J \wedge p\,(X, v), \qquad \ldots \qquad J \wedge p\,(X, v')$$

but no child with intension

$$J \wedge p\,(X, v'').$$

Also, suppose that all the nodes $N, \ldots N'$ and $M$ have practical conditionals for $A$, all with non-null expected outcomes. In that case, the hierarchy will appear more aesthetically pleasing if it includes another child $N''$ of $M$ with the missing intension.

Say the practical conditional for $A$ in $M$ is $P$. If ever Dido chooses $P$ for a round of experiments and all these experiments are successes and if the partition $N, \ldots N'$, $N''$ is not useless, then Dido creates $N''$ and gives it the practical conditional $P$.

### 5.6.5 RESULTS

#### *Computer game*

In the computer game of pirates, magicians, and dwarves, Dido produced a correct classification with the smallest possible number of nodes. After 125 applications of actions to entities, its problem solver worked perfectly. However, at that stage, many of its practical conditionals were still uncertain. All told, Dido made about 400 applications of actions before all practical conditionals were fixated.

#### *Classification*

Scott and Markovitch made some other tests with Dido. One involved a domain in which Dido mimicked a classifier which is fed a training set containing both positive and negative examples. The domain had a single operation, which asked a user what was the class of its argument. This operation does not actually change its argument. Perhaps the way to understand the test is to suppose that each probability in a practical conditional has the form

$$P(\text{class of } X \text{ is } ...) = ....$$

The test was performed with both conjunctive and disjunctive concepts. Dido seems good at learning conjunctive concepts. These can be described by a single node in its hierarchy. It is less good at learning disjunctive concepts, perhaps because these can only be described by a parent node with several children. Decision tree learners appear to work better than Dido on disjunctive concepts.

#### *Noise*

A third test applied Dido to noisy data. In this situation, Dido may never stop because some of its practical conditionals may not fixate for a very long time. The critical parameter is $\delta$. If $\delta$ is small then outcomes with low estimated probability remain in a practical conditional. Random noise in the training data may keep this estimate above $\delta$ for many rounds. On the other hand, if $\delta$ is larger, the wrong outcomes may be removed.

Dido's results were quite creditable. It could recognize when to stop learning with a noise level of 5 per cent, but when the noise level was 10 per cent some of its practical conditionals did not fixate. Even when the noise level was 15 per cent, its performance at predicting outcomes of actions was almost perfect. Beyond this noise level, its performance fell, but remained better than the inherent uncertainty in the training data.

## 5.7 SURVEY OF CLUSTERING METHODS

Each of the programs presented so far uses a particular technique to invent a classification scheme, or similar data structure, of a particular kind. There has been no great attempt to justify one technique over and above all others. The techniques chosen do in fact appear to work quite well, for some training sets, but it would be helpful if we could find a way to decide how to select one learning technique from among many possibilities.

### 5.7.1 COMPARISONS

Here is a selection of criteria which can be applied to the algorithms studied above.

- What is the form of each *input* instance?

- What is the form of the *output*?

- Does the algorithm require its training instances *preclassified*? If so, does it require both positive examples and *counter-examples*?

- Is the algorithm *incremental*?

- What *statistic* does it *optimize*?

- Does it *optimize globally* over the whole classification scheme or just *locally*?

- Is each test made by the output classifier *sensitive* to many features of a datum or is it a crude test of just one feature?

The answers to all these questions are presented in Table 5.10, for all the algorithms described in this chapter and also for Kohonen nets.

**Table 5.10**

| | *Kohonen nets* | *Wolff's algorithm* | *Decision trees* | *Al-Mathami* | *Fung and Crawford* | *Scott–Markovitch* |
|---|---|---|---|---|---|---|
| *Input* | $\mathbb{R}^n$ | words | a/v pairs | a/v pairs | a/v pairs | a/v pairs |
| *Output* | subset of $\mathbb{R}^n$ | C.F.G. | tree | tree | graph | partial order |
| *Preclassified?* | no | +ve | ± | no | +ve | no |
| *Incremental?* | no | yes | no | yes | no | yes |
| *Statistic* | ? | ? | entropy | success rate | backing | entropy |
| *Global?* | broad | broad | no | no | no | no |
| *Sensitive* | yes | yes | no | yes | yes | yes |

In addition, any incremental algorithm can be compared with the simple little prototype incremental clusterer of Section 5.1.1 and one can ask of it

- What *unbalanced test* does it make?

- When it detects an unbalanced cluster, what is its *subdivide* operation?

Of the three incremental algorithms,

1. Wolff's grammar learner detects unbalanced clusters by testing for missing alternatives in any context. Its subdivide operator is *rebuild*.

2.  Al-Mathami's tree-building program does not appear to detect or subdivide unbalanced clusters at all.

3.  The Scott–Markovitch method detects an unbalanced cluster when similar experiments yield different outcomes. Its major subdividing operator is *conjectural specialization*, but it has others.

The decision-tree algorithm is the only one which requires both positive and negative examples of a predefined concept. Al-Mathami's method can also learn from such a training set: the property of being a positive example is just treated as an extra attribute.

The fastest learners are the ones which only optimize a small portion of their output classifiers at a time.

- The decision tree learner and Al-Mathami's method both only optimize in the neighbourhood of a single node of the tree.

- The Scott–Markovitch algorithm might in principle scan a rather larger portion of its hierarchy, but the implementation keeps records of features in the hierarchy which save it from frequent general scans.

- The Fung–Crawford method only optimizes each Markov boundary separately, but for each such boundary it makes a broad search through subsets of all the graph's nodes and this is likely to be a slow process.

- When a Kohonen net begins to learn, it is sensitive to all its nodes, but it becomes progressively less so as its parameter $K$ is reduced.

- Wolff's algorithm is slow because, in each cycle, it surveys perhaps many large parse trees and it considers many possible contexts in each. The whole of its current grammar may occur in a single parse tree. On the other hand, Wolff's method is the only one presented here which can form recursive structures.

### Strength of clustering methods

- They can handle noisy data.

- Some of them are incremental learners.

- Given a confidence level, some methods can stop learning when their output reaches that level.

- Some are not very heavy computationally.

- They discover clusters in forms which can be named.

- They discover intensional descriptions of clusters.

- They produce hierarchical classifications. These are essential for some other learning algorithms, particularly the so-called similarity-based method for learning rules (Chapter 7).

- Many of these methods can invent disjunctive concepts.

### *Limitations*

- Many such methods are only reliable when they are supplied with large training sets. This is true in some cases even when the data are clean.

- They often depend on arbitrary parameters.

- Some are computationally heavy.

- There is not always a simple way of deciding when to stop learning.

### 5.7.2 HOW TO DESIGN A CONCEPTUAL CLUSTERER

Buntine has suggested the outline of a procedure for designing algorithms which construct classifiers. The resultant algorithms will learn from training data how to classify further data, so the procedure invents learning algorithms. Buntine starts from the premiss that the learner should invent a data structure with a *discrete* component called $T$ and a *continuous* part called $\vartheta$.

*Examples:*

1. This learner is intended to invent a tree which classifies examples. Any example $x$ may belong to just one of a few classes $c, c', \ldots c''$.

   $T$ might be a tree whose internal nodes consist of questions which can be asked of $x$ and whose answers are *yes* or *no*. $\vartheta$ is a set of probability distributions over the leaves $l$ of $T$, one called $\theta(c)$ for each class $c$. If the classifier $(T, \vartheta)$ is perfect then $\vartheta_l(c)$ is the probability that, if $x$ is assigned by all the questions to the leaf $l$, then $x$ will belong to class $c$.

   The trees invented by the algorithm of Section 5.3 are more naive. If such a tree assigns $x$ to a leaf $l$, then it predicts that $x$ will always belong to the class of $c_l$ associated with $l$. This is the special case when $\vartheta_l(c_l) = 1$ and all other probabilities $\vartheta_l(c')$ are zero.

2. This learner invents a way to predict the frequencies of words in a language. $T$ might be a context-free grammar. $\vartheta$ is again a set of probability distributions. This time, there is one probability distribution $\vartheta_A$ for each variable symbol $A$ of $T$. If $P$ is a production in $T$ expanding $A$, then $\vartheta_A(P)$ is meant to be the probability that some occurrence of $A$ in a parse tree will be expanded by production $P$.

   We have not seen any structure like this before. With it, one can assign a probability to each possible parse tree:

   The probability of a parse tree is the product

   $$\vartheta_A(P) \times \vartheta_B(Q) \times \ldots \vartheta_C(R)$$

   taken over all internal nodes in the tree. The internal nodes are $A, B, \ldots C$, and $P, Q, \ldots R$ are the productions expanding them.

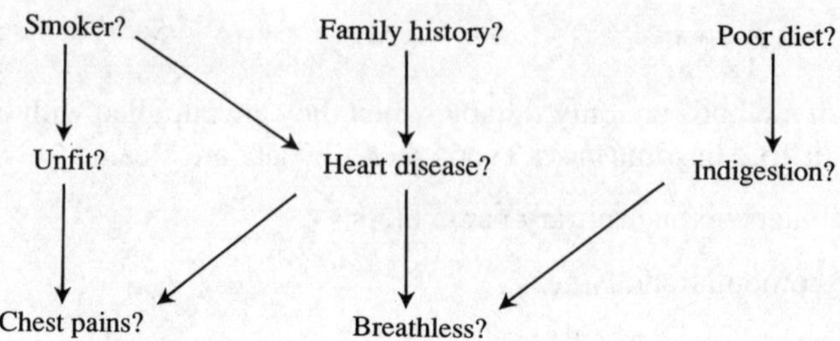



**Fig. 5.18**   An example of a discrete component in a classifier.

Then one can assign a probability to each word in the grammar's language.

> The probability of a word $w$ is the sum of the probabilities of all parse trees for $w$. One might say that the learner is good if the probability predicted by $(T, \vartheta)$ for $w$ is close to the frequency of $w$ in the training set, for most words $w$.

3.   $T$ might be an acyclic directed graph of questions which can be asked of an instance $x$. $\vartheta$ is a set of probabilities, one for each node (question).

> If $Q$ is one such question and $R, \ldots S$ are the questions with edges pointing to $Q$, then $\vartheta(Q)$ should be the probability that if $R(x)$ and $\ldots S(x)$ are all true then $Q(x)$ will be true too.
>
> Buntine presents an example whose $T$ is the graph shown in Fig. 5.18.
> Its $\vartheta$ consists of probabilities
>> $P$(unfit | smoker)
>> $P$(heart disease | smoker, family history)
>> *etc.*

To start with, assume that there is a certain set of all possible structures $(T, \vartheta)$ that the learner might invent. We also assume that, before the training set is known, for each $T$ there is a known 'prior' probability $P(T)$ that $T$ is the right structure and, for each $T$, there is some known prior density $p(\vartheta \mid T)$ for the possible values of $\vartheta$ on $T$. For instance, $T$ might be the set of all trees in which no question is asked more than once down any branch and $P(T)$ might be the same for all trees and $p(\vartheta \mid T)$ might be the uniform distribution at each leaf.

After the training set $TS$ has been revealed, $P(T \mid TS)$ is the revised probability that $T$ is the right structure and $p(\vartheta \mid T, TS)$ is the revised density for the continuous component $\vartheta$ on $T$. When $x$ is any further instance, the system predicts that the probability that $x$ belongs to class $c$,

$$P(c \mid x, TS),$$

is

$$\sum_T \int_\vartheta P(T \mid TS)\, p(\vartheta \mid T, TS)\, P(c \mid x, T, \vartheta, TS)\ d\vartheta .$$

In the case of a tree, this is

$$\sum_T \int_\vartheta p(\vartheta \mid T,TS) \; P(T \mid TS) \; \vartheta_l(c) \; \mathrm{d}\vartheta \qquad \text{when } T \text{ assigns } x \text{ to leaf } l.$$

In practice, in most circumstances, no system could record all the information embodied in the probabilities $P(T \mid TS)$ and $p(\vartheta \mid T, TS)$. The objective of learning is to discover a particular $T$ and $\vartheta$ so that $\vartheta_l(c)$ is a good approximation to $P(c \mid x, TS)$.

The learner works more simply if $T$ and $\vartheta$ are treated separately. Therefore, say

$$E_{x,T,TS}(c) \;\; = \;\; \int p(\vartheta \mid T,TS) \;\; P(c \mid x,T,\vartheta,TS) \;\; d\vartheta$$

so that

$$P(c \mid x,TS) \;\; = \;\; \sum_T P(T \mid TS) \;\; E_{x,T,TS}(c).$$

In such a situation, Buntine suggests the following outline procedure for inventing a learning algorithm.

*Step 1* Choose a particular class of possible structures $T$ and continuous components $\vartheta$.

*Step 2* Choose particular prior distributions $P(T)$ and $p(\vartheta \mid T)$.

*Step 3* Choose a way to estimate the revised distribution $P(T \mid TS)$ for each $T$ in the light of any new training data $TS$.

*Step 4* Invent a heuristic search algorithm in the space of all structures $T$ for one which maximizes the estimate of $P(T \mid TS)$.

*Step 5* Invent a way to find approximations for the revised estimates $E_{x, T, TS}(c)$.

All the specific clusterers described above fit this pattern. Their major subtleties usually lie in step 4. The various probability distributions over $T$ and $\vartheta$ have not been made explicit.

Buntine shows in outline how his procedure can be used to derive learning algorithms for decision trees and for graphs of dependencies between predicates, and he justifies his choices of probability distributions by Bayesian statistics, but his account is abbreviated. He also considers a more general form of step 4. Rather than search for a single $T$ which maximizes the estimate of $P(T \mid i)$, he advocates searching for a small set of structures with large values of this estimate. Then step 5 involves finding $E_{x, T, TS}(c)$ for each $T$ in this small set and then the procedure involves the following extra step.

*Step 6* Choose a way to find an average of the values $E_{x, T, TS}(c)$ over the various structures $T$.

The results using this extension were better, but at a significant increase in computing space and time.


## 5.8   APPENDIX: MORE ON EXPERIMENTS AND ENTROPY

In the discussion of decision trees, we met entropy. Most of that discussion is sound. Entropy is what you think it is and it can be used for building decision trees and in other circumstances  too. However, that is not quite the whole story. Here are a few more details.


### 5.8.1   A SHORT NOTE ON EXPERIMENTS

So far, we have been assuming a particular kind of experiment. It consists of $N$ trials, where $N$ is chosen before the trials start. For instance, the experiment might be to test whether it rains and you might choose in advance to keep records every day for a whole month. Then $N$ is the number of days in the month. The interesting observation will then be the number of days, say $k$ days, on which it actually rains.

This sort of experiment can be depicted thus:

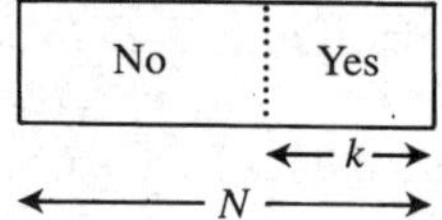


The objective is to discover the position of the dotted line. The whole box is drawn solid because its size is fixed in advance.

There is a second common kind of experiment. In this, the experimenter decides what $k$ will be before he starts and then he continues repeating his tests until he has just $k$ successes, whereupon he stops at once. The interesting number is $N$, the total number of observations he has to make. This kind can be depicted as

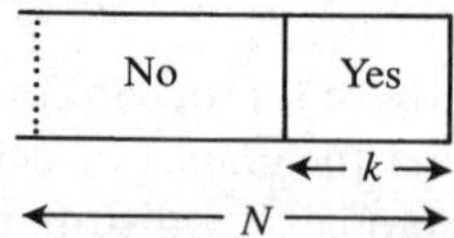


Again, the experimenter is trying to find the position of the dotted line. The important point is, the statistical analysis of such an experiment is *different* from the first kind. This is not the place to explain just what the analysis should be, but you should at least be aware that there is a difference.

Other variants are possible. For instance, suppose you want to estimate the probability that a coin will come up heads when tossed. You opt for the second kind of experiment with $k$ set to 3, so you intend to keep tossing until it comes up heads for the third time. Then what actually happens is, you toss it ten times and it always comes up tails, whereupon you give up and find another coin. Such an experimental situation could be drawn as

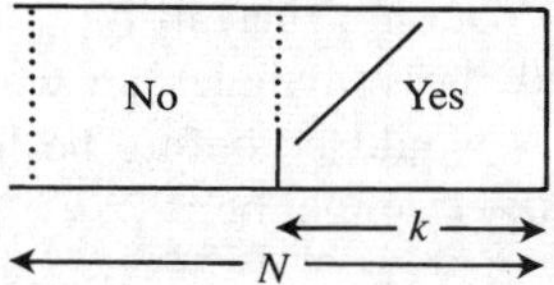


The analysis of any experiment depends subtly on its design and there are many possible designs.

### 5.8.2 A MORE ACCURATE ESTIMATE OF ENTROPY

Recall that in Section 5.3.3 we considered an experiment called $X_1$ which was performed 253 times and of these tests, 249 were successful. From these figures, we estimated its probability $\pi_1$ of success and the entropy $H(X_1)$. The values we chose then were 249/253 for $\pi_1$ and, for $H(X_1)$,

$$-\left(\frac{249}{253}\log_2\frac{249}{253} + \frac{4}{253}\log_2\frac{4}{253}\right).$$

These value are inaccurate. For a start, the best estimate of $\pi_1$ is a little less than 249/253. Also, $H(X_1)$ does not depend so simply on $\pi_1$. We shall now see how to estimate them better. We use the notation of Section 5.3.

If there are only a few results then, whatever the question $Q_j$, one of its two sets TS$Y$ or TS$N$ is bound to be very small. Any conclusion drawn from it would be worthless. The approximate calculation above does not take account of this fact. If TS is small and almost all values in it have the property $P$, then maybe it would be better not to split TS any further but treat this set as a leaf of the decision tree. There is a corresponding experiment.

$X_0$: $x$ will have property $P$, regardless.

We can estimate the entropy of $X_0$, just as we can that of $X_j$ for any question $Q_j$. If $H(X_0)$ is less than $H(X_j)$ for every question $Q_j$, then there is no good question and the current node of the decision tree is best left as a leaf.

Practical examples suggest that this is important. Quinlan has tested one case by building a decision tree using the above approximate entropy method. He then converted the tree into a nearly equivalent set of rules for an expert system. He found that the expert system worked better if some rules based on few results were discarded in favour of others with coarser conditions but more supporting evidence.

How might we estimate the entropy for each question $Q_j$? Imagine that we are faced with some experiment $X$, which can answer *yes* with some probability $\pi$ or no with probability $(1 - \pi)$, and we want to estimate its entropy. To do that, we really want to estimate not $\pi$ but the *probability distribution* of $\pi$. This is a rather odd notion. $\pi$ itself is a probability. The point is, $\pi$ is not known. Experimental evidence may suggest some most likely value for $\pi$, but in principle, whatever the data, $\pi$ could be any number between 0 and 1.

Suppose our experimental data are that, in $N$ samples of the experiment $X$, $k$ produce the answer *yes* and the remainder answer *no*. (We assume that the experimenter decided what $N$ would be before he began testing.) A reasonable guess at $\pi$'s probability density, $\varphi$, might be

$$\varphi(x) \;=\; \frac{(N+1)!}{k!(N-k)!} x^k (1-x)^{N-k}.$$

This function seems to have all the right characteristics:

- It is continuous;

- it is non-negative;

- its integral over the unit interval is 1;

- its form is symmetric in $k$ and $(N-k)$, so its predictions about $X$ are symmetric between *yes* and *no*;

- if any experiment produces *yes* then $\varphi(0)$ vanishes, and the more often we obtain *yes*, the smaller $\varphi$ is near 0;

- it has a single maximum at the point $k/N$, which is a good estimate for $\pi$.

In fact, Bayesian arguments give some more detailed justification for this choice of $\varphi$. To all intents and purposes, it is the same as a function called the $\beta$ distribution. If, before the experiment, we assume that $\pi$ is equally likely to lie anywhere between 0 and 1, then afterwards $\varphi$ is the best estimate we can make for the probability density of $\pi$. Even if there were some better possible assumption about $\pi$ before the experiment, then $\varphi$ would still be quite a good estimate afterwards.

If the probability density of $\pi$ is $\varphi(x)$ then the expected value of $H(X)$ will be

$$-\int_0^1 \left( x \log_2 x + (1-x)\log_2(1-x) \right) \varphi(x)\, dx.$$

One can calculate this integral. It is

$$\frac{\log_2 e}{N+2} \left( (k+1) \sum_{n=0}^{N-k} \frac{1}{k+n+2} + (N-k+1) \sum_{n=0}^{k} \frac{1}{N-k+n+2} \right).$$

Some expected values for $H(X)$, for various $N$ and $k$, are given in Table 5.11.

**Table 5.11**

| | $N=1$ | 2 | 3 | 4 | 5 | 6 | 7 | 8 | 9 |
|---|---|---|---|---|---|---|---|---|---|
| $k=0$ | 0.7214 | 0.6612 | 0.6011 | 0.5490 | 0.5049 | 0.4676 | 0.4357 | 0.4081 | 0.3842 |
| 1 | 0.7214 | 0.8416 | 0.8416 | 0.8095 | 0.7694 | 0.7291 | 0.6910 | 0.6560 | 0.6240 |
| 2 | | 0.6612 | 0.8416 | 0.8897 | 0.8897 | 0.8703 | 0.8433 | 0.8136 | 0.7838 |
| 3 | | | 0.6011 | 0.8095 | 0.8897 | 0.9154 | 0.9154 | 0.9026 | 0.8834 |
| 4 | | | | 0.5490 | 0.7694 | 0.8703 | 0.9154 | 0.9315 | 0.9315 |

*Example:*

From Table 5.11, we can see easily that sometimes it is not worth asking an extra question. For instance, suppose that TS contains seven instances ($N=7$); of these, two have property $P$ ($k=2$), so the estimated entropy of experiment $X_0$ is 0.8433. If there is one question $Q$ which can be asked and the outcomes are

| $P$ | | *yes* | | *no* | |
|---|---|---|---|---|---|
| $Q$ | *yes* | *no* | *yes* | *no* | |
| | 2 | 0 | 3 | 2 | |

then the experiment $X$ for $Q$ has four answers *yes* out of the seven instances, so its entropy is 0.9154, which is larger. Hence, in this case, it is actually better to ignore $Q$ completely and predict that (with probability about 5/7) every example does not have property $P$.

In this example, the Hamming distance between $P$ and $Q$ is 3. There is another 'question' $F$ whose answer is always *no* and the Hamming distance between $P$ and $F$ is 2. Really, when we say that it is not worth asking $Q$, we mean that $P$ is closer to $F$ (or to the 'question' $T$ whose answer is always *yes*) than it is to $Q$.

### 5.8.3 COMPOSITE QUESTIONS

You should not always feel bound by the natural questions. There are occasions when it may be worth while to use a composite test which involves several simple ones. Also, sometimes it is worth trying a test which recognizes circumstances when a property $P$ is true, even though the test makes no prediction about other circumstances when $P$ might be true or false.

Suppose for instance that there are two natural questions, $Q_1$ and $Q_2$. For a change, imagine that $Q_2$ does not yield *yes* or *no*, but answers one of the three values $a$, $b$, or $c$. Out of nine tests, imagine they produce the results:

| $P$ | yes | | no | |
|---|---|---|---|---|
|  | yes | no | yes | no |
| $Q_1$ | 2 | 1 | 3 | 3 |

and

| $P$ | yes | | | no | | |
|---|---|---|---|---|---|---|
|  | $a$ | $b$ | $c$ | $a$ | $b$ | $c$ |
| $Q_2$ | 1 | 2 | 0 | 0 | 4 | 2 |

These suggest the estimates shown in Table 5.12, so the best predictor of $P$ appears to be whether $Q_2$ is $a$.

**Table 5.12**

| Experiment | $N$ | $k$ | Expected entropy |
|---|---|---|---|
| Always $P$ | 9 | 3 | 0.8834 |
| If $Q_1$ then $P$ else $\neg P$ | 9 | 5 | 0.9315 |
|  |  |  | (same as 4 of 9) |
| If $Q_2 = a$ then P else $\neg P$ | 9 | 7 | 0.7838 |
| If $Q_2 = b$ then P else $\neg P$ | 9 | 4 | 0.9315 |
| If $Q_2 = c$ then P else $\neg P$ | 9 | 4 | 0.9315 |

Let us suppose that the actual occurrences of $Q_1$, $Q_2$, and $P$ are as portrayed in Table 5.13. This table is meant to signify that, for instance, there were two occasions out of the nine tests when $Q_1$ was true, $Q_2$ answered $c$, and $P$ was false. From it, we can guess the expected entropy of more subtle composite tests, but note that the estimates of entropy given in Table 5.14 are not quite right, because they are based on a total number of experiments $N$ which was *not* fixed in advance.

**Table 5.13**

| | $Q_1$ | | $\neg Q_1$ | |
|---|---|---|---|---|
| $Q_2 = a$ | $P : 0$ | $\neg P : 0$ | $P : 1$ | $\neg P : 0$ |
| $Q_2 = b$ | $P : 2$ | $\neg P : 1$ | $P : 0$ | $\neg P : 3$ |
| $Q_2 = c$ | $P : 0$ | $\neg P : 2$ | $P : 0$ | $\neg P : 0$ |

**Table 5.14**

| Experiment | $N$ | $k$ | *Estimated entropy* |
|---|---|---|---|
| *If* $\neg Q_1 \wedge Q_2 = b$ *then* $\neg P$ | 3 | 3 | 0.6011 |
| *If* $Q_1 \wedge Q_2 = c$ *then* $\neg P$ | 2 | 2 | 0.6612 |
| *If* $\neg Q_1 \wedge Q_2 = a$ *then* $P$ | 1 | 1 | 0.7214 |
| *If* $(\neg Q_1 \wedge Q_2 = b) \vee (Q_1 \wedge Q_2 = c)$ *then* $\neg P$ | 5 | 5 | 0.5049 |
| *If* $(\neg Q_1 \wedge Q_2 = b) \vee (Q_1 \wedge Q_2 = c)$ *then* $\neg P$ *else* $P$ | 9 | 8 | 0.6240 |

All these tests appear to give more reliable predictions of $P$ than just the simple experiments $Q_1$ or $Q_2$ alone. The drawback with this method is that there are six boxes in Table 5.13, so there are 64 distinct combinations of them which could be used to predict $P$. This number, 64, is quite large. There is a fair chance that, whatever the data, *some* such combination will seem to predict $P$ (or $\neg P$) better than a simple test. Are we prepared to consider them all? Probably not. Then which should we exclude?

The key issue is the Kolmogorov complexity of the test. The least complex tests are the simple ones which just involve $Q_1$ or $Q_2$. There may be a rigorous criterion for deciding which tests are acceptable, at what entropies, but I have not seen one. The decision may always be somewhat arbitrary, since complexity depends on the language in which the tests are described. It appears to be a matter of taste.

### 5.8.4 EXPECTED SUCCESS RATES

Entropy is indeed a beautiful and valuable concept, but perhaps there has been a craze for it over the past few years. There is an alternative criterion for selecting a good question which appears to be simpler and quite as useful. It is this: Choose the test which is most likely, on available evidence, to predict the property $P$ correctly. In most cases, this criterion will suggest the same test as minimizing entropy, but just occasionally, the two criteria differ.

Say $X$ is any experiment which can be true or false, as before. If $X$ is true $k$ times out of $N$ trials, then as before, the success rate $\pi$ of $X$ is an unknown quantity. Our knowledge of it is best described by the probability density

$$\varphi(x) = \frac{(N+1)!}{k!(N-k)!} x^k (1-x)^{N-k}$$

and the expected proportion of occasions when $X$ will be true is

$$\int_0^1 x\,\varphi(x)\,\mathrm{d}x$$

which is $\dfrac{(k+1)}{(N+2)}$.

*Example*

If as yet we have made no tests, then $N$ and $k$ are both 0, and so without further knowledge the expected success rate of $X$ is $\frac{1}{2}$.

*Example: How to estimate $P_I(E)$ when discovering Markov networks*

Recall the example of the weather, from observations of pressure, wind and rain. In it, there was a calculation of a probability $P_I(O)$ of an observation $O$. The value found then was

$$\frac{(4/20)\times(9/20)}{11/20} = \frac{36}{220} \simeq 0.164$$

We can now make a better estimate. It is

$$\frac{(5/22)\times(10/22)}{12/22} = \frac{50}{264} \simeq 0.189$$

which is substantially different.

Let us see how this approach applies to an example in which there are two questions, $Q_1$ and $Q_2$, which each can answer *yes* or *no*. The assertion $P$ which we want to predict is

This bird has forward vision.

$Q_1$ might be

Does this bird have a curved beak?

and $Q_2$ might be

Is this bird large?

Beaks are not very conspicuous and so there might be just one occasion when we can see both a bird's eyes and its beak properly. On this one occasion, the bird's beak is curved and it has forward vision, so for $Q_1$, $N = 1$ and $k = 1$. The question $Q_2$ is much easier to test, so there are nine occasions when we can observe both it and $P$. Say, $Q_2$ corresponds with $P$ seven times. The numbers involved are shown in Table 5.15.

**Table 5.15**

| Experiment | $N$ | $k$ | Expected success rate | Expected entropy |
|---|---|---|---|---|
| $Q_1 \Leftrightarrow P$ | 1 | 1 | 0.667 | 0.7214 |
| $Q_2 \Leftrightarrow P$ | 9 | 7 | 0.727 | 0.7838 |

The test with $Q_1$ has lower entropy, but the one with $Q_2$ is more likely to predict $P$ correctly. Thus, in general, an experiment with least expected entropy is not necessarily the best predictor.

## FURTHER READING

For details of birds' features actually used for classification by ornithologists, see Fitter (1963). Pankhurst (1991) explains how classification is applied in practical biology.

Everitt (1980) and Hartigan (1975) give more thorough discussions of the basic principles of clustering. The distinctions between optimizing, clumping, and hierarchical methods are made by Fisher and Langley (1985). The analysis by Langley and Iba (1993) throws more light on prototypes.

A range of learning algorithms, many of them clusterers, is incorporated in the 'Machine Learning Toolbox' developed under the ESPRIT project no. 2154. An overview appears in Craw, Sleeman, Graner, Rissakis, and Sharma (1992). The functions of the algorithms are summarized in the project flyer, issued by this group at the University of Aberdeen. Brazdil, Gama and Henery (1994) have made a detailed comparative study of more than twenty algorithms.

Rendell (1988), and Rendell and Cho (1990), discuss languages for describing training sets and related issues. Bergadano, Giordana, and Saitta (1991) give an account of non-classical intensional descriptions and provide more references. Ontological clusters and hierarchies containing them are described by Al-Mathami (1990).

The principal advocates of the term *conceptual clustering* are Michalski and Stepp (1983, 1986). Fisher and Langley (1985) give a careful discussion of several such algorithms, including CLUSTER/2 (invented by Michalski) and UNIMEM (due to Lebowitz).

The clustering approach to learning grammars was investigated by Wolff (1982); see also Hutchinson (1988) and Wolff (1990). There is a good introduction to parsers in Gazdar and Mellish (1989).

Decision trees have been investigated particularly by Quinlan (1979, 1986) who has published several papers on them. There is another pleasant account in Cestnik, Kononenko, and Bratko (1987) and a discussion of variant algorithms in Utgoff (1989). An experiment which translates a decision tree into rules and compares the predictive power of the tree with that of the rules was summarized by Quinlan (1987).

Shannon and Weaver (1949) first introduced entropy. Parry (1981) gives a thorough treatment. For a very readable introduction to the subject, see pages 276–87 of Ross (1976).

The account of the algorithm for finding graphs of interdependent properties is based on Fung and Crawford (1990). Steels and Van de Velde (1989) show how expert system rules can be recovered from such graphs. Geiger, Paz, and Pearl (1990) suggest how to infer causal relationships without temporal data. Their graphs are directed with an arrow from each cause to its effect, but, as their title implies, their method only discovers trees. Buntine (1991) proposes another method which is capable of inventing directed acyclic graphs and he remarks that it is similar to one developed by Cooper and Herskovits (1991) which shows good results. This is one of the examples which Buntine develops to demonstrate his method for inventing clustering algorithms.

Accounts of Dido appear in Scott and Markovitch (1989, 1991, 1992).

See DeGroot (1970) for an analysis of expected distributions of probabilities before and after experiments. He shows, on pages 198–201, that our function $\varphi$ is a good approximation to a best estimate for the density of $\pi$, even when the initial given distribution for $\pi$ is not quite uniform.

Composite questions have been investigated by Ragavan, Rendell, Shaw, and Tessmer (1993) and by Rendell and Ragavan (1993), with success.

# EXERCISES

1.  A door-to-door salesman wants to have the best possible chance of selling a life insurance policy whenever he rings a doorbell. At each house he visits, he notes:

    - the time of day;

    - the size of the house;

    - the age of the house;

    - whether houses in the area have gardens;

    - the state of the garden, if the house has one;

    - the states of the neighbours' gardens;

    - whether there is a car outside;

    - whether there is a bicycle outside;

    - the colour of the paintwork;

    - whether he made a sale there.

    Suggest an algorithm which he might use to learn what sort of house is worth visiting. Describe the data structures which it uses. Will the basic form of the algorithm suffice? If not, what elaborations will it require?

2.  A child can choose to go out and join the neighbours. The factors affecting his decision may include whether

    - it is wet;

    - he has finished his homework;

    - his mother is cheerful;

    - his mother is busy;

    - he can see the boy next door through the window.

    Construct a decision tree which decides if he will go out, from the following eight cases:

| *He goes out* | N | Y | N | N | N | N | Y | Y |
|---|---|---|---|---|---|---|---|---|
| *It is wet* | Y | N | Y | Y | Y | N | Y | N |
| *He has finished his homework* | Y | Y | Y | Y | Y | Y | N | N |
| *His mother is cheerful* | N | N | Y | Y | N | Y | Y | Y |
| *His mother is busy* | Y | Y | Y | Y | Y | Y | Y | Y |
| *He can see the boy next door* | N | Y | Y | N | N | Y | N | N |

3. A decision tree should predict whether

   $P$: someone will feel spry in the morning.

   The available questions are

   1. $Q_1$: What did he do the previous evening?
      to which there are three possible answers:
      - $l$: write a letter
      - $p$: play badminton
      - $g$: go out to dinner

   and

   2. $Q_2$: What is the weather?
      with possible answers
      - $s$: warm and sunny
      - $c$: cool and bright
      - $w$: wet.

   After a month, the observed results are

| $P$: | yes | | | no | | |
|---|---|---|---|---|---|---|
| | $l$ | $p$ | $g$ | $l$ | $p$ | $g$ |
| $Q_1$: | 6 | 6 | 7 | 3 | 0 | 8 |
| | $s$ | $c$ | $w$ | $s$ | $c$ | $w$ |
| $Q_2$: | 10 | 6 | 3 | 3 | 2 | 6 |

   From these data, estimate the entropies of the experiments

   $$If \quad (Q_1 = l) \lor (Q_1 = p) \quad then \quad P \quad else \quad \neg P$$

   and

   $$If \quad (Q_2 = s) \lor (Q_2 = c) \quad then \quad P \quad else \quad \neg P$$

   First, make naïve estimates by using the most likely values of the probabilities $\pi$. Then calculate the most likely values of their entropies, by assuming the Bayesian estimate for the probability distribution of $\pi$ when the prior estimate of $\pi$ is that it is uniformly distributed between 0 and 1.

4. There is a context-free grammar $G$ with terminal symbols: $a$, $b$, $c$, $d$; variable symbols: $P$, $Q$; and productions:

   $$P \to a \mid Qb \mid cQ$$
   $$Q \to d \mid bP.$$

   (a) Write down all words in $G$'s language, $L$, which can be generated from $P$ or $Q$ by using at most five derivation steps. This is most easily done in stages. Draw two columns, one headed $P$ and the other headed $Q$. In each column, write the words which can be derived from that symbol, first in one step, then in two, and so on. The table begins

| $P$ | $Q$ | Stage |
|---|---|---|
| $a$ | $d$ | 1 |
| $db$ | $ba$ | 2 |
| $cd$ | | |
| $bab$ | $bdb$ | 3 |
| $cba$ | $bcd$ | |

(b)   Using these words as input data, all occurring equally often, simulate the *build* operation of the grammar learner: Find the best correlated adjacent pair of symbols and replace this pair everywhere by a new AND symbol. Repeat this process a few times. This may be done systematically by drawing a square table with rows and columns indexed by symbols:

$$
\begin{array}{c|c|c|c|c}
 & a & b & c & d \\
\hline
a & & & & \\
\hline
b & & & & \\
\hline
c & & & & \\
\hline
d & & & &
\end{array}
$$

The entry in any square is the number of occurrences of the pair $ij$ divided by the numbers of occurrences of $i$ and $j$ individually, where $i$ is the symbol to the square's left and $j$ is above the square. Each time that you invent a new symbol, rewrite the above table to include it.

(c)   Simulate the *fold* operation: draw parse trees of the words, using all the symbols found so far. Make a list of all left contexts and another of all right contexts occurring in the trees. Construct a table, indexed down the left side by the left contexts and along the top by right contexts. The entry in it indexed by contexts *lc* and *rc* is the set of symbols $x$ occurring in these contexts:

      *lc x rc*       occurs in some tree.

(d)   Simulate the *generalize* and *rebuild* operations.

(e)   After iterating steps (b), (c), and (d) until the grammar cannot be extended reliably any more, repeat step (a) with the resulting grammar. Compare the sets of words derivable in five steps from the two grammars, the given one above, and the one you have constructed. How large is the symmetric difference of these two sets?

5.   Continue the calculation begun in the second example of Section 5.5, to determine the graph of interdependencies between pressure, wind, and rain.

(a)   From the given data, estimate all the other conditional probabilities:

$$P_I(w \mid p), \quad P_I(r \mid w), \quad P_I(p \mid w), \quad P_I(w \mid r), \quad P_I(p \mid r).$$

Recall that $I$ is $I(r, p, w)$, the hypothesis that $r$ is independent of $w$ given $p$. $P_I(r \mid p)$ was estimated in the example.

(b)   Write a procedure which takes for inputs:

- all the estimates of probabilities $P_I(x \mid y)$ from part (*a*);
- an observation $O = (v_p, v_w, v_r)$;

and which calculates the probability $P_I(O)$.

(c)   Write another procedure which takes as inputs the same probability estimates $P(x \mid y)$ and a summary $E$ of a set of observations, and which calculates $P_I(E)$. The summary $E$ might be in the form of a table:

| $p$ | $Y$ | $Y$ | $Y$ | $Y$ | $N$ | $N$ | $N$ | $N$ |
|---|---|---|---|---|---|---|---|---|
| $w$ | $Y$ | $Y$ | $N$ | $N$ | $Y$ | $Y$ | $N$ | $N$ |
| $r$ | $Y$ | $N$ | $Y$ | $N$ | $Y$ | $N$ | $Y$ | $N$ |
| *no.* | $n_{pwr}$ | $n_{pw\bar{r}}$ | $n_{p\bar{w}r}$ | $n_{p\bar{w}\bar{r}}$ | $n_{\bar{p}wr}$ | $n_{\bar{p}w\bar{r}}$ | $n_{\bar{p}\bar{w}r}$ | $n_{\bar{p}\bar{w}\bar{r}}$ |

where the number $n$ at the foot of each column is the number of observations $O$ in $E$ matching the values in the column above. You can choose how to represent this table in your code.

(d)   Suppose that the meteorologist's second set consisted of 10 observations and the numbers of each were actually as shown in the table below:

| $p$ | $Y$ | $Y$ | $Y$ | $Y$ | $N$ | $N$ | $N$ | $N$ |
|---|---|---|---|---|---|---|---|---|
| $w$ | $Y$ | $Y$ | $N$ | $N$ | $Y$ | $Y$ | $N$ | $N$ |
| $r$ | $Y$ | $N$ | $Y$ | $N$ | $Y$ | $N$ | $Y$ | $N$ |
| *no.* | 2 | 2 | 0 | 1 | 0 | 1 | 0 | 4 |

Use the procedure which you wrote in part (*c*) to estimate $P_I(E)$ for this set $E$ of observations.

(e)   Write a third procedure. This one takes for input a number $N$ and outputs a list of all possible tables of $N$ observations. Thus, when $N$ is 10, the table shown in (*d*) should be in the list.

(f)   The fourth procedure accepts as input the list generated by procedure (e). Its output is a list of pairs of numbers: if $E'$ is an entry in the input list then the corresponding entry in the output list is

$$(P_I(E'), B_{E'}).$$

Here, $B_{E'}$ is a binomial coefficient. It is the number of possible different experimental outcomes which could produce the table $E'$. The value of $B_{E'}$ is

$$\frac{N!}{n_{pwr}!\, n_{pw\bar{r}}!\, n_{p\bar{w}r}!\, n_{p\bar{w}\bar{r}}!\, n_{\bar{p}wr}!\, n_{\bar{p}w\bar{r}}!\, n_{\bar{p}\bar{w}r}!\, n_{\bar{p}\bar{w}\bar{r}}!}.$$

(g)   With the benefit of these procedures, you can now calculate the backing which the experiment $E$ gives to the hypothesis $I(r, p, w)$. This backing is the sum

$$\sum B_{E'}\, P_I(E')$$

taken over all pairs $(P_I(E'), B_{E'})$ for which $P_I(E') < P_I(E)$. Work it out. To within backing of 10 per cent, does experiment $E$ support the hypothesis $I(r, p, w)$?

> *Note on implementation*: If you are using a lazy functional language with a decent garbage collector, then the method just outlined is sensible. If you are writing in a more conventional language, then it might be wise to amalgamate procedures (*e*) and (*f*) into one which just produces the list of pairs of numbers, $(P_I(E'), B_{E'})$. The lists involved are quite long and there is no need to construct the entire list of tables.

(h)   There are six possible edges which might occur in the Markov network relating $p$, $w$, and $r$. Extend your programs so that they calculate the backings for all the hypotheses $I(x, y, z)$ where $\{x, y, z\} = \{p, w, r\}$. Then implement the Fung–Crawford algorithm and determined this network from the given data.

# 6 Pattern matching and generalization

## SUMMARY

This is a large subject. Here, we present four basic methods.

- The first infers interdependencies between simple facts by means of a technique called association chains.

- The second forms general terms and clauses from examples. It depends on:
  - finding how known facts interrelate;
  - substitutions;
  - the commutative and associative laws for disjunction.

  It can give rise to curious mistaken inferences, particularly when the fact that two distinct objects are distinct is not stated explicitly. This kind of generalization leads to consideration of unification and other varieties of laws.

- The third method is a case of program synthesis. The particular programming language used is called *FP*.

- The fourth is a way of inventing recursive definitions from examples. It is sensitive to bias, and the method includes principles which suggest how to adjust the description language so that it is suitable for inference.

## INTRODUCTION

We have met one form of generalization in the last chapter. Wolff's *fold* operation is followed by a step which generalizes the productions for AND symbols. There are other kinds. In the next chapter, we shall generalize proofs. These two sorts of generalizing are quite different. When you generalize a proof, you are removing any unnecessary assumptions from it. That is a simple mechanical process and, as long as the proof's hypotheses are correct, it involves no risk. The sort of generalizing which takes place in Wolff's algorithm, which we shall study in more detail now, involves guessing; and the guesses may be wrong.

Syntactic methods are valuable because they require very small training sets. Three or four examples are usually enough and sometimes the learner can extrapolate from two or even just one. A lot of the methods associated with rule-based systems are also syntactic, as you will see in Chapter 7. The separation of the methods in this chapter from rule-based methods is a bit arbitrary. They have been divided on the grounds of how they are commonly used, as much as on their innate features.

In this chapter, we shall study four purely syntactic methods. The last three are algorithms which invent generalizations. The first is a trick for discovering associations between facts.

The first generalizer is due to Plotkin. Its examples should all have the same structure and the algorithm replaces terms in them with variables. There are actually two forms of this algorithm. The simpler form generalizes from terms with similar structure. The second generalizes from pairs of clauses.

The second generalizer is more subtle. Its input examples do not have identical structures. This second algorithm generalizes the structures themselves. In fact, it does more than that. This second algorithm invents a program which can construct the given structures. The programs found by this method all have a particular form. They involve **linear recursion**: the program calculates the value $f\,t$ of a function $f$ on some term $t$.

$$f\,t \;=\; g(f\,t')$$
$$f\,t' \;=\; g(f\,t'')$$
$$f\,t'' \;=\; g(f\,t''') \ldots$$

The terms $t, t', t'', \ldots$ get progressively simpler. The formula $g$ is the same for all of them.

The third generalizing algorithm is designed to find recursive definitions. Given any predicate, such as *column* which should be true of all columns and not anything else, it can be used to invent rules which describe the predicate. The predicate occurs in the conclusion of each rule and it can also occur in conditions. One feature of this method is that the recursion it invents may be *non-linear*: the defined predicate can occur more than once in a condition for itself.

The data available to a learner often consist of no more than simple raw facts. Some generalizing algorithms cannot handle data in this form. Their training sets have to contain assertions of *cause and effect*. Each training instance expected by the learner is typically a clause, such as

good_at (Edward, chess) $\wedge$ good_at (Rachel, chess) $\Rightarrow$
like_playing (Edward, Rachel, chess)

whereas in practice the learner might only be told that

good_at (Edward, chess)

and

good_at (Rachel, chess)

and

like_playing (Edward, Rachel, chess).

It has to guess for itself the relation between these basic facts. You have already come across one method of inferring such dependencies. The Fung–Crawford statistical technique can form them. However, statistical calculations are cumbersome. They need lots of examples. Therefore, we begin with another method suggested by Vere which lets a learner construe such relations from simple syntax.

Here is an example of how these methods fit together. Binary trees are data structures common in computer science. There are two kinds of binary tree. The interesting ones have branches. They are terms of the form

$$node(l, r)$$

where *node* is a function symbol and the two branches, $l$ and $r$, are themselves binary trees. The other kind of tree is a leaf. We suppose that there is a predicate

$$leaf(x)$$

which is true whenever $x$ is a leaf and not otherwise. There is a predicate called *bt* which should be true of all these trees and of nothing else. We want to find a general definition of *bt*.

*Example: inventing the notion of binary trees*

The learner is given the assertions

$$bt(\mathbf{A}) \quad \text{and} \quad bt(node\ (\mathbf{B}, \mathbf{C}))$$

and various other facts, including

$$leaf(\mathbf{A}) \quad leaf(\mathbf{B}) \quad leaf(\mathbf{C}).$$

From these, Vere's method shows how to construct the ground clauses

$$leaf(\mathbf{A}) \quad\quad\quad \Rightarrow \quad bt(\mathbf{A})$$
$$leaf(\mathbf{B}) \wedge leaf(\mathbf{C}) \quad \Rightarrow \quad bt(node\ (\mathbf{B}, \mathbf{C})).$$

Another similar set of data would yield the ground clauses

$$leaf(\mathbf{D}) \quad\quad\quad \Rightarrow \quad bt(\mathbf{D})$$
$$leaf(\mathbf{E}) \wedge leaf(\mathbf{F}) \quad \Rightarrow \quad bt(node\ (\mathbf{E}, \mathbf{F})).$$

From these, Plotkin's method allows us to conclude that

$$leaf(a) \quad\quad\quad \Rightarrow \quad bt(a)$$
$$leaf(b) \wedge leaf(c) \quad \Rightarrow \quad b(node\ (b, c))$$

for any $a$, $b$, and $c$. Then the last generalizing algorithm will suggest that, from these, we can infer a recursive rule:

$$bt(b) \wedge bt(c) \quad\quad \Rightarrow \quad bt(node\ (b, c)).$$

This together with the base case

$$leaf(a) \quad\quad\quad \Rightarrow \quad bt(a)$$

is the desired definition of *bt*.

## 6.1   INCORPORATING BACKGROUND INFORMATION

Vere starts from the notion that there is a distinction between the observed features which are used to describe a training *instance* and the known facts about the background *situation* in which it occurs. The description of each instance omits lots of background detail, so the learner has to decide what aspects of the background are relevant. Vere showed how one can build rules whose conditions include just the right facts about the background.

Vere's original examples mostly involve operators with side effects. The method is simpler when it is applied to situations with no side effect, so we shall investigate a couple of these first.

### 6.1.1   ASSOCIATION CHAINS

*Example*

Suppose that Edward is good at chess, Rachel is good at chess and tennis, Fiona is good at tennis, Edward and Rachel like playing chess together, and Rachel and Fiona like playing tennis together. The objective is to find conditions for liking to play together. To a human, the natural course of inference is

$$\text{good\_at (Ed, chs)} \quad \wedge \quad \text{good\_at (Ra, chs)} \quad \Rightarrow \quad \text{like\_ playing (Ed, Ra, chs)}$$
$$\text{good\_at (Fi, tns)} \quad \wedge \quad \text{good\_at (Ra, tns)} \quad \Rightarrow \quad \text{like\_playing (Ra, Fi, tns)}.$$

We shall soon study an algorithm due to Gordon Plotkin which will let us infer

$$\text{good\_at } (a, g) \wedge \text{good\_at } (b, g) \wedge \text{good\_at (Ra, } g) \Rightarrow \text{like\_playing } (a, b, g)$$

from these two premises, but before doing that we must see how the premises can be constructed from the raw facts.

The background consists of the four facts

$$\text{good\_at (Ed, chs)} \quad \text{good\_at (Ra, chs)} \quad \text{good\_at (Fi, tns)} \quad \text{good\_ at (Ra, tns)}$$

and the foreground assertions are

$$\text{like\_playing (Ed, Ra, chs)} \quad \text{like\_playing (Ra, Fi, tns)}.$$

The objective is to find background conditions for the two foreground conclusions.

Consider the first: like_playing (Ed, Ra, chs). This contains three constants. Vere argues that there should be a reason why they occur together, so the condition for like_playing (Ed, Ra, chs) should contain facts linking Ed and Ra and chess. Any such facts may form part of a plausible condition. If these facts in turn contain any other constants, there should be further literals in the condition relating them. The argument can apply to more elaborate terms too, not just to constants.

Vere writes that terms in the conclusion should be linked together by 'association chains' of literals in the conditions. Two literals are associated if they are different and they contain a common term. An association chain is a sequence of literals in which adjacent literals are associated.

Say two terms $T_1$ and $T_2$ are **associated** if there is a sequence of literals

$$L_1, \quad L_2, \quad \ldots \quad L_k$$

and there is a sequence of terms

$$T_1 = t_0, \quad t_1, \quad t_2, \quad \ldots \quad t_k = T_2$$

so that, for each $j \geq 1$, the terms $t_{j-1}$ and $t_j$ both occur in $L_j$. This sequence of literals is called an **association chain** between $T_1$ and $T_2$.

In the case above, Ed and chs are two terms occurring in the conclusion.

like_playing (Ed, Ra, chs)   good_at (Ed, chs)   like_playing (Ed, Ra, chs)

is an association chain linking them. Vere says that all the intermediate literals in the chains are likely conditions, so

good_at (Ed, chs)

is a condition for

like_playing (Ed, Ra, chs).

Similarly, Ra and chs are two terms in the conclusion and

like_playing (Ed, Ra, chs)   good_at (Ra, chs)   like_playing (Ed, Ra, chs)

is another association chain linking them, so

good_at (Ra, chs)

is another such condition. Vere's principle leads to precisely the ground clauses above.

Vere's original principle seems to miss the case of facts whose predicates have just one argument. He writes:

We are interested in forming association chains in which the first and last literals in the chain are foreground literals and the rest are background literals.

Such a chain can never contain a background predicate with just one argument. Such predicates do occur in real life. One such is *leaf*, the predicate which provides the base case for the predicate *bt* which is true of binary trees. Therefore, let us extend his method to allow chains with one end a foreground literal and the other background of arity 1. The clauses

$$leaf(\mathbf{A}) \qquad \Rightarrow \quad bt\,(\mathbf{A})$$

and

$$leaf(\mathbf{B}) \;\wedge\; leaf(\mathbf{C}) \Rightarrow bt\,(node(\mathbf{B}, \mathbf{C}))$$

are built from such chains, as shown.

### 6.1.2  BACKGROUND CONDITIONS FOR OPERATORS WITH SIDE EFFECTS

Vere's rules for actions are a little more elaborate than ordinary clauses, because they make a distinction between conditions which the rule's action changes and other conditions which remain true after the action. Such a rule is written thus:

$$[Z]\, X \to Y$$

where $X$, $Y$, and $Z$ are conjunctions of atomic formulas.

$Z$ consists of conditions which remain true after the action.
$X$ is the conjunction of all conditions which cease to be true after the action.
$Y$ is a conjunction of facts which describe the result of the action.

The action is only described by its effects.

*Example: the 'blocks world'*
The two rules in this case will appear as
*Unstack*:
$$[y \neq \textbf{table} \wedge \mathrm{clear}(x)\,]\ \mathrm{on}(x, y)\ \to\ \mathrm{on}\,(x, \textbf{table}) \wedge \mathrm{clear}\,(y)$$

and
*Stack*:
$$[\mathrm{clear}(x) \wedge y \neq \textbf{table} \wedge x \neq y\,]\ (\mathrm{on}\,(x, \textbf{table}) \wedge \mathrm{clear}\,(y)\,)\ \to\ \mathrm{on}\,(x, y).$$

*Example: finding a forced win, in noughts and crosses*
Recall, the board is a $3 \times 3$ array of squares, such as

$$
\begin{array}{c|c|c|c}
 & a & b & c \\
\hline
1 & & \times & \\
\hline
2 & & \bigcirc & \\
\hline
3 & & & \\
\end{array}
$$

The structure of the board can be described by eight facts stating which triples of squares form lines:

*linear* $(a_1, a_2, a_3)$, ..., *linear* $(a_3, b_2, c_1)$, ...

One plausible rule is
    IF   *to-move* $(C) \wedge$ *linear* $(x, y, z) \wedge$
        *filled* $(x, C) \wedge$ *empty* $(y) \wedge$ *empty* $(z) \wedge C \neq D$
THEN
        *play* $(y, C)$; *play* $(z, D)$
where the operation play $(x, C)$ deletes the fact

*empty* $(x)$

and adds

*filled* $(x, C)$    and    *filled*$(z, D)$.

In Vere's form, this rule appears as

$$[\textit{to-move}\ (C) \land \textit{linear}\ (x, y, z) \land \textit{filled}\ (x, C) \land C \neq D]$$
$$(\textit{empty}\ (y) \land \textit{empty}\ (z)) \rightarrow$$
$$(\textit{filled}\ (y, C) \land \textit{filled}\ (z, D)).$$

Any rule can be written in this form. If its set of preconditions is $\alpha$ and after it has been applied the resulting state is described by a set of facts called $\beta$, then

$Z$ is the conjunction of $\alpha \cap \beta$
$X$ is the conjunction of $\alpha \setminus \beta$
$Y$ is the conjunction of $\beta \setminus \alpha$.

Vere makes a further distinction between the part of $Z$ which is special to the example (**foreground** information, $Zf$) and the part which is always assumed true (**background** information, $Zb$). In the last example,

the instance of $Zf$ is   $\textit{to-move}\ (\bigcirc) \land \textit{filled}\ (b_2, \bigcirc)$
the instance of $Zb$ is $\textit{linear}\ (b_2, c_3, a_1) \land \bigcirc \neq \times$.

The whole rule may be written as

$$[Zb, Zf]\ X \rightarrow Y.$$

If background information is omitted, then generalization can easily form a rule with a variable on the right side (in $Y$) which is unrelated with anything on the left (in $X$ or $Z$). This is a bad thing. Vere's approach is designed to prevent it.

A rule

$$[Zb, Zf]\ X \rightarrow Y$$

is **deterministic** if every variable occurring in $Y$ is associated with some variable occurring in $X$ or $Zf$. There must be an association chain linking these variables whose literals are all in the rule, but some of them may be in $Zb$.

In the case of noughts and crosses,

$y$ and $z$ occur in $X$

$C$ occurs in $Zf$, in the condition $\textit{to-move}\ (C)$

$D$ is associated with $C$ by a chain which includes the condition $C \neq D$

so this rule is deterministic.

Vere's **first principle** is that *any learned rule should be deterministic*. If Plotkin generalization produces a rule from the original training set which is not deterministic, then more assumed facts should be included in the descriptions of training instances so that the outcome is deterministic.

His **second principle** is that *association chains should be short*. This principle will suggest which are the most relevant background facts.

Vere wrote a program, which he called 'Thoth-pb', embodying these principles, but I have not seen an account of it nor even an explanation of how it embodied them. Thoth-pb was applied to four cases which had already been studied.

1.  How to predict the next of a sequence of letters, such as

    A B M C D M E F M ...

    The background information was the order of letters in the alphabet:

    *next* (A, B)    *next* (B, C)    *next* (C, D)    ...

    The sequence had to have a fixed period, unlike

    A B B C C C D D D D E ...

2.  Finding visual analogies. If

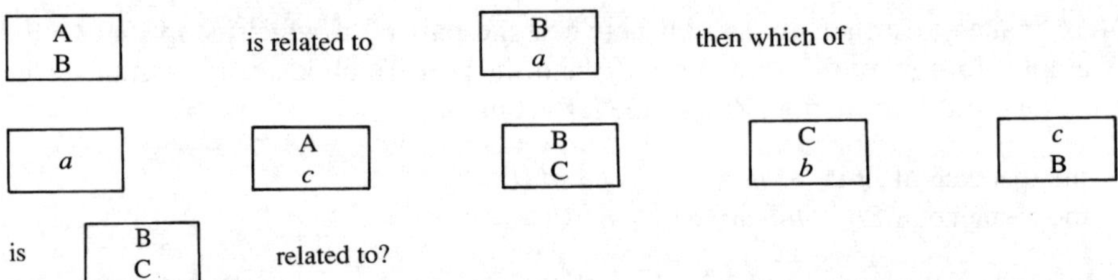

3.  Learning of legal pawn moves in chess. The board was described as a set of 64
    named squares. The background information consisted of adjacencies between
    squares. In this case, Thoth-pb was tested with one training set which included
    counter-examples and another consisting only of examples.

4.  *The Towers of Hanoi* problem.

    In most cases, it found satisfactory general rules. When it did not, there was
    usually some quirk of the training set which made Thoth-pb's output
    reasonable.

## 6.2   FINDING PATTERNS OF CAUSE AND EFFECT

Patterns are formed by generalizing from examples with strict syntax. The
prototypical algorithm which learns such patterns and which we shall present here
was studied by Gordon Plotkin. It is very useful for learning in many
circumstances.

The natural setting for generalization is a formal language. First-order predicate
calculus (FOPC) is a prime example, the one used by Plotkin. He studied two
cases. The first generalizes terms and literals. Recall, a **term** may be either a
constant:

    Peter    Jane

or a variable symbol:

   $x$   $y$

or a composite, formed from a function symbol and its arguments which are other terms:

   father_of (Jane).

Terms represent things which we can make assertions about. A **literal** represents some such assertion. The literals are the simplest possible assertions. They involve no connectives like $\wedge$ or $\vee$ or $\Rightarrow$ and no quantifiers. A literal may be either an atomic formula, consisting of a predicate symbol and its arguments:

   likes (Peter, father_of (Jane))

which you can interpret as

   'Peter likes Jane's father'

or it may be a negated atomic formula:

   $\neg$likes (Peter, father_of (Jane))

which may be read as

   'Peter doesn't like Jane's father'.

Each term and each literal can only be written one way (bar variants in notation).

*Example: simple generalizations*

   Suppose that we are told

      likes (Peter, father_of (Jane)).

   and

      likes (Edward, father_of (Jane)).

   The naïve generalization of these two formulas is

      $\forall x$ likes $(x,$ father_of (Jane)).

   Suppose instead that we were told that

      likes (Peter, father_of (Jane))

   and

      likes (Peter, Jane).

   The generalization of these two formulas which the following theory will provide is

      $\forall x$ likes (Peter, $x$).

In both cases, these generalizations are a bit too crude. In the first case, it might make more sense to suggest that *all young men* like Jane's father. It may well be that Peter's father and Jane's father don't get on too well together. In the second example, it might be wiser to conjecture that Peter likes *anyone connected with Jane*. The assertion that Peter likes absolutely everybody and everything is rather too sweeping. There are ways of modifying the theory, but be warned: as it stands, these are the generalizations which it will produce.

Plotkin's second case generalizes disjunctive clauses. A **clause** is a disjunction of literals in which all variables are universally quantified; for example,

$$\forall x \ (\text{likes } (x, \text{father\_of } (\text{Jane})) \lor \neg \text{likes } (\text{Jane}, x)).$$

Such a clause depicts a statement which may be true or false. This one says that

'Whoever $x$ may be, either $x$ likes Jane's father or Jane doesn't like $x$'

or equivalently

'Anyone Jane likes likes her father'

or equivalently

'Jane dislikes anyone who doesn't like her father'.

Note that a clause need not contain any variables. The expression

$$\text{likes } (\text{Peter}, \text{father\_of } (\text{Jane})) \lor \neg \text{likes } (\text{Jane}, \text{Peter})$$

is also a clause. When a clause or term contains no variables, it is said to be **ground**.

A clause can be re-expressed with other connectives, such as $\land$ or $\Rightarrow$, but the conventional way to write clauses only uses the two connectives $\lor$ and $\neg$. These two suffice. Since all variables are universally quantified, workers in this subject often don't bother to say so. They simply leave off all the $\forall$ symbols:

$$\text{likes } (x, \text{father\_of } (\text{Jane})) \lor \neg \text{likes } (\text{Jane}, x).$$

(Plotkin and some others take this simplification of notation even further. They write a clause as the set of its literals.)

### 6.2.1 UNDERLYING ASSUMPTIONS AND RESTRICTIONS

*Expressing the relevant facts as a clause*

Before the algorithms can be applied, we must translate all relevant knowledge of the real world into clauses.

*Example: generalizations of cause and effect*

Suppose the given knowledge is

- Peter likes Jane's father;

- Jane likes Peter;

- Edward does not like Jane's father;

- Jane does not like Edward.

If these bald statements are translated directly into symbolic form, they will appear as

likes (Peter, father_of (Jane))
likes (Jane, Peter)
¬likes (Edward, father_of (Jane))
¬likes (Jane, Edward).

The result of generalizing these would probably be

likes $(x, y)$ ∨ ¬likes $(x, y)$

which is not much use.

The point is, the bald statements in the example do not capture our intuition that Jane's preferences depend on other peoples' attitudes. Besides the individual facts about Jane, Peter, and Edward, we have general knowledge about fathers and about liking. This extra knowledge is missing from the basic literals. They embody no notion of cause and effect. The statements which are really suitable for generalization are

- Jane likes Peter, *but if Peter didn't* like Jane's father *then she wouldn't;*

- Edward does not like Jane's father, *and so* Jane does not like Edward.

In symbolic form, these appear as

likes (Jane, Peter) ⇒ likes (Peter, father_of (Jane))
¬likes (Edward, father_of (Jane)) ⇒ ¬likes (Jane, Edward)

or, as clauses,

likes (Peter, father_of (Jane)) ∨ ¬likes (Jane, Peter)

likes (Edward, father_of (Jane)) ∨ ¬likes (Jane, Edward).

Generalizing these will give us what we want:

likes $(x$, father_of (Jane)) ∨ ¬likes (Jane, $x)$.

If the information supplied is just in the form of basic facts, then Vere's method can be used to assemble them into clauses. Henceforth, we shall take it for granted that we always know which facts imply which others.

### *The commutative and associative laws*

Generalizing clauses takes a little more effort than generalizing terms and literals, because a clause can be written in different ways by rearranging its disjuncts:

$\neg$likes (Jane, $x$) $\vee$ likes ( $x$, father_of (Jane)).

The formal way to say this is that disjunction ($\vee$) satisfied the **commutative** law:

$P \vee Q$       means the same as       $Q \vee P$.

Furthermore, if there are three or more disjuncts, we can calculate and combine their individual truth values in any order. Thus,

(likes ($x$, father_of (Jane)) $\vee$ $\neg$likes (Jane, $x$)) $\vee$ $x$ = Peter

means the same as

likes ($x$, father_of (Jane)) $\vee$ ($\neg$likes (Jane, $x$) $\vee$ $x$ = Peter).

In jargon, one says that disjunction satisfies the **associative** law:

$(P \vee Q) \vee R$       means the same as       $P \vee (Q \vee R)$.

Plotkin arrived at two separate though related algorithms. The first generalizes pure formulas, not subject to any laws. The second generalizes formulas containing a connective which is subject to both the associative and commutative laws. This is the beginning of a deep subject. There are many possible laws which a connective might obey. (Remember, we are not limited to formulas and connectives solely from the predicate calculus.) Furthermore, a single formula may contain more than one non-trivial connective and there may be subtle laws which interrelate two or more connectives. We shall touch on these ideas later, but they are too much for us to explore in detail.

### 6.2.2 SUBSTITUTIONS

Generalization is most easily expressed by means of substitutions. A **substitution** is a function from formulas to formulas. It leaves the framework of a formula unchanged—the function symbols and constants, parentheses, and commas all stay where they were to begin with. It only changes variables. For instance, a substitution called $\vartheta$ might change each occurrence of the variable $x$ to the constant *David*. Its result when applied to

likes ($x$, father_of (Jane)) $\vee$ $\neg$likes (Jane, $x$)

will be

likes (David, father_of (Jane)) $\vee$ $\neg$ likes (Jane, David).

Traditionally, a substitution is written on the right of the formula, so the above is

(likes $(x$, father_of (Jane)) $\vee \neg$ likes (Jane, $x$)) $\vartheta$.

If this substitution only changes $x$ and never any other variable, then it is written

$\vartheta = [$David$/x]$.

There might be another substitution $\varphi$ which has this effect on $x$ and which also changes some other variable, say $y$:

$\varphi = [$David$/x$, father_of $(z)/y]$.

This will have exactly the same result as $\vartheta$ when applied to the above formula. $y$ does not occur in this formula, so the fact that $\varphi$ substitutes for $y$ doesn't matter. However, the difference between them becomes apparent when they are applied to the formula

likes $(x, y) \vee \neg$likes $(z, x)$.

On this, $\vartheta$ produces

likes (David, $y) \vee \neg$likes $(z$, David)

but $\varphi$ produces

likes (David, father_of$(z)) \vee \neg$likes $(z$, David).

When a substitution changes several variables, then they are all replaced simultaneously. This matters for a substitution like the following:

$\rho = [$David$/x$, father_of $(x)/y]$

which substitutes for the variable $x$ and also for $y$ and when the value which replaces $y$ contains $x$. The result of applying $\rho$ to the formula above

likes $(x, y) \vee \neg$likes $(z, x)$

will be

likes (David, father_of $(x)) \vee \neg$likes $(z$, David).

Two substitutions such as $\vartheta$ and $\rho$ can be applied successively to a formula. The result is the same as applying a single substitution, $\vartheta\rho$, called the **product** of $\vartheta$ and $\rho$. In this case,

$x\vartheta \ = \ $ David $\quad$ so $\quad x\,\vartheta\rho \ = \ $ David $\quad \rho \ = \ $ David

and

$y\,\vartheta \ = \ y \quad$ so $\quad y\,\vartheta\rho \ = \ y \quad \rho \ = \ $ father_of $(x)$.

Neither $\vartheta$ nor $\rho$ alter any other variables and so

$\vartheta\rho = [$David$/x$, father_of $(x)/y]$.

Note that the order matters in such a product. If we follow the corresponding calculation for $\vartheta$ and $\rho$ in the other order, we find that

$$x\,\rho = \text{David} \qquad \text{so} \qquad x\,\rho\vartheta = \text{David } \vartheta = \text{David}$$

but

$$y\,\rho = \text{father\_of }(x) \qquad \text{so} \qquad y\,\rho\vartheta = \text{father\_of (David)}.$$

Thus, the product of $\rho$ and $\vartheta$ is

$$\rho\vartheta = [\text{David}/x, \text{father\_of (David)}/y]$$

which is not the same as $\vartheta\rho$.

We say that one substitution, say $\sigma$, is **more general** than another, $\tau$, if there is a substitution $\upsilon$ so that

$$\tau = \sigma\upsilon.$$

We also say that $\tau$ is **more specific** than $\sigma$. In the cases above, $\vartheta$ is more general than both $\varphi$ and $\rho$ because

$$\varphi = \vartheta[\text{father of }(z)/y]$$

and

$$\rho = \vartheta[\text{father\_of }(x)/y].$$

(You should check this!) The most general substitution is the one which changes absolutely nothing. It is sometimes called $\varepsilon$.

### 6.2.3   GENERALIZING TERMS

The basic idea is simple. In order to generalize from two terms, $F$ and $G$, you scan along them both from their left-hand ends up to the first place where they differ. At this place, $F$ will contain a subterm, say $f$, and $G$ will contain a subterm $g$. You replace both $f$ in $F$ and $g$ in $G$ with a new variable, say $x$. Then you also replace $f$ and $g$ with $x$ at any other places where they both occur. The whole process is repeated until the two terms $F$ and $G$ are identical.

For instance, say we are finding a common least generalization of the two formulas

$$F = \text{likes (Edward, Jane)}$$

and

$$G = \text{likes (Peter, Peter)}.$$

In the first cycle, $f$ is *Edward* and $g$ is *Peter* and the result of replacing them is

$$F_1 = \text{likes }(x, \text{Jane})$$
$$G_1 = \text{likes }(x, \text{Peter}).$$

In the second cycle, $f$ is *Jane* and $g$ is *Peter*. The algorithm's outcome is

$$\text{likes }(x, y).$$

The first occurrence of *Peter* in $G$ is changed to $x$ along with *Edward* in $F$. The point is, we do not change the second occurrence of *Peter* to $x$ too. We only change *Edward* in $F$ and *Peter* in $G$ to $x$ where they occur together.

Here is another slightly more elaborate example. Suppose that

$F$ = likes_playing (Ed, Ra, common-game (Ed, Ra))
$G$ = likes_playing (Pe, Ra, chess).

The first cycle yields

$F$ = likes_playing ($x$, Ra, common_game (Ed, Ra))
$G$ = likes_playing ($x$, Ra, chess).

The second occurrence of *Ed* is not changed because, at the second place where *Ed* occurs in $F$, *Pe* does not occur in $G$. The result of generalizing from these two terms is

likes_playing ($x$, Ra, $y$).

For this reason, we have to be rather careful about how we describe where each subterm occurs in a formula. In the first case above it is easy. In the first $F$, we say that *Edward* occurs at place $\langle 1 \rangle$ and *Jane* occurs at place $\langle 2 \rangle$. Matters become subler when the subterms themselves have sub-subterms. In general, a **place** is a sequence of integers in diamond brackets, like $\langle 3, 1 \rangle$. This particular place is where you will find the *first* subterm of the *third* subterm of the formula. Thus, in

likes_playing (Ed, Ra, common_game (Ed, Ra))

the subterm at place $\langle 3 \rangle$ is

common_game (Ed, Ra)

and the one in place $\langle 3, 1 \rangle$ is

Ed.

We shall say that the $\langle 3, 1 \rangle$th subterm of this formula is *Ed*.

If $t$ is any term, the subterm of $t$ at place $\langle \ \ \rangle$ is just $t$.
If $t = f(t_1, t_2, ..., t_n)$ and $n \geq i$
then the subterm at place $\langle i, j, ..., k \rangle$ in $t$ is
the subterm of $t_i$ at place $\langle j, ... k \rangle$, if it exists.

Plotkin's first generalization algorithm takes as inputs two formulas, $F$ and $G$. Its outputs are a formula $W$ which is a generalization of both $F$ and $G$, and two substitutions $\vartheta$ and $\varphi$. These give $F$ and $G$ when applied to $W$:

$$W\theta = F \qquad \text{and} \qquad W\varphi = G.$$

They satisfy the extra condition that any other generalization $V$ of $F$ and $G$ also generalizes $W$: if

$$V\rho = F \qquad \text{and} \qquad V\sigma = G$$

then there is another substitution $\tau$ such that

$$V\tau = W.$$

See Fig. 6.1

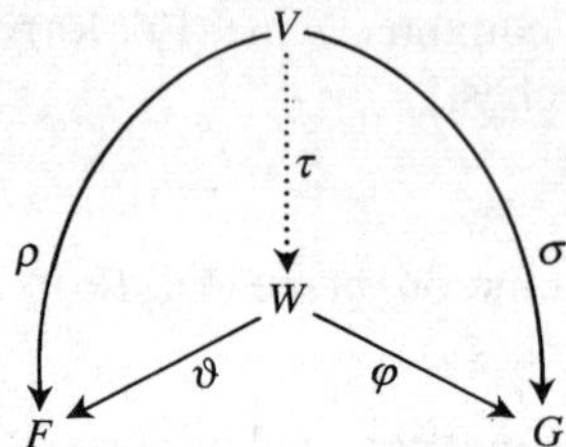

**Fig. 6.1**   The least general generalization $W$ of $F$ and $G$.

If the solid arrows can be filled in, then so can the dotted line too. However, note that $\tau\vartheta$ may not be the same as $\rho$ and $\tau\varphi$ may not be the same as $\sigma$ because $\rho$ and $\sigma$ may affect other variables which do not occur in $V$ or $W$.

The algorithms goes:

$$W_1 := F$$
$$W_2 := G$$
$$\vartheta_1 := \vartheta_2 := \varepsilon$$

WHILE $W_1 \neq W_2$

   $I$ := a shortest place where $W_1$ and $W_2$ differ
   $T_1$ := the $I$th subterm of $W_1$
   $T_2$ := the $I$th subterm of $W_2$
   Choose the place $I$ so that $T_1$ and $T_2$ have different first symbols, so that they are not subterms of any other two such terms.

   If you like, you can take the leftmost place where $W_1$ and $W_2$ differ. Remember, $I$ is a sequence of numbers.

   Choose a new variable, which we call $x$, which does not occur in either $W_1$ or $W_2$ or $\theta_1$ or $\theta_2$.

   FOR every position $J$
       IF the $J$th subterm of $W_1$ is $T_1$
       AND the $J$th subterm of $W_2$ is $T_2$
       THEN at place $J$ in $W_1$, replace $T_1$ by $x$
               at place $J$ in $W_2$, replace $T_2$ by $x$

   $$\vartheta_1 := \vartheta_1 \, [T_1 / x]$$
   $$\vartheta_2 := \vartheta_2 \, [T_2 / x]$$

$$W := W_1$$
$$\vartheta := \vartheta_1$$
$$\varphi := \theta_2.$$

This algorithm always succeeds. When it has finished, $W$ is the common value of $W_1$ and $W_2$.

*Example*

Suppose that

$$F = \text{likes (Edward, Jane)}$$

and

$$G = \text{likes (Peter, Peter)}.$$

In the first cycle of the WHILE loop, $I$ could be either $\langle 1 \rangle$ or $\langle 2 \rangle$. Let us choose

$$I = \langle 1 \rangle.$$

At this place,

$$T_1 = \text{Edward} \qquad \text{and} \qquad T_2 = \text{Peter}.$$

Say our chosen new variable is $x$.

There are just two possible values of $J$: $\langle 1 \rangle$ and $\langle 2 \rangle$. The $\langle 2 \rangle$th subterm of $G$ is $T_2$ but the $\langle 2 \rangle$th subterm of $F$ is not $T_1$. Hence, $\langle 1 \rangle$ is the only place where we change subterms to $x$. At the end of this cycle,

$$W_1 = \text{likes } (x, \text{Jane})$$
$$W_2 = \text{likes } (x, \text{Peter})$$
$$\vartheta_1 = [\text{Edward}/x]$$
$$\vartheta_2 = [\text{Peter}/x].$$

During the second cycle, $I$ can only be $\langle 2 \rangle$.

$$T_1 = \text{Jane}$$
$$T_2 = \text{Peter}$$

If the new variable we choose is $y$, then at the end,

$$W = W_1 = W_2 = \text{likes } (x, y)$$
$$\vartheta_1 = [\text{Edward}/x] \, [\text{Jane}/y] = [\text{Edward}/x, \text{Jane}/y]$$
$$\vartheta_2 = [\text{Peter}/x] \, [\text{Peter}/y] = [\text{Peter}/x, \text{Peter}/y].$$

This algorithm works well for terms. In FOPC, it should not really be applied as it stands to literals, because it might generalize two literals to a variable and in FOPC, a variable cannot take the place of a literal. Hence, Plotkin added an extra restriction:

Either $W_1$ and $W_2$ must both be terms
    or they must both be atomic formulas with the same predicate symbol
    or they must both be negated atomic formulas with the same predicate symbol.

If $W_1$ and $W_2$ satisfy this condition, then Gordon Plotkin says that they are compatible. The algorithm only applies to two literals if they are **compatible.**

### 6.2.4   GENERALIZING CLAUSES

This step is rather more subtle than the straightforward first algorithm for terms. To start with, we have to make up our minds about just what sort of generalization we are looking for. If we are given two clauses such as

likes (Peter, father_of (Jane)) $\vee$ $\neg$likes (Jane, Peter)

and

likes (David, father_of (Jane)) $\vee$ $\neg$likes (Jane, David)

and we also have a considerable body of knowledge about the characters of Peter and David and about Jane's powers of deduction and sense of propriety, then we might be able to arrive at some profound conclusion such as

sensible $(x)$ $\wedge$ sympathetic $(x)$ $\vee$ $\neg$likes (Jane, $x$)

but this is not the generalization we expect. When we speak of generalization, we are *not* supposing any underlying knowledge or deduction. The sort of generalizing which we are doing is purely syntactic.

Suppose that $C_1$ and $C_2$ are any two clauses. We seek another clause which implies them both. In general, it is too much effort to search through the space of all clauses $C$ for which

$$C \Rightarrow C_1 \qquad \text{and} \qquad C \Rightarrow C_2.$$

Implication, '$\Rightarrow$', is too complex. Instead, Plotkin opted for search in the space of clauses $C$ which subsume both $C_1$ and $C_2$.

If $C$ and $D$ are two clauses, we say that $C$ **subsumes** $D$ if there is a substitution $\sigma$ for which every literal disjunct in $C\sigma$ occurs as a disjunct in $D$. Thus,

$\neg$likes (Jane, $x$)

subsumes

likes (Peter, father_of (Jane)) $\vee$ $\neg$likes (Jane, Peter).

The substitution

[Peter/$x$]

will do for $\sigma$. If $C$ subsumes $D$ then also $C \Rightarrow D$, so the search space for subsumption is smaller than for general implication.

Plotkin's second algorithm takes as it input two clauses, $C_1$ and $C_2$, and either fails, or produces another clause $C$ which subsumes both $C_1$ and $C_2$ and which is subsumed by any other clause $D$ which subsumes both $C_1$ and $C_2$:

$C \sigma_1 = C_1$ for some substitution $\sigma_1$
$C \sigma_2 = C_2$ for some substitution $\sigma_2$.
If $D$ subsumes $C_1$ and $D$ subsumes $C_2$ then $D$ subsumes $C$.

A **selection** $K$ is a pair of compatible literals

$$\{L_1, L_2\}$$

where $L_1$ is a disjunct in $C_1$ and $L_2$ is a disjunct in $C_2$. The algorithm will succeed if and only if $C_1$ and $C_2$ have a selection. If $K$ is any such selection, then say $L$ is the generalization of $K$ produced by the first algorithm. ($L$ exists because $L_1$ and $L_2$ are compatible.) $L$ actually subsumes $C_1$ and $C_2$; so $L$ is a good approximation to the generalization $C$ which we are looking for. If $K$ is the only selection for $C_1$ and $C_2$ then $C$ is in fact $L$. All the new work of the algorithm goes into coping with all possible selections simultaneously.

Suppose that all the possible selections from $C_1$ and $C_2$ are

$$K_1, K_2, \ldots, K_n.$$

The least generalization $C$ has one disjunct $L_j$ for each such selection $K_j$. Furthermore, if

$$K_j = \{L_{j1}, L_{j2}\}$$

then the substitution $\sigma_1$ is such that

$$L_j \sigma_1 = L_{j1}$$

and similarly for $\sigma_2$. The algorithm achieves this by generalizing all the selections simultaneously. The technical way to describe this is, you write down two large (!) terms. Each has the same function symbol and $n$ arguments. Any function symbol '*fun*' will do. The arguments of the first are the literals $L_{j1}$ and those of the second are $L_{j2}$. (Never mind that these are literals rather than terms—we imagine that their leading symbols are functions, not predicates or negated predicates.) Thus, these big terms are

$$\text{fun }(L_{11}, L_{21}, L_{31}, \ldots, L_{n1})$$

and

$$\text{fun }(L_{21}, L_{22}, L_{32}, \ldots, L_{n2}).$$

You then find the least common generalization of these two terms, by Plotkin's first algorithm. Suppose it is

$$\text{fun }(L_1, L_2, L_3, \ldots, L_n).$$

The clause we seek, $C$, is

$$L_1 \vee L_2 \vee L_3 \vee \ldots \vee L_n.$$

In practice, it is easier on paper not to write out the whole terms. Instead, Plotkin suggested writing out the selections in a column. Each cycle of the first algorithm generalizes the column a little. For each cycle, he wrote another column, showing the states of the selections after that cycle.

*Example*

Suppose that

$$C_1 = \text{likes (Peter, father\_of (Jane))} \lor \neg\text{likes (Jane, Peter)}$$

and

$$C_2 = \text{likes (David, father\_of (Jane))} \lor \neg\text{likes (Jane, David)}.$$

The columns are shown in Table 6.1.

**Table 6.1**

| Initial selections | Selections after one cycle |
|---|---|
| likes (Peter, father_of (Jane)) | likes$(x,$ father_of (Jane)) |
| likes (David, father_of (Jane)) | likes $(x,$ father_of (Jane)) |
| $\neg$likes (Jane, Peter) | $\neg$likes (Jane, $x$) |
| $\neg$likes (Jane, David) | $\neg$likes (Jane, $x$) |

In this simple case, the final generalization appears after a single step:

$$C = \text{likes } (x, \text{ father\_of( Jane))} \lor \neg\text{likes(Jane, } x).$$

*Example: (Plotkin)*

Consider a very simple board game for two players, $\times$ and $\circ$. The rules are obscure, rather like those of *Mornington Crescent*, but at least we know that the board has just two squares and that the following positions are both wins for $\times$:

The objective is to guess the general form of a winning position for $\times$.

*Note*: at this point, write down your guess.

In order to apply the algorithm, we must describe the positions in a first-order language. Say

$$\text{Occ}(N, T, P)$$

means that in position $P$, a token $T$ occurs in square $N$. The left square is numbered 1 and the other is square 2.

$$\text{Win}(P)$$

means that the position $P$ is won for $\times$. The two statements are

$$C_1: \neg\text{Occ}(1, \times, p_1) \vee \neg\text{Occ}(2, \circ, p_1) \vee \text{Win}(p_1)$$

and

$$C_2: \neg\text{Occ}(1, \times, p_2) \vee \neg\text{Occ}(2, \times, p_2) \vee \text{Win}(p_2).$$

The columns of selections appear in Table 6.2. As soon as the two literals in a selection become identical, we stop writing them in further columns.

**Table 6.2**

| | | | | |
|---|---|---|---|---|
| $\neg\text{Occ}(1, \times, p_1)$ | $\neg\text{Occ}(1, \times, P)$ | | | |
| $\neg\text{Occ}(1, \times, p_2)$ | $\neg\text{Occ}(1, \times, P)$ | | | |
| | | | | |
| $\neg\text{Occ}(1, \times, p_1)$ | $\neg\text{Occ}(1, \times, P)$ | $\neg\text{Occ}(N, \times, P)$ | | |
| $\neg\text{Occ}(2, \times, p_2)$ | $\neg\text{Occ}(2, \times, P)$ | $\neg\text{Occ}(N, \times, P)$ | | |
| | | | | |
| $\neg\text{Occ}(2, \circ, p_1)$ | $\neg\text{Occ}(2, \circ, P)$ | $\neg\text{Occ}(2, \circ, P)$ | $\neg\text{Occ}(M, \circ, P)$ | $\neg\text{Occ}(M, T, P)$ |
| $\neg\text{Occ}(1, \times, p_2)$ | $\neg\text{Occ}(1, \times, P)$ | $\neg\text{Occ}(1, \times, P)$ | $\neg\text{Occ}(M, \times, P)$ | $\neg\text{Occ}(M, T, P)$ |
| | | | | |
| $\neg\text{Occ}(2, \circ, p_1)$ | $\neg\text{Occ}(2, \circ, P)$ | $\neg\text{Occ}(2, \circ, P)$ | $\neg\text{Occ}(2, \circ, P)$ | $\neg\text{Occ}(2, T, P)$ |
| $\neg\text{Occ}(2, \times, p_2)$ | $\neg\text{Occ}(2, \times, P)$ | $\neg\text{Occ}(2, \times, P)$ | $\neg\text{Occ}(2, \times, P)$ | $\neg\text{Occ}(2, T, P)$ |
| | | | | |
| $\text{Win}(p_1)$ | $\text{Win}(P)$ | | | |
| $\text{Win}(p_2)$ | $\text{Win}(P)$ | | | |

The resulting generalization, $C$, is

$$\neg\text{Occ}(1, \times, P) \vee \neg\text{Occ}(N, \times, P) \vee \neg\text{Occ}(M, T, P) \vee \neg\text{Occ}(2, T, P) \vee \text{Win}(P)$$

This can be interpreted as:
    If there is a $\times$ in square 1
    and there is a $\times$ somewhere on the board
    and there is something somewhere on the board
    and this something is in square 2
    then the position is a win for $\times$.

### 6.2.5   REDUCING A CLAUSE

The clause $C$ is actually a correct solution, but it is not what we want. The point which has been overlooked so far is, there are *many* clauses which are all 'least' generalizations of $C_1$ and $C_2$. All these generalizations subsume each other. They are logically equivalent. What we really want is a **reduced** least generalization— that is to say, some $C$ as above with the additional property that

    If $D$ is any disjunction of literals chosen from $C$
    and $C$ subsumes $D$
    then $D$ is actually the whole of $C$.

Any such $D$ naturally subsumes $C$; for $\sigma$, take the empty substitution $\varepsilon$. The condition '$C$ subsumes $D$' says that $C$ and $D$ are equivalent. $C$ is reduced if it is not equivalent to any proper subset of itself.

In Plotkin's example, the $C$ produced is not reduced. We can find a reduced version easily. It is

$$D: \quad \neg\mathrm{Occ}(1, \times, P) \quad \vee \quad \neg\mathrm{Occ}(2, T, P) \quad \vee \quad \mathrm{Win}(P).$$

The substitution $\sigma$ by which $C$ subsumes $D$ is

$$[1/N, 2/M].$$

You can always find a reduced version $D$ of $C$. All that is required is a little procedure to determine whether one clause subsumes another, and some backtracking search.


### 6.2.6    DISTINGUISHING MURDER FROM SUICIDE

Although Plotkin's algorithms do all that is claimed for them, they sometimes produce unforseen results. This is because we know much more about Peter, David, Jane, and her father than the data supplied to the algorithm. For instance, we know that anyone called Peter is male and anyone called Jane isn't; hence, Peter is not the same person as Jane. P. H. Winston experimented with an example which makes this phenomenon particularly vivid. He encoded a representation of the various motives and other factors which led to all the murders in *Macbeth* and then looked for common features. His process of generalization went a bit further than Plotkin's. All constants in the story, such as Banquo, Macbeth, and Lady Macbeth, were converted to variables. The kind of generalization which his analysis produced resembled:

> If $x$ is a nobleman
> and $x$ has a greedy wife
> and $x$ is weak-willed
> and $y$ wants to be king
> and $y$ is a nobleman
> then $x$ will murder $y$.

This is a classic slander produced by a dominant nation in order to denigrate a small neighbour. In the Tudor period, the English spread many such silly fables about the Scots and the Irish.

The interesting feature of Winston's naïve generalization is that it predicts that Macbeth will kill himself! If

$$x = \mathrm{Macbeth} \qquad \text{and} \qquad y = \mathrm{Macbeth}$$

then all the rule's conditions are fulfilled. Clearly something has gone awry. It is the following: in all examples from which the generalization was formed, two terms (Jane and Peter, or Macbeth and Banquo) are distinct:

$$\neg(\text{Jane} = \text{Peter}) \quad \text{or} \quad \neg(\text{Macbeth} = \text{Banquo})$$

so these literals can be added to the training instances as extra disjuncts. They are invariably true, so for most purposes one does not bother. However, when generalizing, they actually play a significant role.

*Example*

Suppose the input clauses are

likes (Peter, father_of (Jane)) $\vee \neg$likes (Jane, Peter)

and

likes (David, father_of (Jane)) $\vee \neg$likes (Jane, David).

The naïve generalization of these is, as before,

likes ($x$, father_of (Jane)) $\vee \neg$likes (Jane, $x$).

Suppose instead that the input clauses were

likes (Peter, father_of (Jane))
$\vee \neg$likes (Jane, Peter)
$\vee$ (Peter = Jane)

and

likes (David, father_of(Jane))
$\vee \neg$likes (Jane, David)
$\vee$ (David = Jane).

In this case, Plotkin's algorithm yields

likes ($x$, father_of (Jane)) $\vee \neg$likes (Jane, $x$) $\vee$ ($x$ = Jane).

These two outcomes are not equivalent. They do not mean the same thing. As in the *Macbeth* story, they may have radically different consequences. The moral is: whenever generalizing a set of clauses, if there is any tacit assumption $P$ in any training instance then add an extra disjunct

$$\ldots \vee \neg P$$

to that instance.

The inequality predicate is the most common tacit assumption, but other predicates may occur similarly.

### 6.2.7 UNIFICATION

Generalization is dual to another process called **unification**. Unification has many applications in its own right. It is one of the two basic features of Prolog. (The other is depth-first backtracking search.) It is necessary for another generalizing technique which we shall meet later in this chapter and it also has various applications in rule-based systems. This is a convenient point to introduce it.

There is a family of algorithms called *unification algorithms*. The first and simplest, **string unification**, was invented by Alan Robinson. It has applications in many branches of computing. Its input is a pair of formulas, $F$ and $G$. If it succeeds, then its output is a substitution $\sigma$ which produces the same thing when applied to both $F$ and to $G$:

$$F\sigma = G\sigma.$$

It runs as follows.

IF there is a variable which occurs in both $F$ and $G$
THEN the algorithm *fails*.
$\quad$ $F$ and $G$ cannot be unified.
Let us assume that $F$ and $G$ have no variable in common.
$F_1 := F$
$G_1 := G$
$\sigma \;\; := \varepsilon$ (the most general substitution).

WHILE $F_1 \neq G_1$
$\quad$ say $\;p\;\;$ is the position, counting from the left end, where they first differ;
$\qquad\quad t\;\;$ is the subterm in $F_1$ beginning at position $p$;
$\qquad\quad u\;\;$ is the subterm in $G_1$ beginning at position $p$.
$\quad$ IF neither $t$ nor $u$ is a variable THEN the algorithm *fails*.
$\quad$ Say the subterm $t$ in $F_1$ at position $p$ is a variable, $x$.
$\quad$ It may happen that
$\qquad$ (a)$\quad t$ is not a variable, but $u$ is a variable. In that case, just interchange $t$
$\qquad\qquad$ and $F_1$ with $u$ and $G_1$ in what follows.
$\qquad$ (b)$\quad u$ is a variable $y$ too. You can take either $x$ or $y$ as the variable in what
$\qquad\qquad$ follows. The algorithm is symmetric in $F$ and $G$.
(*)$\quad$ IF the variable $x$ occurs in the subterm $u$
$\qquad$ THEN
$\qquad\qquad$ the algorithm *fails*
$\qquad$ ELSE
$\qquad\qquad F_1 := F_1\,[u/x]$
$\qquad\qquad G_1 := G_1\,[u/x]$
$\qquad\qquad \sigma \;:= \sigma\,[u/x].$

At the end of the loop, if the algorithm did not fail, then

$$F\sigma = F_1 = G_1 = G\sigma.$$

Robinson proved that $\sigma$ is a **most general** unifier of $F$ and $G$: if there is any other substitution $\tau$ for which

$$F\tau = G\tau$$

then there is another substitution $\vartheta$ for which $\tau = \sigma\vartheta$.

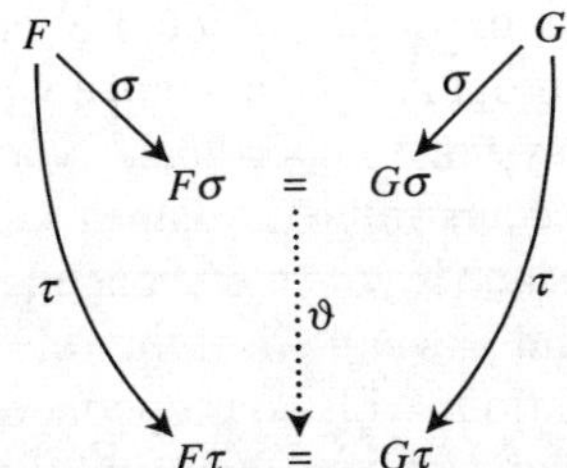

**Fig. 6.2** The most general unifier $\sigma$ of $F$ and $G$.

This is illustrated in Fig. 6.2. If ever the arrows for $\sigma$ and $\tau$ can be filled in, then so can the dotted arrow as well.

The algorithm given here is the proper unification algorithm. The one embodied in Prolog is a slight perversion. Prolog omits the test marked ($^*$), because it is too heavy computationally—and also, it is sometimes useful to be able to 'unify' terms when this test would prevent it. The result is a data structure which loops back and includes itself as a subterm. First-order predicate calculus cannot allow such structures, but computing people often use them. This test is called the **occurs check**.

### 6.2.8 GENERALIZING AND UNIFYING SUBJECT TO LAWS

As we have seen, Plotkin invented two algorithms, one which generalizes pairs of terms and the other for pairs of clauses. The reason we need two is that clauses contain the disjunction symbol, $\vee$, which obeys the

- *commutative law*: $A \vee B$ is the same as $B \vee A$;

- *associative law*: $(A \vee B) \vee C$ is the same as $A \vee (B \vee C)$;

and so a single clause can be written in different ways. We are so used to these laws that we are almost never aware of them. It does not occur to us to put parentheses into a clause, but strictly, they should be there. The associative law says that they don't matter.

Any law has a similar form. It says that two syntactically different expressions have identical interpretations. Generalization works from syntax. Whatever the output, the choice of the way that inputs are expressed should not matter. Equivalent inputs should yield equivalent outputs.

There are many possible laws. One which occurs often in algebra is the **distributive law**:

$(A + B)C$ is the same as $AC + BC$.

Another is the **homomorphism law**:

$f(A) \cdot f(B)$ is the same as $f(A \cdot B)$.

These are not so common in the fields which are used as test beds for artificial intelligence. More often, AI programs have to cope with **geometric symmetry**: the world is the same, whatever system of coordinates you choose to describe it.

A particular case of this occurs in board games such as chess, go, and noughts and crosses: the board is the same if you interchange left and right or front and back (except in chess, for kings and pawns). Also, the rules and tactics of the game are unchanged if you switch positions with your opponent (except for the first move). Each set of laws (note—each *set*, not each individual law) poses a different problem and requires its own generalizing algorithm.

In the dual situation of unification, algorithms have been invented for many sets of laws. It transpires that, usually, a single unifying substitution is not enough. There is no one substitution $\sigma$ with which you can always fill in the diagram of Fig. 6.3, so that, for any $\tau$, there is a $\vartheta$. If there is such a $\sigma$, it is called a **most general unifier** of $F$ and $G$.

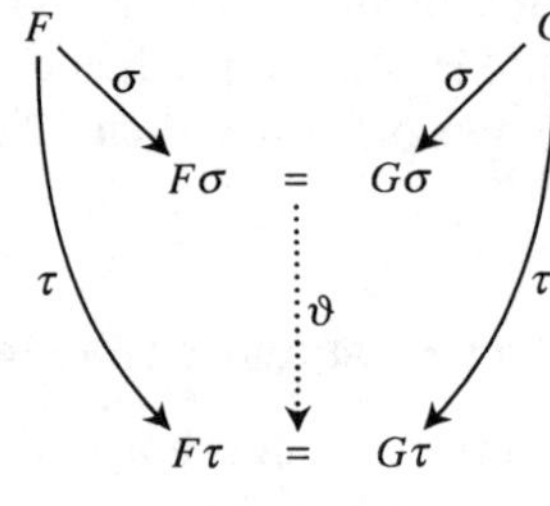

**Fig. 6.3**

When there is no unique $\sigma$, the next best thing is a **complete set of most general unifiers** for $F$ and $G$. This is a set of substitutions with the property that, given $F$, $G$, and $\tau$, there is some $\sigma$ in the set for which you can fill in the diagram with a $\vartheta$. Clearly, we would like a minimal complete set of most general unifiers, but finding any such set may be hard.

Straightforward unification, subject to no laws, is the easiest case. At the other extreme, there is the problem of unification subject to both the distributive and the associative laws. Then, for given $F$ and $G$, there may be no finite complete set. What is more, there may be no systematic way to decide whether $F$ and $G$ can be unified at all. Generalizing subject to laws is a younger subject, still open to research.

## 6.3   PROGRAM SYNTHESIS

One bad way to 'specify' a program is to give a few examples of possible inputs for it and, for each input, say what the output should be. Of course, such a finite set of input/output pairs does not specify the program at all. Even so, this is often an effective way of telling someone about the program. A competent programmer can often guess what the intended program should do, from just such data. The algorithm which we shall investigate next, which was invented by Zhu and Jin, does the same. It constructs a program with a loop in it, from a finite set of input/output pairs.

In this case, the term 'algorithm' is something of a misnomer. Zhu and Jin describe a procedure which involves some arbitrary choices. They actually show how to find a set of programs which meet the given conditions. The set may contain several programs, in which case the user can choose any one of them. Alternatively, the set may be empty, in which case the 'algorithm' fails. Zhu and Jin call their approach **knowledge-based** because a fully automatic implementation of it would require an expert system to make the choices.

The sort of program which this algorithm can invent has a strictly limited form. Say its input is called *in* and its output is *out*, then you can imagine that the program might be something like

$D := in$;

WHILE NOT *test* $(D)$ DO
    $D := step\ (D)$;

$out := D$.

The algorithm can invent programs which are a bit more elaborate than this, but not much. In particular, the program can only contain one WHILE loop. This particular example depends on two subsidiary functions which here are called *test* and *step*. The biggest task of the algorithm is to construct these two functions.

Be careful not to confuse the input and output of the program, *in* and *out*, with the input and output of the algorithm. The algorithm's input is a set of *in/out* pairs and its output is the entire program.

The above little program is written in a style which (I hope) readers find easy to understand. Unfortunately, this style involves several features which do not lend themselves to program synthesis. Particularly, synthesis is a lot harder for programs which contain

- variables, like $D$;

- assignment statements, containing ': =';

- control structures like WHILE which depend for their meaning on time: as the program runs, events in it have to happen in a certain order.

Therefore, the synthesized programs will look quite different. They will be written in a language called FP, due to John Backus. There is a correspondence between programs written in FP and programs written with WHILE loops and assignments, but it is quite obscure. In fact, even the following extension of the little program above cannot be rendered simply into a single FP program:

$$D := in;$$

WHILE NOT *test* (*D*) DO
  $$D := step\ 2\ (in, D);$$

$$out := tidy\ (D).$$

This does not mean to say that FP code is more complex than WHILE-type code. They are just different.

### 6.3.1  THE PROGRAMMING LANGUAGE FP

In FP, a program is an expression defining a function. Let us call the whole program, *prog*. Then the behaviour of *prog* is given by the statement

$$out = prog{:}in \qquad \text{for every possible pair } in/out.$$

The notation *prog*:*in* is the FP way of writing the value of the function *prog* when its argument is *in*. Thus, the FP way of writing

$$test(D)$$

is

$$test{:}D$$

It is not just *prog* that is a function. Everything within it is a function too. We began on the right tack in the original form of the program, since its subsidiary routines *test* and *step* are functions. The difference between the first form and the FP version is that all the control structures, such as WHILE, are represented with functions too. Here is what a synthesized program will look like:

$$prog = test \rightarrow base;\ h \circ [e, prog \circ d]$$

Programs in this form are known as **linear recursive** programs. Perhaps this cryptic form is a little more comprehensible if you read it as

$$prog{:}in\ =\ \ if\ test{:}\ in$$
$$then\ base{:}\ in$$
$$else\ h{:}\ \langle e{:}in, prog{:}\ (d{:}in)\rangle$$

The code for *prog* includes a call to *prog* itself. This is how one encodes a loop. The internal call is nested inside calls to various other functions: *h*, *e*, and *d*. Never mind them for the moment. The program will run thus:

- *test* returns *true* or *false*. The program begins by *test*ing its input.

- If *test* of the input is *true* then the program's output is *base* of the input.

- If not, the program first calculates *d* of the input and then *prog* of that. This is where the loop comes in. The recursive call to *prog* may itself call *prog* many times. When the inner call to *prog* eventually returns its value, then the value of the whole program is calculated by the functions *e* and *h*.

Every user-defined function in an FP program takes just one argument. This is sometimes inconvenient. The function *h* in this code depends on two quantities, *e:in* and *prog*: (*d:in*). In languages with WHILE loops, functions and procedures can often take several arguments. In FP, Backus got round that by making the argument to a function such as *h* be a sequence of all the things it depends on. Thus, *h* takes one argument, which is a sequence of all the quantities which we think of as being its arguments. The explanatory version shows this approach.

Another curious feature of FP is the complete lack of any mention of the input. The whole program *prog* is put together by composing other functions without mentioning their arguments. FP adopts the convention that any function in an expression takes the whole program's input as its argument, unless it is part of a construction which dictates otherwise. Here are some more details.

$$t \rightarrow a \; ; b \qquad \text{means} \qquad \textit{'if t then return the value of a}$$
$$\textit{else return the value of b'}$$

*t* is a predicate and *a* and *b* are functions. On input *in*, the value of the whole expression

$$(t \rightarrow a \; ; b) : in$$

will be *a : in* if *t: in* is *true*;
      *b : in* if *t: in* is *false*.

  *a ∘ b* means *return the value of a of b of the input.*

As before, *a* and *b* are functions. The value of
  (*a ∘ b*): in
is *a*: ( *b*: *in* )

Using just these forms, we can now write the FP equivalent of the first little program. It is *prog*, where

$$prog = test \rightarrow id; prog \circ step$$

*id* is a built-in function, defined as part of the FP language. Its value is always the same as its argument:

$$id : in = in \qquad \text{for every possible input } in.$$

As mentioned above, FP uses a sequence of arguments for any function which depends on several inputs. Hence, sequences play a big role in FP code. Whenever *a, b, c, … d* are functions, then there is another function

$$[a, b, c, \ldots d]$$

which Backus calls a 'construction'. It is a function whose value is a sequence. Its value on input *in* is

$$[a, b, c, \ldots d] : in = \langle a:in, b:in, c:in, \ldots d:in \rangle.$$

FP also has a mechanism for taking sequences apart. For each positive number $j$, there is a function on sequences which extracts the $j$th entry from the sequence. Perhaps perversely, this function is actually written as $j$. Thus, the composite function

$$2 \circ [a, b, c, \ldots, d]$$

is equivalent to the function

$$b$$

because, whatever the input *in*,

$$2 : \langle a{:}in, b{:}in, c{:}in, \ldots, d{:}in \rangle = b{:}\ in$$

We can therefore write

$$2 \circ [a, b, c, \ldots, d] = b$$

which states that the two expressions represent the same function. (However, be warned that this equality is not quite true. Backus chose 'eager' semantics for FP. This means that if any of the functions $a$ or $c$ or $\ldots d$ enters an unending loop then the expression on the left will loop but $b$ may not.)

*Example*

The second version of the little WHILE program, which depends on *step2* and *tidy* as well as *test*, corresponds to a more elaborate FP program which can be called *prog2*:

$$prog2 = tidy \circ 2 \circ loop \circ [id, id]$$

where

$$loop = (test \circ 2) \rightarrow id;\ loop \circ [1, step2]$$

(I have used the same name for the function called *step2* in the WHILE version, which takes two arguments, and the corresponding function in the FP version which takes a sequence of two entries.)

*prog2* involves the constructions $[id, id]$ and $[1, step2]$ with two entries because the code involves sequences $\langle in, D \rangle$ of two data, *in* and $D$. Initially, $D$ is given the value *in* and so the first value of this sequence is $\langle in, in \rangle$ which is

$$[id, id] :in$$

The *test* is applied to $D$ alone, so the test which the *loop* makes on the sequence is $(test \circ 2)$. If this is satisfied, the output of the looping stage will be just the sequence $\langle in, D \rangle$. Otherwise, the loop continues. Each cycle of the loop takes $\langle in, D \rangle$ for input and returns $\langle in, step2{:}\langle in, D \rangle \rangle$.

Since '2' denotes a function, FP needs some other notation whenever a program involves the number *two*. Note that the constant number *two* itself will never occur in an FP program, since everything in an FP program is a function with one argument. What a program may contain is a function whose value is always the number *two*, whatever its input. This function is written as

$$\overline{2}$$

Similarly, the function whose value is always the empty sequence is written

$$\overline{<>}$$

In general, if $c$ is any constant such as 2 or *true* or the ASCII character 'A', the function whose value is always this constant is written as

$$\overline{c}$$

FP has several built-in functions. One such is the identity function, *id*, mentioned above. Another is *addition*, '+'. This expects two numbers. If it were written in strict FP style, then its input should really be a sequence of two numbers, so

$$+ \circ [\overline{3}, \overline{5}] = \overline{8}$$

but as a condescension, you are allowed to write it in the normal way:

$$\overline{3} + \overline{5} = \overline{8}$$

Just as '+' is a natural function on numbers, there are natural functions on sequences and truth values. One such is 'and', which in FP is written '&'. It can also appear between its arguments:

$$\overline{true} \ \& \ \overline{true} \ = \ \overline{true}$$
$$\overline{true} \ \& \ \overline{false} \ = \ \overline{false}$$

and so on.
On sequences, the function *tail* returns the sub-sequence of all entries except the first. Thus,

$$1 \circ tail = 2$$

Another very useful function is the test for equality, *eq*. This also expects two inputs. It returns *true* if they are the same and *false* otherwise. Thus, you can check easily that

$$eq \circ [id, id] \ = \ \overline{true}$$

For ease of reading, we shall sometimes write *eq* as '=' and put it in between its arguments too.

*Examples of FP functions*

   1.   The length of a sequence:

$$length \; = \; eq \circ [id, \overline{<>}] \; \to \; \overline{0} \; ; \; + \circ [\,\overline{1}, \; (length \circ tail)]$$

   *or*

$$id = \overline{<>} \; \to \; \overline{0} \; ; \; \overline{1} \; + \; (length \circ tail)$$

   2.   A function which returns *true* if its argument is a sequence and all entries in
        the sequence are *true*:

$$map\text{-}and \; = \; eq \circ [id, \overline{<>}] \; \to \; \overline{true} \; ; \; \& \circ [1, \; (map\text{-}and \circ tail)]$$

   *or*

$$id = \overline{<>} \; \to \; \overline{true} \; ; \; 1 \; \& \; (map\text{-}and \circ tail)$$

At various stages, we shall write statements of the form

  $prog1 = prog2$

This means that the functions represented by the two expressions, $prog1$ and $prog2$,
are the same. The expressions themselves may be different. For instance, you can
convince yourself that

$$map\text{-}and \; \circ \left[\overline{true}, \; \overline{true}, \dots \; \overline{true}\right] \; = \; \overline{true}$$

Whenever one discusses a function, one faces the issue of its domain of definition.
The conventional approach is to say that each function $f$ is defined for all arguments
$x$ in a certain set called the **domain** of $f$. If $y$ is not in this set then $f(y)$ would be
meaningless. This approach is sometimes inconvenient. For instance, when we
discuss programs and regard them as functions, we have to consider the possibility
that a program may go into an endless loop. Life would be easy if we could always
decide in advance whether $prog(in)$ will loop. Unfortunately, we cannot. Put
another way, we cannot decide what is the domain of $prog$.

   Those who study computer science have opted for another approach. The domain
of every program is *everything*. Consequently, there must be a 'value' which a
program 'outputs' even when it goes into a loop or when it cannot return a 'normal'
value for any other reason. This special value is called **bottom** and is written '$\perp$'.
*Bottom* is also known as the **totally undefined** value. You can easily write a
program with this output:

  $\perp$ is the output of *forever*, where $forever = forever + \overline{1}$

*Bottom* can also be the output of a program for other reasons. For instance, if it
expects a character as input and it is given a sequence, the output is $\perp$. In an
expression of any of the forms

  $a \; \& \; b \;$   or   $\; a + b \;$   or   $\; p \to .. \; ; \; ..$

if $a$ or $b$ or $p$ returns $\perp$ then the whole expression will have value $\perp$ too. Some people have suggested using different varieties of $\perp$ for different purposes, but for us now, one $\perp$ is enough.

By this trick, we can extend the domain of every program to include all possible data. It is still convenient to distinguish the class of those data on which a program $p$ does not take the value $\perp$. This is written

$$Dom(p) \;=\; \{x \mid p(x) \neq \perp\}$$

Thus,

$$Dom(tail)$$

is the class of all non-empty sequences.

It often happens that a conditional expression always returns $\perp$ if its test fails, so it is written

$$p \to a; \overline{\perp}$$

Writing $\overline{\perp}$ all the time is tedious, so we adopt the convention that it and the semicolon before it are left off. Thus,

$$p \to a$$

on its own means the same.


### 6.3.2   SUMMARY OF PROGRAM SYNTHESIS

The synthesis process is quite involved. Here is a summary of it. The following six sections give full details and you are encouraged to read this summary in conjunction with them.

1.   Input: A sequence of *input/output* pairs

$$in_1/out_1 \quad in_2/out_2 \quad \ldots \quad in_k/out_k$$

Restrictions on input:

- Each $in_j$ and $out_j$ should be in the syntax of Section 6.2.3;

- $in_j$ cannot contain any function symbol;

- each variable can occur at most once in $in_j$;

- if $i \neq j$ then $in_i$ and $in_j$ cannot be unified;

- for each function symbol $f$ occurring in any $out_j$ the system should be provided with an FP function $f'$ interpreting $f$.

Output: an FP program

$$prog = test \to base; \; h \circ [e, prog \circ d]$$

such that, for all pairs in the input sequence,

$$prog: in_j = out_j$$

2.  Simple program for one *in/out* pair:

$$SimProg(in, out) = P(in) \rightarrow F(in, out)$$

where

If *in* is a variable then $P(in) = \overline{true}$

If *in* is a constant, $c$, then $P(in) = id = \overline{c})$
If *in* is a sequence $\langle t_1, t_2, t_3, \dots t_n \rangle$ of length $n$ then
$$P(in) = (length = \overline{n}) \,\&\, (map\text{-}and \circ [P(t_1) \circ 1, P(t_2) \circ 2, \dots P(t_n) \circ n])$$

and

$$
\begin{aligned}
F(in, out) &= pp_{in}(out) \quad \text{if } out \text{ is a variable}\\
&= \overline{out} \qquad\quad\ \text{if } out \text{ is a constant}\\
&= [F(in, t_1), \dots F(in, t_n)]\\
&\qquad\qquad \text{if out is a sequence } \langle t_1, \dots t_n \rangle\\
&= f' \circ F(in, \langle t_1, \dots t_n \rangle)\\
&\qquad\qquad \text{if } out \text{ is } f(t_1, \dots t_n)
\end{aligned}
$$

and

If $y$ occurs at just one place in $t$ then
$\quad pp_t(y)$ is projection onto the place where $y$ occurs in $t$
otherwise
$\quad pp_t(y)$ is $\overline{\perp}$

3.  $test\ = P(in_1)$
$base = F(in_1, out_1)$

4.  The conditions on $d$ are

$$P(in_1) \circ d = P(in_2)$$
$$\dots$$
$$P(in_{k-1}) \circ d = P(in_k)$$

5.  The program $e$ is chosen so that, for some other program $p$, for any datum $t$, either

$$Dom(p \circ [e, d]) = Dom([e, d]) = Dom(d)$$

and

$$p \circ [e, d] : t = t$$

or

$$p \circ [e, d] : t = \perp$$

6. The condition on $h$ is

$$h \circ [e, F(in_j, out_j) \circ d] = F(in_{j+1}, out_{j+1})$$

for all $j < k$.

### 6.3.3 PERMITTED INPUT/OUTPUT PAIRS

Each training instance is an input/output pair. The learning algorithm will not work if it is given a pair in some unexpected format. The permitted format is very flexible, but there must be some restriction on it. Each input or output of a pair is called a *term*. The permitted syntax of a term is

$$term :: = constant \,|$$
$$variable \,|$$
$$\langle term \,\{, term\}\rangle \,|$$
$$f(term \,\{, term\})$$

Here, $f$ can be any function symbol. This is just the syntax of terms in predicate calculus, except that there is an explicit way of writing sequences of terms:

$$\langle t_1, t_2, t_3, \ldots t_n \rangle$$

by enclosing them in diamond brackets. Note that in FP this sequence is not the same as

$$\langle t_1, \langle t_2, \langle t_3, \ldots \langle t_n \rangle \ldots \rangle\rangle\rangle.$$

The first has $n$ entries whereas the second has just two. Note also that a term can contain variables. This fact means that the algorithm for synthesizing programs is quite versatile.

The algorithm will only be reliable if we place two restrictions on each term *in*:

(1) *in* cannot contain any function symbol;

(2) each variable can occur at most once in *in*
(we shall see how to get round this, later).

Furthermore, the entire training set

$$T = \{in_1/out_1, in_2/out_2, \ldots, in_k/out_k\}$$

must satisfy the restriction:

If $i \neq j$ then $in_i$ and $in_j$ cannot be unified.

### 6.3.4 THE SIMPLE PROGRAM FOR A SINGLE INPUT/OUTPUT PAIR

To begin with, the algorithm forms a 'simple' program for each input/output pair *in/out*. This simple program will be of the form

$$SimProg(in, out) = P(in) \to F(in, out)$$

Here, $P(in)$ is an expression in FP which tests whether its argument matches *in*. $F(in, out)$ is another expression. Given any input $t$, this simple program will yield $\perp$ if *in* does not match $t$. If they do match, it will yield an instance of *out*.

If there is a most general substitution $\vartheta$ so that
$$in\ \vartheta = t$$
then
$$SimProg(in,out){:}t = out\ \vartheta$$
otherwise
$$SimProg(in,out)\ {:}t = \perp$$

The expressions $P(in)$ and $F(in,out)$ are defined by induction over the structures of *in* and *out*. The form of $P(in)$ is quite simple, since the form of *in* is restricted:

If *in* is a variable then

$$P(in) = \overline{true}$$

If *in* is a constant, $c$, then

$$P(in) = (id = \overline{c})$$

If *in* is a sequence of length $n$, $\langle t_1, t_2, t_3, \ldots t_n \rangle$, then

$$P(in) = (length = \overline{n})\ \&\ (map\text{-}and \circ [P(t_1) \circ 1, P(t_2) \circ 2, \ldots P(t_n) \circ n])$$

$F(in,out)$ is more elaborate. To start with, for each function symbol $f$ occurring in *out*, we have to have an FP program $f'$. Whenever its input is a sequence, $f'$ will return the term consisting of $f$ of the entries in the sequence:

$$f'{:}\ \langle t_1, t_2, t_3, \ldots, t_n \rangle = f(t_1, t_2, t_3, \ldots t_n)$$

Otherwise, $f'$ returns $\perp$. We shall just assume that $f'$ is given.

The other complication in the definition of $F(in, out)$ has to do with variables. If *out* is a variable, say $y$, which does not occur in *in*, then $F(in, y)$ is $\overline{\perp}$. If $y$ occurs in *in* then it occurs at some unique place, by the restriction on input/output pairs. In that case, $F(in, y)$ is the function which projects any term $t$ onto $t$'s subterm at that place. If $t$'s structure has no such place, then $F(in, y){:}t$ is $\perp$. Thus, if $a$ and $b$ and $c$ are constants and $x$ and $y$ are distinct variables, then

$$F(x, y) = \overline{\perp}$$
$$F(x, x) = id$$
$$F(\langle a, x \rangle, y) = \overline{\perp}$$
$$F(\langle a, x \rangle, x) = 2$$
$$F(\langle a, \langle x, a \rangle \rangle, x) = 1 \circ 2$$

so

$$F(\langle a, x \rangle, x) : b = \perp$$
$$F(\langle a, x \rangle, x) : \langle b, c \rangle = c$$

Let us say that, for any term $t$ and any variable $y$,

> If $y$ occurs at just one place in $t$ then
>> $pp_t(y)$ is projection onto the place where $y$ occurs in $t$
> otherwise
>> $pp_t(y)$ is $\underline{\bot}$

Now we are in a position to define $F$ by induction on the structure of *out*:

$$
\begin{aligned}
F(in, out) &= pp_{in}(out) \quad \text{if out is a variable} \\
&= \overline{out} \qquad\quad \text{if } out \text{ is a constant} \\
&= [F(in, t_1), \ldots F(in, t_n)] \\
&\quad\; \text{if } out \text{ is a sequence } \langle t_1, \ldots t_n \rangle \\
&= f' \circ F(in, \langle t_1, \ldots t_n \rangle) \text{ if } out \text{ is } f(t_1, \ldots t_n)
\end{aligned}
$$

This completes the definition of the simple function for one training instance *in/out*.

*Example*

Suppose that *in* is a 1-sequence, $\langle x \rangle$ and *out* is the pair $\langle c, x \rangle$ where $x$ is a variable and $c$ is a constant. Then

$$
\begin{aligned}
P(\langle x \rangle) &= \left( length = \overline{1} \right) \,\&\, \left( map\text{-}and \circ [P(x) \circ 1] \right) \\
&= \left( length = \overline{1} \right) \,\&\, \left( map\text{-}and \circ [\overline{true} \circ 1] \right)
\end{aligned}
$$

and

$$
\begin{aligned}
F(\langle x \rangle, \langle c, x \rangle) &= [F(\langle x \rangle, c), F(\langle x \rangle, x)] \\
&= [\overline{c}, pp_{\langle x \rangle}(x)] \\
&= [\overline{c}, 1]
\end{aligned}
$$

so

$$
\begin{aligned}
&SimProg(\langle x \rangle, \langle c, x \rangle) \\
&\quad = \left( length = \overline{1} \right) \,\&\, \left( map\text{-}and \circ [\overline{true} \circ 1] \right) \rightarrow [\overline{c}, 1]
\end{aligned}
$$

This expression for $SimProg(\langle x \rangle, \langle c, x \rangle)$ can be simplified to

$$
(length = \overline{1}) \rightarrow [\overline{c}, 1]
$$

### 6.3.5 GENERALIZING A SEQUENCE OF SIMPLE PROGRAMS

Suppose the training set for the algorithm is

$$
T = \{in_1/out_1, in_2/out_2, \ldots in_k/out_k\}
$$

The algorithm can synthesize one simple program for each instance:

$$
SimProg(in_j, out_j)
$$

for $j \leq k$. The desired program $Prog(T)$ should subsume all these simple programs. That is to say, for any term $t$ and for all $j \leq k$,

if $SimProg(in_j, out_j) : t$ is not $\perp$
then $Prog(T) : t = SimProg(in_j, out_j) : t$

All this is sometimes summarized by writing

$$
\begin{aligned}
Prog(T) \geq \quad & SimProg\ (in_1, out_1) \\
+ \ & SimProg\ (in_2, out_2) \\
+ \ & \ldots \\
+ \ & SimProg\ (in_k, out_k)
\end{aligned}
$$

Since $in_i$ and $in_j$ cannot be unified when $i \neq j$, there are never two conflicting conditions on $Prog(T)$ for different instances in $T$. (One sometimes says that the set of simple programs forms an **orthogonal system**.) The most basic program satisfying these conditions is the one which coincides with each $SimProg$ whenever it is not $\perp$ and which is $\perp$ for every other input $t$. The essential objective is to find a program $Prog(T)$ which is more general than the most basic one. $Prog(T)$ should return something other than $\perp$ for infinitely many inputs $t$.

Recall, the form of $Prog(T)$ is going to be

$$
test \rightarrow base;\ h \circ [e, prog \circ d]
$$

If you work through the evaluation of this expression for some input $t$, you will find that it calculates

$$
test(t), \quad test(d(t)), \quad test(d(d(t))), \quad \ldots
$$

successively until one of them is *true*, whereupon it finds the answer by evaluating *base* and then repeatedly evaluating $e$ and $h$. Hence,

$$
\begin{aligned}
Prog(T) \ = \quad & test \rightarrow base \\
+ \ & test \circ d \rightarrow h \circ [e, base \circ d] \\
+ \ & test \circ d \circ d \rightarrow h \circ [e, h \circ [e, base \circ d] \circ d] \\
+ \ & test \circ d \circ d \circ d \rightarrow h \circ [e, h \circ [e, base \circ d] \circ d] \circ d]
\end{aligned}
$$

on condition that these sub-programs are orthogonal:

If $i \neq j$ then
    for all terms $t$,
        $test \circ d^i : t$ and $test \circ d^j : t$
    are not both *true*.

Now the essential ingredient of the algorithm becomes apparent. If the training set consists of the first few simplest instances of possible inputs and it is ordered in increasing order of complexity, then these two expressions for $Prog(T)$ can be compared, summand by summand:

$$
\begin{aligned}
test \rightarrow base &= P(in_1) \rightarrow F(in_1, out_1) \\
test \circ d \rightarrow h \circ [e, base \circ d] &= P(in_2) \rightarrow F(in_2, out_2) \\
test \circ d \circ d \rightarrow h \circ [e, h \circ [e, base \circ d] \circ d] &= P(in_3) \rightarrow F(in_3, out_3)
\end{aligned}
$$

We can read off expressions for the functions *test* and *base*:

$$test = P(in_1)$$
$$base = F(in_1, out_1)$$

and we can put tight restrictions on $d$, $e$, and $h$. The rest of the algorithm consists of guessing suitable values for $d$, $e$, and $h$.

### 6.3.6   CHOOSING A SUITABLE EXPRESSION FOR $d$

The conditions on $d$ are

$$P(in_1) \circ d = P(in_2)$$
$$\ldots$$
$$P(in_{k-1}) \circ d = P(in_k)$$

These all have the form

$$R \circ d = S$$

Zhu and Jin invented a procedure for finding *many* solutions for such an equation. They write

$$\partial_r\left(\frac{S}{R}\right)$$

for the set of all its solutions. They do not find the whole of this set, but they find a large part of it, usually enough to contain a solution $d$. The required $d$ will be a program in the intersection

$$\partial_r\left(\frac{P(in_2)}{P(in_1)}\right) \ \cap \ \partial_r\left(\frac{P(in_3)}{P(in_2)}\right) \ \cap \ldots \partial_r\left(\frac{P(in_k)}{P(in_{k-1})}\right)$$

If this intersection is empty, then of course there is no possible such program $d$. If the intersection has more than one element, then the equations for $d$ have several solutions.

$\partial_r(S/R)$ is found by induction on the structures of $S$ and $R$. There are several base cases. Here are some examples:

$$id \ \in \ \partial_r(S/S) \qquad \text{for any program } S;$$
$$S \ \in \ \partial_r(R \circ S)/R) \quad \text{for any programs } S \text{ and } R;$$
$$\overline{b} \ \in \ \partial_r(\overline{a}/R) \qquad \text{whenever } a \text{ and } b \text{ and constants and } R(b) = a;$$

$$tail^{n-m} \in \partial_r\left(\frac{length = \overline{n}}{length = \overline{m}}\right) \quad \text{whenever } n \geq m;$$

$$[g_1, \cdots g_{n-1}, S, g_{n+1}, \cdots g_m] \in \partial_r(S/n) \quad \text{whenever}$$
$$m \geq n \text{ and } Dom(g_j) \supseteq Dom(S) \text{ for all } j.$$

When $R$ and $S$ are composite, we use the following rules:

$$\partial_r\left(\frac{S_1 \circ S_2}{R}\right) \supseteq \partial_r\left(\frac{S_1}{R}\right) \circ S_2$$

$$\partial_r\left(\frac{S}{R_1 \circ R_2}\right) = \partial_r\left(\frac{\partial_r(S/R_1)}{R_2}\right)$$

$$\partial_r\left(\frac{[S_1, \cdots S_n]}{[R_1, \cdots R_n]}\right) \supseteq \partial_r\left(\frac{S_1}{R_1}\right) \cap \cdots \partial_r\left(\frac{S_n}{R_n}\right)$$

$$\partial_r\left(\frac{[S_1, \cdots S_m]}{[R_1, \cdots R_n]}\right) \quad \text{is empty if } m \neq n.$$

We are being a bit careless with notation and sometimes writing a set of programs where you would expect just one. Thus,

$$\partial_r\left(\frac{S_1}{R}\right) \circ S_2 \quad \text{stands for} \quad \left\{p \circ S_2 \mid p \in \partial_r\left(\frac{S_1}{R}\right)\right\}$$

and

$$\partial_r\left(\frac{\partial_r(S/R_1)}{R_2}\right) \quad \text{stands for} \quad \bigcup \left\{\partial_r(p/R_2) \mid p \in \partial_r(S/R_1)\right\}.$$

From these, we can derive some useful subsidiary rules:

$$\partial_r\left(\frac{T \circ S}{T \circ R}\right) = \partial_r\left(\frac{\partial_r(T \circ S/T)}{R}\right) \supseteq \partial_r\left(\frac{S}{R}\right)$$

and

$$\partial_r\left(\frac{[S_1, \cdots S_n]}{[R_1 \circ 1, \ldots R_n \circ n]}\right) \supseteq \left[\partial_r\left(\frac{S_1}{R_1}\right), \cdots \partial_r\left(\frac{S_n}{R_n}\right)\right].$$

*Example*

Suppose that

$$in_1 = \langle x \rangle \qquad \text{and } out_1 = \langle x, c \rangle$$
$$in_2 = \langle x, y \rangle \qquad \text{and } out_2 = \langle x, y, c \rangle$$
$$in_3 = \langle x, y, z \rangle \quad \text{and } out_3 = \langle x, y, z, c \rangle$$

where $x$, $y$, and $z$ are variables and $c$ is a constant. Then

$$P(in_1) = \left(length = \overline{1}\right) \,\&\, \left(map\text{-}and \circ [\overline{true \circ 1}]\right)$$

$$P(in_2) = \left(length = \overline{2}\right) \,\&\, \left(map\text{-}and \circ [\overline{true \circ 1}, \overline{true \circ 2}]\right)$$

$$P(in_3) = \left(length = \overline{3}\right) \,\&\, \left(map\text{-}and \circ [\overline{true \circ 1}, \overline{true \circ 2}, \overline{true \circ 3}]\right)$$

If these are written in strict FP style, then $P(in_1)$ becomes

$$\& \circ \left[\ eq \circ [length, \overline{1}], map\text{-}and\ \circ [\overline{true} \circ 1]\ \right]$$

and $P(in_2)$ appears similar. By the rules above,

$$\partial_r(P(in_2)/P(in_1))$$

$$\supseteq\ \partial_r\left(\frac{[eq \circ [length, \overline{2}],\ \ map\text{-}and\ \circ [\overline{true} \circ 1, \overline{true} \circ 2]\ ]}{[eq \circ [length, \overline{1}],\ \ map\text{-}and\ \circ [\overline{true} \circ 1]\ ]}\right)$$

$$\supseteq\ \partial_r(\overline{2}/\overline{1})\ \cap\ \partial_r\left(\frac{[\overline{true} \circ 1,\ \overline{true} \circ 2]}{[\overline{true} \circ 1]}\right)$$

but this last set is empty. In this case, it pays to simplify the expressions $P(in)$ first. Since

$$\overline{true} = \overline{true} \circ n \qquad \text{for any } n$$

and

$$map\text{-}and\ \circ \left[\overline{true}, \overline{true}, \ldots, \overline{true} = \overline{true}\right]$$

and

$$\& \circ [f, \overline{true}] = f \qquad \text{for any function } f$$

we have that

$$P(in_j) = (length = \overline{j}) \text{ for all } j$$

and so

$$tail \in\ \partial_r(P(in_2))/(P(in_1)) \cap\ \partial_r(P(in_3)/P(in_2))$$

Thus, the function *tail* is a suitable candidate for $d$.


### 6.3.7  CHOOSING A SUITABLE EXPRESSION FOR $h$

The restrictions on $h$ and $e$ are

$$h \circ [e, base \circ d] = F(in_2, out_2)$$
$$h \circ [e, h \circ [e, base \circ d] \circ d] = F(in_3, out_3)$$
$$h \circ [e, h \circ [e, h \circ [e, base \circ d] \circ d] \circ d] = F(in_4, out_4)$$

and so on,
so

$$h \circ [e, F(in_j, out_j) \circ d] = F(in_j + 1, out_j + 1) \text{ for any } j < k.$$

One possible solution for $e$ is *id*. This will always work, if there is any solution, but the corresponding $h$ is often more complex than necessary. Zhu and Jin suggest a subtler choice of $e$, which we shall explore below. For the moment, let us assume some value for $e$ and use it to calculate $h$.

The procedure for finding $h$ is very like the one for $d$. All the equations for $h$ have the form

$$h \circ R = S$$

Say

$$\partial_l\left(\frac{S}{R}\right)$$

is the set of all solutions of this equation. Then, much as for $\partial_r$:

$$id \;\in\; \partial_l(S/S) \qquad \text{for any program } S;$$
$$S \;\in\; \partial_l(S \circ R)/R) \quad \text{for any program } S \text{ and } R;$$
$$g \;\in\; \partial_l(\overline{a}/\overline{b}) \qquad\quad \text{whenever } a \text{ and } b \text{ and constants and } g(b) = a.$$

There are many more such base cases. When $R$ and $S$ are composite, we use the following rules:

$$\partial_l\left(\frac{S_1 \circ S_2}{R}\right) \;\supseteq\; S_l \circ \partial_l\left(\frac{S_2}{R}\right)$$

$$\partial_l\left(\frac{S}{R_1 \circ R_2}\right) \;=\; \partial_l\left(\frac{\partial_l(S/R_2)}{R_1}\right)$$

$$\partial_l\left(\frac{S}{[R_1, \cdots R_n]}\right) \;\supseteq\; \bigcup_{j=1}^{n} \partial_l\left(\frac{S}{R_j}\right) \circ j$$

$$\partial_l\left(\frac{[S_1, \cdots S_n]}{R}\right) \;\supseteq\; \left[\partial_l\left(\frac{S_1}{R}\right), \cdots \partial_l\left(\frac{S_n}{R}\right)\right].$$

The possible functions $h$ can be found by induction over $R$ and $S$, just as $d$ was.

### 6.3.8  CHOOSING A SUITABLE EXPRESSION FOR $e$

The function $e$ is something of a luxury. If there is any solution for $h$ and $e$ then there is another solution in which $e$ is the identity function:

$$(h \circ [e \circ 1, 2]) \circ [id, prog \circ d] = h \circ [e, prog \circ d]$$

We could equally well choose $(h \circ [e \circ 1, 2])$ in place of $h$ and $id$ in place of $e$. The advantage of a more elaborate $e$ is that it makes our choice of $h$ simpler. The penalty of an elaborate $e$ is that it may lose information. For any datum $t$, there may be some key property which can be calculated from $t$ itself but not from $e : t$. If the required value

$$h : \langle e : t, prog : d : t \rangle$$

depends on this property, then a bad choice of $e$ will mean that there is no possible choice for $h$. The ideal function $e$ will just capture the information required by $h$.

Zhu and Jin suggest that a good choice for $e$ is likely to be some function which retains the features which are lost by $d$. Then the pair $[e, d]$ will retain all the essential properties of any input. To be precise, they require of $e$ that

$$(1) \quad Dom(e) = Dom(d)$$

and there is some other program, say $p$, for which

$$(2) \quad p \circ [e, d] \leq id,$$

that is, for any datum $t$ either

$$p \circ [e, d] : t = t$$

or

$$p \circ [e, d] : t = \bot$$

$$(3) \quad Dom(p \circ [e, d]) = Dom([e, d]) = Dom(d)$$

If any such program $e$ exists, then it is called a (left) **adjunction** of $d$. Of course, if $d$ has one adjunction, then it probably has many. The chosen $e$ should retain just the information which $d$ loses and no more and the expression for $e$ should be as simple as is practical.

Conditions (2) and (3) can be summarized by saying that the program $[e, d]$ has a **partial left inverse**: if $q$ is any program, then another program $p$ is a partial left inverse for $q$ if

$$p \circ q \leq id$$

and

$$Dom (p \circ q) = Dom(q)$$

If $d$ is any program, let us write $U(d)$ for the set of programs $e$ which satisfy conditions (1), (2) and (3).

*Examples* of adjunctions

$$- \quad \in \quad U(+)$$
$$1 \quad \in \quad U(tail)$$

If $e$ is an adjunction of $d$ then $d$ is an adjunction of $e$.

*Examples* of other programs not in $U(d)$ because of condition (1)

$$\times \quad \in \quad U(\div)$$
$$\bot \quad \in \quad U(g) \quad \text{if } g \text{ has a partial left inverse}$$
$$id \quad \in \quad U(g) \quad \text{for any program } g$$

One can often discover useful programs by investigating adjunctions and inverses. For example, *tail* is an adjunction for the function 1 on non-empty sequences, and there is a function called *cons* which is a partial inverse for the construct [1, *tail*].

$$cons : \langle x, \langle y, z, \ldots \rangle \rangle = \langle x, y, z, \ldots \rangle$$

There is a procedure which can calculate a subset of $U(d)$, for any $d$. These examples are all base cases for it. As with $\partial_r$ and $\partial_l$, the procedure works by induction over the structure of $d$. The subset of $U(d)$ which it finds may be empty, in which case we have to resort to *id* for $e$, but often it contains more helpful alternatives. It employs the rules:

$$U(d_1 \circ d_2) \qquad\qquad\quad \supseteq\; [U(d_2), U(d_1) \circ d_2]$$
$$U(d_1 \circ d_2) \qquad\qquad\quad \supseteq\; U(d_1) \circ d_2 \text{ if } d_2 \qquad \text{has a partial left inverse}$$
$$U([d_1, d_2, \ldots, d_n]) \qquad \supseteq\; \bigcup \{\, U(d_j) \mid Dom(d_j) = Dom\,([d_1, d_2, \ldots, d_n])\}$$
$$U([d_1 \circ 1, d_2 \circ 2, \ldots, d_n \circ n]) \supseteq\; [U(d_1 \circ 1), U(d_2 \circ 2), \ldots, U(d_n \circ n)]$$

One might add the rule

$$U(p \to q\,;\, r) \qquad\qquad\qquad \supseteq\; \{p \to q'\,;\, r' \mid q' \in U(q) \wedge r' \in U(r)\}$$

but conditionals never occur in the programs $d$ synthesized by this algorithm, so it is not needed.

*Example* (Zhu and Jin): inner products

Each input to the desired program is a pair of vectors

$$\langle\langle a_1, a_2, \ldots, a_n\rangle, \quad \langle b_1, b_2, \ldots, b_n\rangle\rangle$$

where the $a_i$s and $b_i$s are numbers. The corresponding output is their inner product

$$a_1 \times b_1 + a_2 \times b_2 + \ldots + a_n \times b_n.$$

Let us suppose that the training set $T$ consists of the first three such cases:

$$\{\langle\langle a_1\rangle, \langle b_1\rangle\rangle \,/\, a_1 \times b_1,$$
$$\langle\langle a_1, a_2\rangle, \langle b_1, b_2\rangle\rangle \,/\, a_1 \times b_1 + a_2 \times b_2,$$
$$\langle\langle a_1, a_2, a_3\rangle, \langle b_1, b_2, b_3\rangle\rangle \,/\, a_1 \times b_1 + a_2 \times b_2 + a_3 \times b_3\}.$$

For the algorithm to work, all entries in these examples must be *variables*. It will not work if any $a_i$ or $b_i$ is actually a number.

For these three cases,

$$P(\langle\langle a_1\rangle, \langle b_1\rangle\rangle)$$
$$= (length = \bar{2})\; \&\; \textit{map-and} \circ [(length = \bar{1}) \circ 1, (length = \bar{1}) \circ 2]$$

$$P(\langle\langle a_1, a_2\rangle, \langle b_1, b_2\rangle\rangle)$$
$$= (length = \bar{2})\; \&\; \textit{map-and} \circ [(length = \bar{2}) \circ 1, (length = \bar{2}) \circ 2]$$

$$P(\langle\langle a_1, a_2, a_3\rangle, \langle b_1, b_2, b_3\rangle\rangle)$$
$$= (length = \bar{2})\; \&\; \textit{map-and} \circ [(length = \bar{3}) \circ 1, (length = \bar{3}) \circ 2]$$

and a straightforward calculation shows that

$$\partial_r \left( \frac{P(\langle\langle a_1, a_2 \rangle, \langle b_1, b_2 \rangle\rangle)}{P(\langle\langle a_1 \rangle, \langle b_1 \rangle\rangle)} \right) \supseteq \{[tail \circ 1,\ tail \circ 2]\}$$

$$\partial_r \left( \frac{P(\langle\langle a_1, a_2, a_3 \rangle, \langle b_1, b_2, b_3 \rangle\rangle)}{P(\langle\langle a_1, a_2 \rangle, \langle b_1, b_2 \rangle\rangle)} \right) \supseteq \{[tail \circ 1,\ tail \circ 2]\}$$

so $d = [tail \circ 1,\ tail \circ 2]$

The choice of $e$ runs thus:

$$
\begin{aligned}
U(d) \ &=\ U([tail \circ 1,\ tail \circ 2]) \\
&\supseteq\ [U(tail) \circ 1,\ U(tail) \circ 2] \\
&\supseteq\ \{[1 \circ 1,\ 1 \circ 2]\}
\end{aligned}
$$

so for $e$ we choose $[1 \circ 1,\ 1 \circ 2]$.

When the products $a_i \times b_i$ have been written in FP style,

$$
\begin{aligned}
F(\langle\langle a_1 \rangle,\ &\langle b_1 \rangle\rangle,\ \times (a_1, b_1)) \\
&=\ \times \circ F(\langle\langle a_1 \rangle,\ \langle b_1 \rangle\rangle,\ \langle a_1, b_1 \rangle) \\
&=\ \times \circ [F(\langle\langle a_1 \rangle,\ \langle b_1 \rangle\rangle,\ a_1),\ F(\langle\langle a_1 \rangle,\ \langle b_1 \rangle\rangle,\ b_1)] \\
&=\ \times \circ [pp_{\langle\langle a_1 \rangle,\ \langle b_1 \rangle\rangle(a_1)},\ pp_{\langle\langle a_1 \rangle,\ \langle b_1 \rangle\rangle(b_1)}] \\
&=\ \times \circ [1 \circ 1,\ 1 \circ 2]
\end{aligned}
$$

Note: this is the step which requires that the $a_i$s and $b_i$s are all variables. If they were not, then $F(in, out)$ would contain constant functions instead of projections.

Similarly,

$$
\begin{aligned}
F(\ \langle\langle a_1, a_2 \rangle,\ &\langle b_1, b_2 \rangle\rangle,\ + (\times(a_1, b_1), \times(a_2, b_2))\ ) \\
&=\ + \circ [\times \circ [1 \circ 1,\ 1 \circ 2],\ \times \circ [2 \circ 1,\ 2 \circ 2]\ ]
\end{aligned}
$$

and

$$
\begin{aligned}
F(\ \langle\langle a_1, a_2, a_3 \rangle,\ &\langle b_1, b_2, b_3 \rangle\rangle,\ +(\times(a_1, b_1)\ ,\ +(\times (a_2, b_2)\ ,\ \times(a_3, b_3)\ )\ )\ ) \\
&=\ + \circ [\times \circ [1 \circ 1,\ 1 \circ 2]\ ,\ + \circ [\times \circ [2 \circ 1,\ 2 \circ 2]\ ,\ \times \circ [3 \circ 1,\ 3 \circ 2]\ ]\ ]
\end{aligned}
$$

Then

$$
\begin{aligned}
[e,\ F(in_1,\ out_1) \circ d] \\
&=\ [\ [1 \circ 1,\ 1 \circ 2],\ \times \circ [1 \circ 1,\ 1 \circ 2] \circ [tail \circ 1,\ tail \circ 2]\ ] \\
&=\ [\ [1 \circ 1,\ 1 \circ 2],\ \times \circ [1 \circ tail \circ 1,\ 1 \circ tail \circ 2]\ ] \\
&=\ [\ [1 \circ 1,\ 1 \circ 2],\ \times \circ [2 \circ 1,\ 2 \circ 2]\ ]
\end{aligned}
$$

and so

$$\partial_l(F(in_2, out_2) \,/\, [e, F(in_1, out_1) \circ d])$$

$$= \partial_l\left( \frac{+ \circ [\times \circ [1 \circ 1, 1 \circ 2], \times \circ [2 \circ 1, 2 \circ 2]]}{[[1 \circ 1, 1 \circ 2], \times \circ [2 \circ 1, 2 \circ 2]]} \right)$$

$$\supseteq\ + \circ \partial_l\left( \frac{[\times \circ [1 \circ 1, 1 \circ 2], \times \circ [2 \circ 1, 2 \circ 2]]}{[[1 \circ 1, 1 \circ 2], \times \circ [2 \circ 1, 2 \circ 2]]} \right)$$

$$\supseteq\ + \circ \left[ \partial_l\left( \frac{\times \circ [1 \circ 1, 1 \circ 2]}{[1 \circ 1, 1 \circ 2]} \right) \circ 1,\ \partial_l\left( \frac{\times \circ [2 \circ 1, 2 \circ 2]}{\times \circ [2 \circ 1, 2 \circ 2]} \right) \circ 2 \right]$$

$$\supseteq\ + \circ \left[ \times \circ \partial_l\left( \frac{[1 \circ 1, 1 \circ 2]}{[1 \circ 1, 1 \circ 2]} \right) \circ 1, id \circ 2 \right]$$

so one possible choice for $h$ is

$$h = + \circ [\times \circ 1, 2]$$

One can check that this choice of $h$ is also a solution of the other condition

$$h \circ [e, F(in_2, out_2) \circ d] = F(in_3, out_3)$$

Hence, the complete program for calculating inner products

$$ip = test \to base;\ h \circ [e, ip \circ d]$$

is

$$\begin{aligned}
ip = &\ (length = \overline{2})\ \&\ map\text{-}and \circ [(length = \overline{1}) \circ 1, (length = \overline{1}) \circ 2] \\
&\to\ \times \circ [1 \circ 1, 1 \circ 2]\, ; \\
&\quad + \circ [\times \circ 1, 2] \circ [\, [1 \circ 1, 1 \circ 2], ip \circ [tail \circ 1, tail \circ 2]\, ]
\end{aligned}$$

This can be simplified to

$$\begin{aligned}
ip = &\ (length = \overline{2})\ \&\ map\text{-}and \circ [(length = \overline{1}) \circ 1, (length = \overline{1}) \circ 2] \\
&\to\ \times \circ [1 \circ 1, 1 \circ 2]\, ; \\
&\quad + \circ [\times \circ [1 \circ 1, 1 \circ 2], ip \circ [tail \circ 1, tail \circ 2]]
\end{aligned}$$

This program is in fact the FP version of a natural procedure for working out inner products. Zhu and Jin give other examples.

### 6.3.9 EXTENDED FORMS OF SYNTHESIS

The assumptions which Zhu and Jin make on their training set are quite stringent.

- The synthesis process will not work if some instance's input *in* contains a function symbol, because then the unification used to describe the test for that case should be subject to a law describing the values of that function.

- It cannot be guaranteed to work when some *in* contains a variable $x$ in more than one place, because then the projection $pp_t(x)$ used to define $F$ is itself not well defined.

- It may also not work if any two distinct *in*s can be unified, since then the desired program would be over-specified.

- It is based on the assumption that the training sequence contains just the first few instances, ordered correctly.

- It is also based on the assumption that there is a unique base case. Whatever the program's input, if the function $d$ is applied to it often enough then eventually *test* will become true.

- The function *test* performs a simple syntactic matching test on *in*.

The first of these restrictions could only be overcome by an analysis of unification and generalization subject to laws, which we are not equipped for. If there is more than one base case, then really the training set should be divided because the learning task involves synthesizing more than one program. The last restriction seems inevitable in any syntactic process. However, the others are not such great obstacles.

### Repeated occurrences of a variable

If *in* contains $x$ more than once, *in* will be a tuple

$$\langle t_1, t_2, t_3, \ldots t_n \rangle.$$

It may have several repeated variables: $x_1, x_2, \ldots x_m$. Say $\pi_{ij}$ is the projection onto the place in *in* containing the $j$th occurrence of $x_i$. The modified definition of the simple function for this example is

$$
\begin{aligned}
P(in) \ &= \\
&(length = \bar{n}) \ \& \\
&\pi_{11} = \pi_{12} \ \& \ \pi_{12} = \pi_{13} \ \& \ \pi_{13} = \pi_{14} \ldots \& \\
&\pi_{21} = \pi_{22} \ \& \ \pi_{22} = \pi_{23} \ \& \ \pi_{23} = \pi_{24} \ldots \& \\
&\quad \ldots \\
&\pi_{m1} = \pi_{m2} \ \& \ \pi_{m2} = \pi_{m3} \ \& \ \pi_{m3} = \pi_{m4} \ldots \& \\
&P(t_1) \circ 1 \ \& \ P(t_2) \circ 2 \ \& \ P(t_3) \circ 3 \ \& \ \ldots P(t_n) \circ n
\end{aligned}
$$

and

$$
\begin{aligned}
&F(in, x_i) = \pi_{i1} \text{ for any variable } x_i \text{ occurring in } in \\
&F(in, y) = \bar{\perp} \text{ for any other variable } y.
\end{aligned}
$$

The other cases for $P$ and $F$ are as before.

There is essentially no change in the description of $F$. The change in $P$ guarantees that $F$ is well defined.

The new definition of $P$ is ambiguous. It depends on a choice of order for the variables $x_1, x_2, \ldots$ in *in* and also on a choice of order for the occurrences of each variable. Different choices will lead to permutations of the terms $\pi_{ij}$ in $P(in)$. To a human reader, this may not appear significant, but it could easily upset the synthesis process. For instance, when calculating $d$, the set

$$\partial_r \left( \frac{P(in_2)}{P(in_1)} \right)$$

depends on the precise forms of $P(in_2)$ and $P(in_1)$. The process may work for one choice of orderings, but not for others.

### When training instances are not orthogonal

Program synthesis works much better when the training set consists of instances containing may variables and preferably no constants. Any constant will appear in $P(in)$ and $F(in, out)$ and will probably find its way through into the final program, which is not satisfactory.

A likely scenario is the following: a training set $T$ consists of many instances, each an input/output pair and each containing constants.

Perhaps the best approach will be to:

1. Partition $T$ into subsets $T_0, T_1, T_2, T_3, \ldots$ which each have a common generalization, formed by replacing constants with variables. Thus, for each $j$, there is a term $t_j$ so that each instance in $T_j$ is matched by $t_j$. The terms $t_j$ can be found by Plotkin's method. Of course, any two instances in $T$ will have a trivial common generalization—a variable, $x$. The subsets $T_j$ should be chosen so that each $t_j$ is as elaborate as possible. In particular, if $i \neq j$ then it should not be possible to unify $t_i$ and $t_j$.

2. Form a second training set

   $$T' = \{t_0, t_1, t_2, t_3, \ldots\}$$

   and use $T'$ as input to the process of program synthesis.

Since each $t_j$ contains variables rather than constants, the resulting program is likely to be much more general and useful.

### Gaps in the training set

The program synthesis method described here depends on nice features of its training set. One which we have not discussed much is that the instances can be ranged in an order

$$in_1/out_1, \quad in_2/out_2, \quad in_3/out_3 \quad \ldots$$

so that all the sets

$$\partial_r \left( \frac{P(in_{j+1})}{P(in_j)} \right)$$

contain a common program $d$. If the sequence has gaps, then this won't be so. If some instance $in_j/out_j$ is missing, then some set

$$\partial_r\left(\frac{P(in')}{P(in)}\right)$$

will contain $d \circ d$ rather than $d$. If $T$ only contains a few widely spaced examples, then all the known sets

$$\partial_r\left(\frac{P(in_i)}{P(in_j)}\right)$$

will contain high powers $d^{i-j}$. The right program $d$ may still be found, by a process analogous to Euclid's algorithm for finding the highest common factor of two whole numbers: repeatedly, use the rule

$$d^{a-b} \in \partial_l\left(\frac{d^a}{d^b}\right) \cap \partial_r\left(\frac{d^a}{d^b}\right)$$

but it may be hard to find $d$ if the $\partial_l$ and $\partial_r$ sets are large.

If the very first instance in $T$ is missing, then the synthesis algorithm will not find the right base case. It will invent the program

$$p' = test' \rightarrow base'; h \circ [e, p' \circ d]$$

where

$$test' = test \circ d$$

and

$$base' = h \circ [e, base \circ d]$$

instead. The sub-programs $test'$ and $base'$ will be known, but the ones which we really want are $test$ and $base$. Suppose that there is a program $test$ in

$$\partial_l\left(\frac{test'}{d}\right)$$

and also programs $u$ and $base$ such that

$$[e,u] \in \partial_r\left(\frac{base'}{h}\right)$$

and

$$base \in \partial_l\left(\frac{u}{d}\right)$$

In that case, there is a more general solution $p$ of the original synthesis task:

$$p = test \rightarrow base ; h \circ [e, p \circ d]$$

which covers a lower base case than the base case of $p'$.

### More than one base case

Consider the function $s$ which adds 1 to a number. If numbers are represented as decimals, then this is trivially

$$s = + \circ [id, \overline{1}]$$

Numbers can also be represented as nested sequences of units, so 3 is represented as $\langle 1, \langle 1, \langle 1 \rangle \rangle \rangle$. In this notation, $s$ is even simpler:

$$s = [\overline{1}, id]$$

However, $s$ is considerably more complicated when numbers are represented in a binary form:

$$
\begin{array}{lll}
0 & \text{as} & \langle 0 \rangle \\
1 & \text{as} & \langle 1 \rangle \\
2 & \text{as} & \langle 0, \langle 1 \rangle \rangle \\
3 & \text{as} & \langle 1, \langle 1 \rangle \rangle \\
4 & \text{as} & \langle 0, \langle 0, \langle 1 \rangle \rangle \rangle,
\end{array}
$$

and so on. Then

$$
\begin{aligned}
s \quad = \quad & length \; = \; \overline{1} \; \& \; 1 = \overline{0} \rightarrow [\overline{1}]; \\
& length \; = \; \overline{1} \; \& \; 1 = \overline{1} \rightarrow [\overline{0}, [\overline{1}]\,]; \\
& \qquad\qquad\qquad 1 = \overline{0} \rightarrow [\overline{1}, tail]; \\
& \qquad\qquad\qquad\qquad [\overline{0}, s \circ tail]
\end{aligned}
$$

This form of $s$ is actually a linear recursive function:

$$
\begin{array}{lll}
d & = & tail \\
e & = & \overline{0} \\
h & = & id
\end{array}
$$

but its *test* is too complex and it has no simple unique *base*. This program $s$ is perhaps easier to understand if it is viewed as one of a recursive pair:

$$s = 1 = \overline{0} \rightarrow s\text{-}eve; \; s\text{-}odd$$

where

$$s\text{-}odd \; = \; length = \overline{1} \; \& \; 1 = \overline{1} \; \rightarrow \; [\overline{0}, [\overline{1}]\,]; \; [\overline{0}, s \circ tail]$$

and

$$s\text{-}eve \; = \; length = \overline{1} \; \& \; 1 = \overline{0} \; \rightarrow \; [\overline{1}]; \; [\overline{1}, tail]$$

It would be very interesting if someone could devise an extension of the method which copes with this example. One approach might be to separate the training sequence into a pair of sequences, the ones with even *in* and the ones with odd *in* as shown in Table 6.3, but there is no obvious way to do this automatically.

**Table 6.3**

| in | out | in | out |
|---|---|---|---|
| ⟨0⟩ | ⟨1⟩ | ⟨1⟩ | ⟨0,⟨1⟩⟩ |
| ⟨0,⟨1⟩⟩ | ⟨1,⟨1⟩⟩ | ⟨1,⟨1⟩⟩ | ⟨0,⟨0,⟨1⟩⟩⟩ |
| ⟨0,⟨0,⟨1⟩⟩⟩ | ⟨1,⟨0,⟨1⟩⟩⟩ | ⟨1,⟨0,⟨1⟩⟩⟩ | ⟨0,⟨1,⟨1⟩⟩⟩ |

# 6.4   OTHER PROGRAM SYNTHESIS METHODS

There is a large body of research on various methods. Work by Shapiro has stimulated several studies of 'logic program synthesis' and ways to invent first-order theories. Another approach synthesizes finite state automata (FSA). Each FSA can recognize the strings in a particular language. The languages which can be recognized are the so-called **regular** languages.

The regular languages are excessively restricted. For example, the language consisting of the strings

$$ab \quad aabb \quad aaabbb \quad aaaabbbb \quad aaaaabbbbb \quad \ldots$$

is not regular. I suspect that the algorithm for synthesizing FP programs will not work if its input is the sequence of pairs

| in | out |
|---|---|
| ⟨x⟩ | ⟨a, b⟩ |
| ⟨x, y⟩ | ⟨a, a, b, b⟩ |
| ⟨x, y, z⟩ | ⟨a, a, a, b, b, b⟩ |

unless it is supplied with a very elaborate built-in function and some subtle knowledge on how to apply it. However, FP program synthesis will work with the training sequence

| in | out |
|---|---|
| ⟨x⟩ | ⟨⟨a⟩, ⟨b⟩⟩ |
| ⟨x, y⟩ | ⟨⟨a, a⟩, ⟨b, b⟩⟩ |
| ⟨x, y, z⟩ | ⟨⟨a, a, a⟩, ⟨b, b, b⟩⟩ |

and it can also generate a sequence of little programs which, when put together, will concatenate two sequences. Hence, FP program synthesis *can* form a sequence of programs which together will generate the non-regular language above (although, of course, the method will not suggest how the sequence should be put together). It does therefore appear that the method of Zhu and Jin may be more powerful than the algorithms which invent finite state automata. On the other hand, the extra power may not lie in the method itself, but rather in the functions such as *length* and *tail* which the method can call on while building its programs.

## 6.5    INVENTING RECURSIVE DEFINITIONS

So far, we have come across two procedures for inventing recursive structures. The first is Wolff's algorithm which invents context-free grammars. Its generalizing step can give rise to recursion. The second is program synthesis, which can construct linear recursive programs. The little algorithm suggested here will invent clauses which form recursive definitions.

Program synthesis in the FP language can only yield linear programs. This algorithm for generating recursive definitions can do more than that. One of our examples will show how to invent a definition of the concept of binary trees, in which the recursion is not linear: the predicate $p$ which is being defined occurs more than once in its own definition.

Observe that any algorithm which invents a recursive structure must involve some kind of guessing. Its input is a finite training set which can be described without any recursion, just by listing its members. The output will be some datum which represents all the inputs or a way to construct them and infinitely many other possible inputs too. That is in the nature of recursion. Any finite set is part of more than one recursively defined infinite set. Hence, there can never be any way to guarantee that the output is correct, in any precise sense. The best that can be asked of the algorithm is that its output should represent a 'reasonable' recursive extension of the input. What 'reasonable' means is up to you. It might mean that the output definition has minimal Kolmogorov complexity.

### 6.5.1    THE SETTING FOR INFERENCE

The algorithms considered here are designed to learn definitions for a certain predicate symbol, $p$. Each such algorithm has inputs:

- the name of this predicate, $p$.

- a set *Exs* of ground clauses. Each clause includes just one literal whose predicate symbol is $p$ and this literal is positive ; so the clause can be written as

$$Q \Rightarrow P$$

  where $Q$ is a conjuction of literals not involving $p$ and $P$ is an atomic formula whose predicate symbol is $p$.

- a set *Cxs* of ground instances of $p$ (the counter-examples).

- *Background*. This is a source of information about other predicates which may occur in conditions $Q$.

Each algorithm outputs another set *RS* of clauses.

The desired definition will be a set of clauses, some subset of the output *RS*. Such clauses are often called **rules**. This method can be used to invent the kind of rule which we shall study further in Chapter 7.

*Exs* is the set of (positive) examples. In practice, each positive example may only be presented as *P*, in which case you have to decide which background facts should be incorporated into *Q*.

*Background* may be a collection of ground facts or it may be an oracle as for contradiction backtracing (see Chapter 7) or it may consist of a complete theory described by axioms and a theorem prover.

The algorithm constructs a set *RS*. Each entry in *RS* is a pair consisting of a rule *R* for *p* and a set $I_R$ of known instances of *R*. It will only keep rules which have known instances. When it ends, some of the rules R may be recursive:

$$Q' \wedge P \Rightarrow P'$$

where *P* and *P'* are atomic formulas whose predicate symbols are both *p*. It is up to you to decide whether such a recursive rule really describes the concept *p* you want. You have the job of deciding which set of rules *R* forms your desired definition, which rules are surplus, and which (if any) are too general and, hence, wrong. A set of rules for *p* might be considered satisfactory if:

(1)   all examples can be derived as instances of them; and

(2)   no counter-example contradicts any of them (see below); and

(3)   the set of ground instances derivable from them contains terms of unbounded size; and

(4)   they have low Kolmogorov complexity.

If there are several such sets of rules, then the one generating the smallest set of ground instances is probably best.

All the algorithms which we shall consider involve two generalizing operations, which we can call *FunG* and *PredG*. *FunG* generalizes terms within a clause by Plotkin's second method. If *C* and *D* are any two clauses, then say *FunG(C, D)* is the result of Plotkin generalization applied to them. In fact, we shall extend Plotkin's basic method a little. For instance, if *C* is

$$on\ (\mathbf{B}, \mathbf{table}) \wedge on\ (\mathbf{C}, \mathbf{B}) \Rightarrow column(\ [\mathbf{C}, \mathbf{B}]\ )$$

and *D* is

$$on\ (\mathbf{E}, \mathbf{table}) \wedge on\ (\mathbf{D}, \mathbf{E}) \Rightarrow column\ (\ [\mathbf{D}, \mathbf{E}]\ )$$

then the result of Plotkin's method applied to them is

$$on\ (b, \mathbf{table}) \wedge on\ (c, b) \wedge on\ (w, x) \wedge on\ (y, z) \Rightarrow column\ ([c, b])$$

where $b$, $c$, $w$, $x$, $y$, and $z$ are new universally quantified variables. The last two conjuncts in the condition are redundant. This generalization is equivalent to the shorter form:

$$on\ (b, \mathbf{table}) \wedge on\ (c, b) \Rightarrow column\ ([c, b]).$$

This is *FunG(C, D)*.

*PredG* generalizes portions of clauses. It is sometimes called *inverse implication* or *abduction.* Say

$$C \quad \text{is} \quad X \Rightarrow P$$

and

$$D \quad \text{is} \quad Q \wedge X' \Rightarrow P'$$

where $X$, $Q$, and $X'$ are literals or conjunctions of literals and all variables are assumed universally quantified as usual. The operation *PredG* can be applied if $X$ and $X'$ can be unified. In that case, say $\vartheta$ is their most general unifier; then

$$PredG(C, D) \quad \text{is} \quad Q\vartheta \wedge P\vartheta \Rightarrow P'\vartheta.$$

There is a sort of approximate correspondence between functions and predicates. Any first-order theory with functions can be interpreted in another without functions. For instance, suppose that the original theory is given by axioms which are all disjunctive clauses, with all variables universally quantified. If you prefer your language without functions and $f$ is a function occurring in axioms, then

(1)   invent a new predicate (say $F$) and an axiom for it:

$$\forall x\ \exists y\ \forall z\ (F(x, z) \Leftrightarrow z = y)$$

   and then in every clause among the axioms, for each term $f(x)$ in it constructed with $f$,

(2)   replace $f(x)$ by a new universally quantified variable, say $y$, and

(3)   introduce a new disjunct $\ldots \vee \neg F(x,y)$.

The new predicate $F$ is interpreted as saying that $y$ is the value of $f(x)$. Observe that the new extra axiom is not in clausal form. It would be nice to imagine that the algorithm's two operations correspond under this correspondence between functions and predicates. They do not exactly.

*Cxs* and the *Background* are used to eliminate over-general rules. Say *Cxs* **contradicts** a rule $R: Q \Rightarrow P$ if there is a substitution $\varphi$ for which $P\varphi \in Cxs$ and *Background* implies that $Q\varphi$ is true.

Conversely, an example $Q' \Rightarrow P'$ in *Exs* **satisfies** $R$ if $P$ can be unified with $P'$, say by $\varphi$, and the inference method in *Background* suffices to show that that $Q\varphi$ is true.

### 6.5.2. A SIMPLE FORM OF THE ALGORITHM

We shall consider two forms of the algorithm in detail. The second form is a seemingly trivial extension of the first, but it does appear to have significantly greater power. Even so, the first form is adequate for some interesting examples, so we start by examining it. It goes:

$$RS := \{(C,\{C\}) \mid C \in Exs\};$$

WHILE $RS$ contains two pairs $(S, I_S)$ and $(T, I_T)$
    AND some operation $Op$, either $FunG$ or $PredG$, can be applied to $S$ and $T$

    $R := Op(S, T)$;
    IF     $Op$  is  $FunG$  THEN  $I_R := I_S \cup I_T$;
    IF     $Op$  is  $PredG$ THEN
        $I_R := \{X \in I_S \cup I_T \mid X$ satisfies both $S\vartheta$ and $T\vartheta$
                where $\vartheta$ is the substitution in $PredG\}$;
    IF     $I_R$   is not empty
    AND  $Cxs$  does not contradict $R$
    THEN

$$RS := RS \cup \{(R, I_R)\};$$

Here are two examples of how it works. They do not include explicit descriptions of $Cxs$ and $Background$ and they do not show the sets $I_R$ of instances, but you can see easily how these details fit in.

The first example shows how to invent a definition of the concept of binary trees, in which the recursion is not linear: the predicate $p$ occurs more than once among a rule's conditions. There are two kinds of binary tree. The interesting ones have branches. They are terms of the form

   $node(l, r)$

where *node* is a function symbol and the two branches, $l$ and $r$, are themselves binary trees. The other kind of tree is a leaf. We suppose that there is a predicate

   $leaf(x)$

which is true of leaves and not of anything else. In this case, the predicate $p$ will be called *bt*. It should be true of all these trees and no others.

*Example*: inventing the notion of binary trees

   $p$ is *bt*. *Exs* consists of

$$leaf(A) \Rightarrow bt(A)$$
$$leaf(B) \wedge leaf(C) \Rightarrow bt(node(B, C))$$
$$leaf(D) \Rightarrow bt(D)$$
$$leaf(E) \wedge leaf(F) \Rightarrow bt(node(E, F))$$

Apply *FunG* twice:

$$leaf(a) \Rightarrow bt(a)$$
$$leaf(b) \wedge leaf(c) \Rightarrow bt(node(b, c))$$

Apply *PredG* twice:

$$bt(b) \wedge leaf(c) \Rightarrow bt(node(b, c))$$
$$leaf(b) \wedge bt(c) \Rightarrow bt(node(b, c))$$

and then, by applying *PredG* again to either of these and

$$leaf(a) \Rightarrow bt(a)$$

we obtain

$$bt(b) \wedge bt(c) \Rightarrow bt(node(b, c))$$

This is the required recursive rule for *bt*.

The second example shows how to invent a definition of the notion of a column, in the 'blocks world' of Chapter 2. A column is a list of blocks, each resting on the one below, except for the bottom one which is resting on a table. If a column consists of three blocks and the top one is **A**, the middle one is **B**, and the bottom one is **C**, then this column is the list [**A**, **B**, **C**].

Lists like this are actually terms formed from the constants, **A**, **B**, **C**, and the empty list, [ ], by means of a function which we shall call *dot* and write as '.'. Thus, when written out fully, the above list should appear as

.(**A**, .(**B**, .(**C**, [ ] ) ) )

The first notation is a shorthand form. The second, with '.', will be familiar to anyone who has studied the theory underlying Lisp. Where convenient, the two notations may be mixed, so this same column may also be written as

.(**A**, .(**B**, [**C**] ) ).

*Example*: definition of the predicate *column*

*p* is *column*. *Exs* consists of six ground clauses describing six columns.

$$on\,(\mathbf{A}, \mathbf{table})\ \Rightarrow\ column\,([\mathbf{A}])$$
$$on\,(\mathbf{B}, \mathbf{table}) \wedge\ on\,(\mathbf{C}, \mathbf{B})\ \Rightarrow\ column\,([\mathbf{C}, \mathbf{B}])$$
$$on\,(\mathbf{D}, \mathbf{table}) \wedge on\,(\mathbf{E}, \mathbf{D}) \wedge\ on\,(\mathbf{F}, \mathbf{E})\ \Rightarrow\ column\,([\mathbf{F}, \mathbf{E}, \mathbf{D}])$$
$$on\,(\mathbf{F}, \mathbf{table})\ \Rightarrow\ column\,([\mathbf{F}])$$
$$on\,(\mathbf{E}, \mathbf{table}) \wedge\ on\,(\mathbf{D}, \mathbf{E})\ \Rightarrow\ column\,([\mathbf{D}, \mathbf{E}])$$
$$on\,(\mathbf{C}, \mathbf{table}) \wedge on\,(\mathbf{B}, \mathbf{C}) \wedge\ on\,(\mathbf{A}, \mathbf{B})\ \Rightarrow\ column\,([\mathbf{A}, \mathbf{B}, \mathbf{C}])$$

On applying *FunG* three times, we obtain

$C1:$                 $on\,(a, \mathbf{table})\ \Rightarrow\ column\,([a])$

$C2:$          $on\,(b, \mathbf{table}) \wedge\ on\,(c, b)\ \Rightarrow\ column\,([c, b])$

$C3:\ on\,(d, \mathbf{table}) \wedge\ on\,(e, d) \wedge\ on\,(f, e)\ \Rightarrow\ column\,([f, e, d])$

where $a$, $b$, $c$, $d$, $e$, and $f$ are new variables.

These three clauses include three pairs to which *PredG* is applicable:

| $X'$ | in $D$ | | $X$ | in $C$ |
|---|---|---|---|---|
| *on* ($b$, **table**) | in $C_2$ | unifies with | *on* ($a$, **table**) | in $C_1$ |
| *on* ($d$, **table**) | in $C_3$ | unifies with | *on* ($a$, **table**) | in $C_1$ |
| *on* (d, **table**) $\wedge$ *on* ($e$, $d$) | in $C_3$ | unifies with | *on* ($b$, **table**) $\wedge$ *on* ($c$, $b$) | in $C_2$ |

The results of these three applications of *PredG* are

$$column\ ([b]) \wedge on\ (c, b) \Rightarrow column\ ([c, b])$$
$$column\ ([d]) \wedge on\ (e, d) \wedge on\ (f, e) \Rightarrow column\ ([f, e, d])$$
$$column\ ([e, d]) \wedge on\ (f, e) \Rightarrow column\ ([f, e, d])$$

Note that these rules do not really deserve to be called recursive. Although each has a conclusion whose predicate is *column* and this same predicate occurs as a condition, the arguments of all these occurrences of *column* are lists of particular lengths. Any column of the form $[c, b]$ has height 2 and a *column* $[b]$ has height 1. In a genuine recursive rule, the *columns* could be as big as you like. The rules derived so far do not fulfil the third condition on a satisfactory definition.

The proper recursive rule for *column* is obtained by applying *FunG* just once more, to the pair

$$column\ ([b]) \wedge on\ (c, b) \Rightarrow column\ ([c, b])$$
$$column\ ([e, d]) \wedge on\ (f, e) \Rightarrow column\ ([f, e, d])$$

which for this purpose are better written as

$$column\ (.(b, [\ ])) \wedge on\ (c, b) \Rightarrow column\ (.(c, .(b, [\ ])))$$
$$column\ (.(e, [d])) \wedge on\ (f, e) \Rightarrow column\ (.(f, .(e, [d])))$$

The result of generalizing from them is

$$column\ (.(x, z)) \wedge on\ (y, x) \Rightarrow column\ (.(y, .(x, z)))$$

This and the base case

$$on\ (a, \textbf{table}) \Rightarrow column\ ([a])$$

together form a definition of *column*. All the other cases can be derived from these two.

### 6.5.3 CONSEQUENCES OF INAPPROPRIATE BIAS

This algorithm is sensitive to the description language. The first example could be described without the predicate *leaf*, using instead a function symbol, say *leaff*. In the example as it stands, anything satisfying the condition leaf is a *leaf*. Instead, we could impose a much stricter condition: something is a leaf if and only if it is a term of the form

$$leaff(x).$$

In this representation, the training set would consist of

*bt(leaff(A))*
*bt(node(leaff(B), leaff(C)))*
*bt(leaff(D))*
*bt(node(leaff(E), leaff(F)))*

If the algorithm is applied to this training set, the outcome is quite different. *PredG* cannot be applied at all because there are no conditions $X$ and $X'$. Hence, there is no possibility of inventing any recursive rule for *bt*.

This example seems to suggest that inference of recursive rules will be easier in a language with predicates rather than function symbols. It seems best to keep the right side of each input rule quite general, with lots of variables, and then impose conditions on the variables.

Might this principle be taken too far the other way? In the second example, the final generalization step using *FunG* occurred between the clauses

$$column\ (.\ (b, [\ ])) \land on\ (c, b) \Rightarrow column\ (.\ (c, .\ (b, [\ ])))$$

and

$$column\ (.\ (e, [d])) \land on\ (f, e) \Rightarrow column\ (.\ (f, .\ (e, [d])))$$

If all function symbols were done away with, then *FunG* would not work. It generalizes the two terms

$$[\ ] \quad \text{and} \quad [d]$$

to a variable. If all function symbols are avoided, then these two terms would not appear and so *FunG* could not apply to them.

Rather than express rules with a function called '.', one could write them with a ternary relation called *co*, where

$$co(a, b, c)$$

holds just when

$$c = .\ (a, b)\ .$$

In the language with the predicate *co*, *FunG* generates the intermediate clauses

$C_1:\ on\ (a, \mathbf{table})\ \land\ co\ (a, [\ ], x)\ \Rightarrow\ column\ (x)$
$C_2:\ on\ (b, \mathbf{table})\ \land\ on\ (c, b) \land co(b, [\ ], y) \land co(c, y, z)$
$$\Rightarrow\ column\ (z)$$
$C_3:\ on\ (d, \mathbf{table})\ \land\ on\ (e, d) \land on\ (f, e)$
$$\land\ co\ (d, [\ ], u) \land co\ (e, u, v) \land co\ (f, v, w)$$
$$\Rightarrow\ column\ (w)$$

These rules contain many matching pairs of conjuncts of conditions. The particularly useful pairs are

$on\,(a, \textbf{table}) \wedge co\,(a, [\ ], x)$ in $C_1$ with
$on\,(b, \textbf{table}) \wedge co\,(b, [\ ], y)$ in $C_2$

and

$on\,(b, \textbf{table}) \wedge on\,(c, b) \wedge co\,(b, [\ ], y) \wedge co\,(c, y, z)$ in $C_2$ with
$on\,(d, \textbf{table}) \wedge on\,(e, d) \wedge co\,(d, [\ ], u) \wedge co\,(e, u, v)$ in $C_3$.

From these pairs, *PredG* will generate

$$column\,(y) \wedge on\,(c, b) \wedge co(c, y, z) \Rightarrow column\,(z)$$

and

$$column\,(v) \wedge on\,(f, e) \wedge co\,(f, v, w) \Rightarrow column(w)$$

These are very nearly the recursive rules we want. The only flaw in them is that they omit anything which says that $b$ is the head of $y$ and that $e$ is the head of *column v*. They have lost the literals

$$co\,(b, [\ ], y) \qquad \text{and} \qquad co\,(e, u, v).$$

### 6.5.4 A CRUDE SOLUTION: BROADENING THE SEARCH

If these two literals are put back in, we obtain

$$column\,(y) \wedge co\,(b, [\ ], y) \wedge on\,(c, b) \wedge co\,(c, y, z) \;\Rightarrow\; column\,(z)$$

and

$$column\,(v) \wedge co\,(e, u, v) \wedge on\,(f, e) \wedge co\,(f, v, w) \;\Rightarrow\; column\,(w)$$

This last is exactly the recursive rule we seek. The algorithm can be extended so that it will produce the desired result in this case. One could introduce an extension of *PredG*, say *PredG+*, which might read:

Given two clauses

$$C: Q \wedge X \;\Rightarrow\; P$$

and

$$D: Q' \wedge X' \;\Rightarrow\; P'$$

such that $X$ and $X'$ can be unified, say by $\vartheta$,

$$PredG{+}(C, D) \quad \text{is} \quad P\vartheta \wedge Q\vartheta \wedge Q'\vartheta \Rightarrow P'\vartheta$$

This extended version will yield the correct recursive rule in the language with *co* in place of '.'. However, it will yield a lot of other misleading rules too. There seems to be no way to stop it from over-generalizing.

### 6.5.5 A BETTER SOLUTION: ADAPTING THE LANGUAGE

The operation *PredG* is the key to discovering recursion. *FunG* prepares training instances for application of *PredG* and then it also extends the range of clauses which already have a suitable form, but it is *PredG* which creates clauses of the form

$$Q \wedge P \Rightarrow P'.$$

Therefore, our effort will be spent on discovering good circumstances for using *PredG*.

The best language seems to be one with just enough predicates. In all other places, function symbols restrict the search more and so make search safer. As long as *PredG* can be applied then functions are preferable. If *PredG* cannot be applied at all, then perhaps some function is better interpreted by a new predicate.

#### *Excessive use of functions*

This manifests itself when the training set includes a pair

$$C: \quad p(f(t_1))$$
$$D: \quad p(g(f(t_2)))$$

where $t_1$ and $t_2$ are two terms which can be unified by some substitution $\vartheta$:

$$t = t_1 \vartheta = t_2 \vartheta.$$

The solution seems to be to invent a new predicate, say $F$, which satisfies the defining condition

$$\forall x \, \forall y \, (F(x, y) \Leftrightarrow y = f(x)).$$

When $C$ and $D$ are rewritten with $F$ in place of $f$, they become

$$F(t_1, y_1) \;\Rightarrow\; p(y_1)$$
$$F(t_2, y_2) \;\Rightarrow\; p(g(y_2))$$

Applied to this pair, *PredG* yields

$$p(y) \;\Rightarrow\; p(g(y))$$

The case of *leaff* was solved by this technique. Observe that neither $f$ nor the new predicate symbol $F$ occurs in the inferred recursive clause.

#### *Excessive use of predicates*

Suppose that the training set contains the pair of clauses

$$C: \qquad\qquad X(t_1) \wedge F(t_1, y_1) \;\Rightarrow\; p(y_1)$$
$$D: \quad Q(t_2) \wedge X(t_2) \wedge F(t_2, y_2) \;\Rightarrow\; p(g(y_2))$$

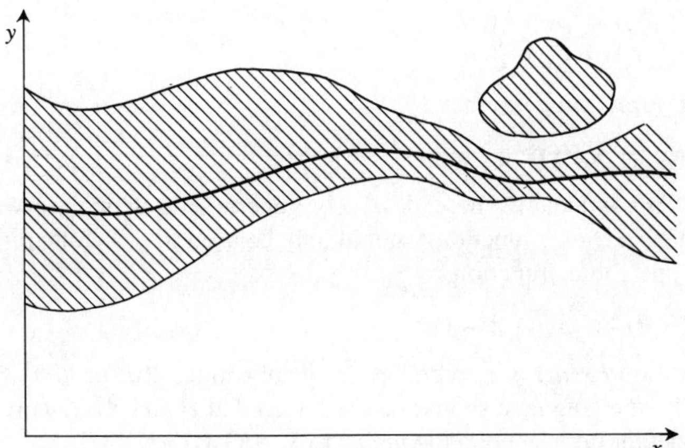

**Fig. 6.4** The graphs of a relation and of a choice function for it. The shaded area represents the coordinates for which $F(x, y)$ is true. The line within the shaded area is a graph of a choice function of $F$.

where, as before $t_1 \vartheta = t_2 \vartheta = t$. When applied to these two, *PredG* will yield

$$Q(t) \wedge p(y) \Rightarrow p(g(y)).$$

This clause is too general because it misses the relationship between the terms $t$ and $y$. This is the case when $F$ is *co*.

Sometimes, there may be no sensible solution. However, suppose that there is an algorithm which, given any term $x$, computes a value which we call $f(x)$ which satisfies the condition

$$\forall x \, F(x, f(x)\,).$$

The algorithm defines a computable choice function $f$ or $F$ (Fig. 6.4). Given some such $f$, the two original clauses imply

$$X(t_1) \Rightarrow p(f(t_1))$$
$$Q(t_2) \wedge X(t_2) \Rightarrow p(g(f(t_2)))$$

From these two, *PredG* yields

$$Q(t) \wedge p(f(t)\,) \Rightarrow p(g(f(t)))$$

This inference looks much more plausible.

### *Criteria for choosing between functions and predicates*

There will always be circumstances when inductive inference is misleading. The case just considered, where we use a computable choice function instead of a relation, can easily lead to errors. Consider the possibility that there may be two such functions, $f_1$ and $f_2$, for $F$; then by this procedure we could infer two new rules

$$Q(t) \wedge p(f_1(t)) \Rightarrow p(g(f_1(t)))$$
$$Q(t) \wedge p(f_2(t)) \Rightarrow p(g(f_2(t))).$$

From these, *FunG* suggests that

$$Q(t) \wedge p(y) \Rightarrow p(g(y))$$

for any $y$. This is exactly the excessively general rule which we were trying to avoid. Perhaps a choice function $f$ should only be used in place of a relation $F$ when $F$ has only one choice function:

$$\forall x \ \exists y \ \forall z \ (F(x, z) \Leftrightarrow z = y).$$

The operation *PredG* is a crude device. It substitutes $P\vartheta$ for $X'\vartheta$. If it is to work satisfactorily, the language should be chosen so that $P$ and $X'$ contain just the right amount of detail. Let us depict the inputs to *PredG* as

$$
\begin{aligned}
C: \quad & X(r) && \Rightarrow p(s) \\
D: \quad & Q(t) \ \wedge \ X(u) && \Rightarrow p(v)
\end{aligned}
$$

where $r$, $s$, $t$, $u$, and $v$ are perhaps composite terms.

When the description language uses functions excessively, then $X$ is too simple and $s$ is too elaborate. Choosing a predicate instead of some function in $s$ will cause the outcome of *PredG* to be more general.

When predicates are used excessively, then $X$ embodies a lot of detail. If $X(u\vartheta)$ is replaced by $p(s\vartheta)$ then this detail may be lost. In particular, if $X(u)$ includes some relation between $t$ and $v$, then this relation will be lost. Choosing a function in place of some part of $X$ will cause the outcome to be more specific.


### Criterion 1

All free variables in $t$ should appear in $v$.

Then all free variables in $t\vartheta$ will also appear in $v\vartheta$ and *PredG* cannot break the interdependence between $Q'\vartheta$ and $P'\vartheta$.


### Criterion 2

As long as Criterion 1 is not violated, $s$ should be as general as possible.

All restrictions on it are better embodied in $X(r)$. Then *PredG* will generalize as far as is reasonable. (One could adopt the opposite approach, but machine learning seems more effective when it over-generalizes and then discovers restrictions.)


### Criterion 3

When replacing a predicate with a computable choice function, only one such function should be used.

If there are several algorithms which compute choice functions, then just one such algorithm should be selected. Otherwise, *FunG* is liable to over-generalize.

### 6.5.6 A SECOND VERSION OF THE ALGORITHM: NEW VARIABLES AND EQUALITY

The classic test of inference is to find a predicate called *sort*. This takes two arguments which should both be lists.

$$sort(a, b)$$

is true when $b$ is a sorted list with the same entries as $a$. There are many algorithms for calculating $b$ from $a$. Each such algorithm can be taken as a definition of *sort*. The literature on machine learning contains several articles on procedures which can be used to invent such algorithms. The best known depend on an operation called *inverse resolution*, embodied in variants of logic program synthesis. Inverse resolution is very like abduction. Accounts of these methods can be found in the references at the end of the chapter.

The first version of our algorithm cannot invent an adequate definition for *sort*, but there is an extension of it which can. The extension is allowed to take a known rule

$$X(a) \Rightarrow P(a)$$

and from it to invent another:

$$X(a) \wedge b = a \Rightarrow P'(a, b)$$

where $P'(a,b)$ is obtained from $P(a)$ by changing some occurrences of $a$ into $b$. This seemingly trivial extension has substantial consequences.

Lists normally appear as $[A, B, C]$. In the example of *column*, we saw how a list can also be written with a function symbol called '.', thus:

$$.(A, .(B, .(C, [\ ])))$$

The dot notation is better for applications of *FunG*. We shall now simplify this notation yet further by leaving out the parentheses and commas, so that this same list appears as

$$.A .B .C [\ ]$$

There is always a unique way to put the commas and parentheses back in again.

*Example*: sorting

    $p$ is *sort*. *Exs* includes

| | |
|---|---|
| $sort [\ ] [\ ]$ | (1) |
| $sort .A [\ ] .A [\ ]$ | (2) |
| $A < B \Rightarrow sort .A .B [\ ] .A .B [\ ]$ | (3) |

$$B < A \Rightarrow sort\ .A\ .B\ [\ ]\ .B\ .A\ [\ ] \tag{4}$$
$$C < D \wedge D < E \Rightarrow sort\ .C\ .D\ .E\ [\ ]\ .C\ .D\ .E\ [\ ] \tag{5}$$
$$C < D \wedge D < E \Rightarrow sort\ .C\ .E\ .D\ [\ ]\ .C\ .D\ .E\ [\ ] \tag{6}$$
$$C < D \wedge D < E \Rightarrow sort\ .D\ .C\ .E\ [\ ]\ .C\ .D\ .E\ [\ ] \tag{7}$$
$$C < D \wedge D < E \Rightarrow sort\ .D\ .E\ .C\ [\ ]\ .C\ .D\ .E\ [\ ] \tag{8}$$
$$C < D \wedge D < E \Rightarrow sort\ .E\ .C\ .D\ [\ ]\ .C\ .D\ .E\ [\ ] \tag{9}$$
$$C < D \wedge D < E \Rightarrow sort\ .E\ .D\ .C\ [\ ]\ .C\ .D\ .E\ [\ ] \tag{10}$$

and perhaps more besides.

The symbols $A$, $B$, and so on are all constants. Let us suppose that the algorithm begins by performing enough cases of *FunG* to invent rules with all these changed to variable: $A$ to $a$, $B$ to $b$, and so on. Thus, the form of example (3) which will be used henceforth is

$$a < b \Rightarrow sort\ .a\ .b\ [\ ]\ .a\ .b\ [\ ] \tag{3}$$

Any further use of *FunG* on these examples produces a rule which is contradicted by some counter-example. For instance, *FunG* with example (1) and any other example yields

$$sort\ x\ x \qquad \text{or} \qquad sort\ x\ y$$

without any conditions.

*FunG* with examples (3) and (5) yields

$$c < d \wedge y < z \Rightarrow sort\ .c\ .d\ x\ .c\ .d\ x$$

which is false if $x$ is not sorted.

*FunG* with examples (3) and (6) yields

$$c < d \wedge y < x \Rightarrow sort\ .c\ .x\ u\ .c\ .d\ v$$

which is false if $v$ is not sorted.

*FunG* with examples (5) and (6) yields

$$c < d \wedge d < e \wedge v < x \wedge w < y \Rightarrow sort\ .c\ .x\ .y\ [\ ]\ .c\ .d\ .e\ [\ ]$$

which is false unless $x$ and $y$ are $d$ and $e$.

Before proceeding, it is worth noting a suitable definition which we hope the algorithm might discover. One such is the set of rules:

$$sort\ [\ ] \qquad [\ ]$$
$$sort\ .a\ [\ ]\ .a\ [\ ]$$
$$c < d \wedge sort\ x\ .d\ y \Rightarrow sort\ .c\ x\ .c\ .d\ y$$
$$d < c \wedge sort\ x\ .d\ y \wedge sort\ .c\ y\ z \Rightarrow sort\ .c\ x\ .d\ z$$

The first two have already emerged easily, just by use of *FunG*. This algorithm will never find the other two, because it cannot construct the literal

*sort x .d y*

It can only form a literal beginning *sort x* ... by generalizing a pair

*sort* [ ] ...     and     *sort .u* ... ...

By induction on the number of generalizing steps, the first of these will have to be *sort* [ ] [ ] and the second must be *sort .u* ... *.v* ... so the generalization of the given pair will be either

*sort x x*     or     *sort x y*

A similar argument applies to the variable $z$ in the fourth rule. Hence, it seems that the best definition for *sort* which we might hope to find with this algorithm is

$$c < d \wedge sort \ .e \ x \ .d \ y \ \Rightarrow \ sort \ .c \ .e \ x \ .c \ .d \ y$$

and

$$d < c \wedge sort \ .e \ x \ .d \ y \wedge sort \ .c \ y \ .f \ z$$
$$\Rightarrow \ sort \ .c \ .e \ x \ .d \ .f \ z$$

These two are actually derivable, but only a trick which lets us introduce new variables via a tautological 'rule'. Here is a derivation of the simpler of them.
From example (3)

$$a < b \Rightarrow sort \ .a \ .b \ [ \ ] \ .a \ .b \ [ \ ]$$

and basic properties of equality, we can conclude that

$$a < b \wedge f = b \Rightarrow sort \ .a \ .f \ [ \ ] \ .a \ .b \ [ \ ]$$

Also, from example (2)

$$f = b \Rightarrow sort \ .f \ [ \ ] \ .b \ [ \ ]$$

From these two, *PredG* lets us infer that

$$a < b \wedge sort \ .f \ [ \ ] \quad .b \ [ \ ] \quad \Rightarrow sort \ .a \ .f \ [ \ ] \quad .a \ .b \ [ \ ]$$

*PredG* with examples (4) and (6), with $d < e$ for $X'$, yields

$$c < d \wedge sort \ .e \ .d \ [ \ ] \ .d \ .e \ [ \ ] \Rightarrow sort \ .c \ .e \ .d \ [ \ ] \ .c \ .d \ .e \ [ \ ]$$

*FunG* with these last two yields the desired recursive rule.

The second version of the algorithm includes the tricks which introduce new variables, as above. It goes:

$RS := \{(C, \{C\}) \mid C \in Exs\};$

REPEAT
WHILE    $RS$ contains two pairs $(S, I_S)$ and $(T, I_T)$
    AND    some operation $Op$, either $FunG$ or $PredG$, can be
           applied to $S$ and $T$

    $R := Op(S, T);$
    IF   $Op$ is $FunG$  THEN $I_R := I_S \cup I_T;$
    IF   $Op$ is $PredG$ THEN
        $I_R := \{X \in I_S \cup I_T \mid X$ satisfies both $S\vartheta$ and $T\vartheta$
             where $\vartheta$ is the substitution in $PredG\};$

    IF   $I_R$ is not empty
    AND   $Cxs$ does not contradict $R$
    THEN
        $RS := RS \cup \{(R, I_R)\};$

IF $RS$ does not yet contain a satisfactory definition of $p$
THEN
   Choose some pair $(S, I)$ from $RS$;
   Choose a variable $a$ which occurs in $P$ more than once
       where $S$ is $X \Rightarrow P$;
   Choose a set of occurrences of $a$ in $P$;
   $RS := RS \cup \{(S', I)\}$
       where  $S'$ is $X \wedge b = a \Rightarrow P'$
          and  $b$ is a new variable
          and  $P'$ is $P$ with the chosen occurrences of $a$ replaced by $b$;

  UNTIL  $RS$ contains a satisfactory definition of $p$
     OR   all possibilities in $RS$ have been chosen.


### 6.5.7   OTHER ASPECTS OF THE ALGORITHM

Note that the new operation which introduces a new variable is *not* equivalent to the process of changing representation. The second form of the algorithm still cannot derive the definition of *bt* from the training data which involves the function *leaff*.

There is a third form of the algorithm which can introduce a new variable into a condition, $X$. I do not know if this will help.

There is a fourth form of the algorithm which *can* invent *bt* from these data. It chooses an arbitrary term $t$ in the conclusion, and a new variable $x$, and replaces $t$ in certain positions in the conclusion by $x$, and adds the new condition

$x = t.$

The second form of the algorithm is the restricted case in which $t$ must be a variable, $a$.

As described, all forms of the algorithm involve non-deterministic choices. In an implementation on a sequential computer, these choices should be made by some deterministic procedure which is probably best encoded as a set of heuristic rules. Thus, the algorithm is really a heuristic search procedure whose goal is a satisfactory set of rules. The derivation of one of the recursive rules for *search*, in the example above, was actually discovered by manual goal-directed search. The algorithm, by contrast, must inevitably be a forward searcher and its quality will depend critically on its heuristic component. It may be possible to discover good general heuristic conditions for it by the method of **rule ordering**, described in Chapter 7. The process of inductive inference itself may be a fertile field for the application of other learning algorithms.

The various forms of the algorithm have different operators, so they explore distinct search spaces. The examples considered above suggest that these search spaces are substantially different. Since all the operators are simple syntactic processes which preserve validity, it should be possible to develop a rigorous theory of these search spaces and how they interrelate.

## 6.6   PROPERTIES OF SYNTACTIC GENERALIZATION

Some of the algorithms presented here may involve generalization subject to laws. We have already seen how this affects Plotkin's method.

Vere's technique can be adapted easily so that it incorporates laws. Any law of the form $X = Y$ can itself be treated as background information. The method can add an instance of the law itself as a condition.

The process of synthesizing programs depends on any basic programs which may be available, such as addition, '+', and *tail*. Both of these functions are subject to laws. Addition is associative and (in FP notation) *tail* obeys the law

$$1 \circ tail = 2$$

The method of Zhu and Jin depends both on these laws and on heuristic knowledge about how to apply them.

***Strengths :***

- These methods are practical ways to extract general forms from specific data.

- They require only small training sets.

- Their outputs are accessible.

- Many of these methods do not need counter-examples.

- The algorithms of Vere and Plotkin do not involve search.

- Plotkin's algorithm supplies a unique 'right' answer. Vere's method may produce more than one answer, but only when its data suggest several answers.

- The two methods which invent recursion involve choices, so they may not yield unique 'right' answers, but there is often an answer which is clearly simplest or easiest to find. The goal conditions on the search for definitions are somewhat arbitrary, but this is in the nature of the task.

- In most of these algorithms, the order of the examples does not matter. It does matter for program synthesis, but an error in the order can usually be noticed and corrected.

- The first three methods stop. The last can be stopped at any point and it is usually easy to tell when to stop it.

- Plotkin's algorithms have good underlying theories. The others can be justified by simple intuition or by arguments invoking complexity.

- The method for inventing recursive definitions is sensitive to the description language. However, there seem to be systematic ways to adapt the description language so that it will usually work.

*Limitations :*

- The data must be clean.

- Learning is not incremental.

- The generalization produced is purely syntactic. It may not yield the answers that a person expects.

- The methods which can invent recursion involve heuristic search.

- Program synthesis and more advanced forms of Plotkin's method may involve searching in spaces subject to laws, which is hard.

- Program synthesis depends on the basic functions known to the synthesis system. If the system does not know of the right basic functions, then it cannot construct the program.

- There are several versions of the method which invents recursive definitions. For any particular task, one has to decide which version to use.


## FURTHER READING

Vere (1977) reports several experiments concerned with synthesizing rules. Other people have used similar approaches, among them van Someren (1984) and the authors of the method called focusing which will be discussed in Chapter 7.

The original form of the first generalization algorithm was published by Plotkin (1969). This reference contains detailed proofs of all the properties we have mentioned. The technique for obtaining a reduced least generalization of two clauses is discussed by Gottlob and Fermüller (1991). Their paper is referred to in OGAI (1992).

Kodratoff, Ganascia, Clavieras, Bollinger, and Tecuci (1984) recognized the significance of the facts that Peter isn't Jane, and that Macbeth is not likely to kill himself.

The full unification algorithm was first expounded by Robinson (1965). Luckham (1967) made a careful analysis of it.

Unification subject to laws is subtle. If you care to follow this topic into its depth, take a look at Siekmann (1986, 1989). An equivalent approach is described by Williams (1991). Generalizing subject to laws is a younger subject. An introduction to it appears in Baader (1990), which exhibits the connection between generalization and unification.

The method of program synthesis described here was published by Zhu and Jin (1991). The original reference for the FP programming language is Backus (1978). Some of the ideas embodied in other methods appear in Biermann and Guiho (1983). The major stimulus for logic program synthesis was Shapiro (1983). This has been extended by, for instance, Huntbach (1986) and Muggleton and Buntine (1988a, b). Muggleton (1992) gives a survey. Muggleton and De Raedt (1994) make a thorough analysis. For an application see Sternberg *et al.* (1994). One algorithm for learning finite state automata appears in Porat and Feldman (1991). Inverse implication has been investigated by Aha *et al.* (1994).

# EXERCISES

1. Write Vere's algorithm in the form of pseudocode. List its inputs and outputs.

2. Say

$$\vartheta = [\text{dinner\_of } (y)/x, z/y]$$
$$\varphi = [\text{cat}/y, \text{sardines}/z].$$

   Write down the expressions

   $$(\text{warm } (y) \Rightarrow \text{likes}(x, y)) \; \vartheta$$
   $$(\text{warm } (y) \Rightarrow \text{likes}(x, y)) \; \varphi$$
   $$(\text{warm } (y) \Rightarrow \text{likes}(x, y)) \; \vartheta\varphi.$$

   What is the product substitution $\vartheta\varphi$? Is it the same as $\varphi\vartheta$?

3. Which of the following pairs of terms can be unified? For those pairs which can be unified, write down a most general unifier. For those which cannot, where does Robinson's unification algorithm fail?

   dinner_from (fish, mother's $(x)$, $y$)
   dinner_from $(u, v, w)$

   dinner_from (fish, mother's $(x)$, cauliflower)
   dinner_from $(x, y, z)$

   grows_in $(x, \text{peat}, \text{sunshine})$
   grows_in (carrots, peat, $y$)

   $x$
   grows_in (potatoes, loam, $y$)

   grows_in $(x, \text{peat}, \text{sunshine})$
   dinner_from $(u, v, w)$.

4.  When it is autumn in the northern hemisphere, oak and beech trees shed their leaves. When it is autumn in the southern hemisphere, *Nothofagus antarctica* trees shed their leaves. Invent a first-order language in which these assertions can be expressed and generalize from them.

5.  Edward is good at chess. Rachel is good at chess and tennis. Fiona is good at tennis. Edward and Rachel like playing chess together and Rachel and Fiona like playing tennis together. What conclusion can be drawn by syntactic generalization? (You can assume that each of these participants likes doing the things that she/he is good at, just because she/he is good at them.)

6.  Keith attends Munty College, where the system of discipline is strict. When Keith plays football on Saturdays, out of school hours, he is rowdy. James attends the Parten School, where discipline is also strict. He is rowdy at the boys' club on Thursday evenings after school. Invent names for the relations between these various participants and attributes. Construct association chains which suggest conditions for a boy to be rowdy. What can you infer from the resulting clauses by syntactic generalization?

7.  A training set for program synthesis consists of the following input/output pairs:

| *input* | *output* |
| --- | --- |
| $\langle x \rangle$ | $x$ |
| $\langle x, y \rangle$ | $x + y$ |
| $\langle x, y, z \rangle$ | $x + y + z$ |

where $x$, $y$, and $z$ are distinct variables. Show how the synthesis algorithm can generate an FP program from this training set. Indicate any simplifications in program fragments which may help the algorithm.

Note: this question is ambiguous, because the sum $x + y + z$ might be $x + (y + z)$ or $(x + y) + z$. It is much easier one way than the other. Which way makes the question easy? Assume that the sum associates so that the question is easy.

8.  Show how the last rule for *sort* can be found by the inference algorithm for recursive rules. *Hint*: Show how to infer

$$d < c \wedge sort\ .e\ [\ ]\ .d\ [\ ]\ \wedge sort\ .c\ [\ ]\ .f\ [\ ]$$
$$\Rightarrow sort\ .c\ .e\ [\ ]\ .d\ .f\ [\ ]$$

and

$$d < c \wedge sort\ .e\ .g\ [\ ]\ .d\ .h\ [\ ]\ \wedge sort\ .c\ .h\ [\ ]\ .f\ .i\ [\ ]$$
$$\Rightarrow sort\ .c\ .e\ .g\ [\ ]\ .d\ .f\ .i\ [\ ]$$

# 7 Learning in rule-based systems

## SUMMARY

Rule-based systems allow the following kinds of learning:

- inventing new conditions;

- inventing new actions;

- inventing new conflict resolution methods;

- discovering (and correcting) errors in the old system.

New conjunctive conditions may be learned by

- inductive methods: focusing; version spaces;

- deductive methods: EBG; lemma generation; chunking.

New actions are formed by composing simpler ones: generalizing plans; macros.
The best way to improve conflict resolution is by rule ordering which can also invent an equivalent of disjunctive conditions.
Flaws in an existing rule base can be found by

- an ideal trace;

- contradiction backtracing.

## INTRODUCTION

The last search method described in Chapter 2 was a form of hill climbing with heuristic rules. It searches for a goal, which is a particular state. It moves from state to state and the rules are what it uses when it chooses where it should move to. We shall now study the kinds of learning associated with rule-based search.

Many researchers would have you believe that any program which is not rule-based is not really anything to do with artificial intelligence. Their view has become less dominant since the rise of neural networks, but it is still influential. Rule-based systems are particularly interesting because:

- very many difficult and intriguing problems can be expressed naturally in terms of searching;

- searching algorithms are well understood. It is possible to compare them and prove theorems about them, independently of any particular set of rules;

- such systems can be given helpful user interfaces. For example, they can 'explain' how they derive results and why they choose particular search paths;

- most of the knowledge within a rule-based system is held explicitly in the rules, which are accessible;

- each rule is in principle independent of all others; as a consequence,

- a user can study the effect of a single rule on the search process;

- each rule embodies just a small fragment of the system's knowledge. With luck, a change in one rule will only produce a small or local change in the system's overall performance; so

- learning is relatively easy, as a learner can add or change just one rule at a time.

Unfortunately, not all these advantages apply to all rule-based systems.

- The search algorithm is *not* encoded in the rules. Hence, it is not accessible. (There are a few search programs which use their rules to simulate other search programs. When this is done properly, the algorithm of the simulated program becomes accessible.)

- In practice, it is rare for one rule to be independent of all others. There is a phenomenon in rule bases analogous to the classic problem of old-fashioned spaghetti logic in programs designed by flow charts. When that occurs, all the advantages of accessibility and local effects are lost.

   Conditions in a rule are of different sorts.

1.  If a *necessary* condition is not satisfied, then the rule's action may not be applicable at all to the current state. The problem solver may crash if it applies this action.

2.  A *heuristic* condition suggests when the action may help. However, a heuristic condition may be fallible and mislead the search.

3.  A *sufficient set* of conditions is of course highly desirable; but if you can find one, then your problem is not really a matter of searching at all.

In principle, the various methods for learning conditions will produce conditions of different sorts. However, all such methods depend on training data and assumptions and the data may be misleading and the assumptions may be wrong; so we can almost never rely totally on the outcome of learning. All learned rules are best regarded as heuristic.


## 7.1   WHAT CAN BE LEARNED WITHIN A RULE BASE

Recall, the rule-based searcher of Chapter 2 takes for inputs:

- a *Start state*;

- *Goal_ conds*;

- a knowledge base of *Rules*.

It produces:

- an *Outcome*, which may be a goal state or the constant '*fail*';

- a *Trace*.

*Goal_conds* define the goals. Any state which satisfies them is a goal.

*Rules* are the IF…THEN… rules which the searcher can call on while searching.
Each has a condition and an action and maybe a name and other details.

Strictly speaking, the *Trace* should be a list of the names of the rules applied
along a successful search path together with details of how each rule was applied.
A rule usually has parameters and the trace should include the values of these
parameters at each search step. I shall oversimplify the description and only
describe traces as if they hold names of rules.

Within its loop, the searcher has variables including a current state called $N$, a
conflict set $CS$, a best rule $R$, and this rule's operator $Op$. Its code is

```
N := start
TRACE := [ ]

WHILE N is not a goal and N ≠ fail
    CS :=conflict-set(N, KB)

    IF CS = { }
    THEN
      N := fail
    ELSE
      R := best-rule (CS, Goal-conditions, N)
      {say R has  action  Op
              name  Name}
      add Name to the back of TRACE
      N := do-action (Op,N)
```

It has three major subfunctions:

- *conflict-set*

- *best-rule*

- *do-action*

Learning within such a searcher can be divided into three parts, which correspond
to these three parts of the searcher. One can learn new conditions, conflict
resolution methods, and actions. However, there is more to learning in a rule-based
system than this simple study suggests.

There are two basic approaches to learning conditions: **deductive** and **inductive**.
Deductive methods assume a substantial reliable amount of knowledge and
extrapolate from it. The assumed knowledge has to include generalizations, such as

rules with variables in them. The outcome is typically an extra rule which simplifies search, although it does not contain any essential new information. The deductive methods explained below are called EBG, lemma generation, and chunking.

Inductive learning usually starts from pragmatic knowledge of specific facts and generalizes to form rules containing variables. Some inductive learning must precede deduction. Plotkin's algorithms are classic forms of induction and they will be invoked in two of the methods described below.

All these methods are good for inventing new knowledge. Inevitably, they sometimes make mistakes, so it is also necessary to trace faults in existing knowledge.

A **factual fault** in the rule base manifests itself when the conflict set contains an error. It occurs when some rule has conditions which do not suit its action. Either the conditions are too restrictive, so the searcher does not consider this action when it should; or the conditions are too lax, so the searcher considers the action when it shouldn't. A factual fault is curable by learning better conditions.

A **control fault** occurs when the conflict set contains the best rule, but *best rule* misses it and selects a worse one. Such faults can be cured either by improving the mechanism of *best rule*, for example by learning better weights for the rules, or by improving the choice of conflict set so that it does not contain the worse rule.

Thus, learning better conditions can cure both control faults and factual faults. Learning an improved conflict resolution method only cures control faults. When one learns a new action, of course one then has to learn conditions for it, and also one must learn how conflict resolution works with it.


## 7.2   TRACING A FAULT

You can try dreaming up new rules off the top of your head. This method works quite well to begin with, but after a while your collection of nearly random rules will become incoherent. At that point, it pays to be more systematic.

If one is going to improve a rule base, one must first find a flaw in it. There is a standard approach. The second output *Trace* of the *search* algorithm above is the **rule trace** of an attempt at search. It is a list of names of all the rules applied, in order:

[$Name_1$, $Name_2$, $Name_3$, ...].

Since the searcher is fallible, there may be a better selection of rules (Fig.7.1). Let us suppose that a perfect searcher would follow some sequence of rules whose names are

[$Iname_1$, $Iname_2$, $Iname_3$, ...].

This sequence is called the **ideal trace**. As far as the rule trace and the ideal trace agree, the searcher worked well. Where they first differ, there is a flaw.

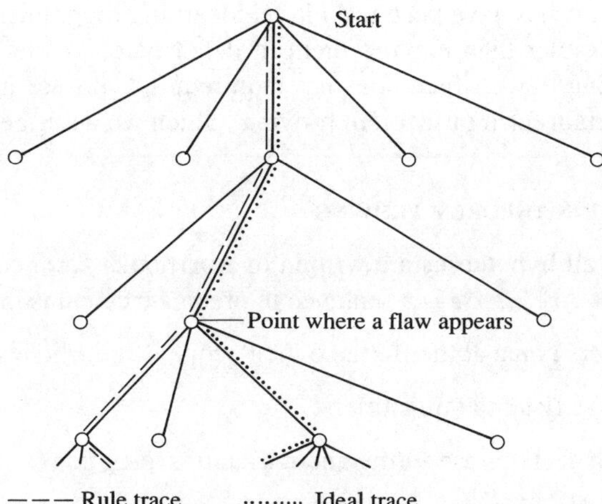



**Fig. 7.1** Finding a fault with an ideal trace. Dashed line: rule trace. Dotted line: ideal trace.

A common technique for tracing flaws runs thus: attempt to solve a problem, with the given rule base. If the solution is not perfect, find a better solution, and compare the traces of the two solutions. The trace of the better solution is used as the ideal trace.

This method does not reveal what kind of flaw occurred. To do that, you examine the searcher's conflict set at the point of the flaw.

1.  If the rule in the ideal trace is in the conflict set, then there was a control fault.

2.  If it is not, but the rule base contains a rule with the right action, then there was a factual fault.

3.  Otherwise, the appropriate action must be learned.

The only way to find a better solution to a problem is to have a better problem solver. You can use whatever solvers are available.

One of the best solvers is you yourself: if the rule-based solver fails, then tell it where you think it went wrong and what it should have done instead. You do not have to provide a complete ideal trace. The flawed solver can help you by displaying relevant parts of its calculation.

In particular cases, there may be some algorithm which solves the problem without search. It may seem that using such an algorithm is cheating. If it exists, then why bother with a searching solver? The answer is, the algorithm may only work in special cases. The learning rule-based solver may be more powerful when it has finished learning.

When the problem is a case of planning, another common technique is to solve it by broadening the search. A full breadth-first search will always find a shortest solution, if the problem can be solved at all. When a full search is not feasible, then some more limited search may still find a better trace.

All these approaches have been used in real learning programs. One such will be studied below. Rather than explore them in detail here, we shall discuss another method for finding flaws which does not require an ideal trace at all. The method depends on a certain form of theorem proving, which we describe first.

### 7.2.1  RESOLUTION THEOREM PROVING

We assume that all hypotheses are written in a particular form called (disjunctive) clauses. A *disjunctive clause* is a sentence in predicate calculus in which:

(1)   all quantifiers come at the front, so their scope is the whole sentence;

(2)   there are no existential quantifiers;

(3)   the only connectives are $\vee$ (disjunction) and $\neg$ (negation);

(4)   negations only apply to atomic formulae, never to disjunctions.

If in addition

(5)   in the sentence, just one literal is not negated

then the sentence is called a **Horn clause**. Thus, a Horn clause looks like

$$\forall x \, \forall y \, ... \, \forall z \, (\neg P \vee \neg Q \vee ... \neg R \vee S)$$

where $P, Q, ... R$, and $S$ are atomic formulas and all variables in them are among $x$, $y, ... z$. The same idea can also be written

$$\forall x \, \forall y \, ... \, \forall z \, (P \wedge Q \wedge ... R \Rightarrow S)$$

or as an IF ... THEN ... rule:

IF $P$ AND $Q$ AND ... $R$
THEN conclude $S$.

Note that there may be no negated atomic formulas in the clause. In that case, it simply asserts an unconditional fact:

$$\forall x \, \forall y \, ... \, \forall z \, S.$$

Since all variables are universally quantified, there is no particular point in saying so for each clause. Hence the quantifiers are usually left off.

Suppose we are given two disjunctive clauses

$$\neg P \vee \neg Q \vee ... R \vee S$$

and

$$\neg S \vee \neg T \vee ... U.$$

From these, we can deduce

$$\neg P \vee \neg Q \vee ... R \vee \neg T \vee ... U.$$

When $S$ is true, this follows from the second hypothesis. When $S$ is false, it follows from the first. This is the basis of resolution. $S$ is called the **resolvent**. Sometimes we say that the two input clauses are **resolved along** $S$.

The full resolution algorithm covers a little more than this. Suppose we are given two hypotheses

$$\neg P \vee \neg Q \vee \dots R \vee S_1$$

and

$$\neg S_2 \vee \neg T \vee \dots U$$

with no variables in common and also suppose that $S_1$ and $S_2$ can be unified by some substitution $\vartheta$. Then from these two hypotheses, we can deduce

$$\neg P\vartheta \vee \neg Q\vartheta \vee \dots R\vartheta \vee \neg T\vartheta \vee \dots U\vartheta$$

That is resolution. One can show that any set of hypotheses in FOPC can be recast as a set of disjunctive clauses and anything which can be proved at all in first-order logic can be proved by resolution.

The next obvious question is, how should one set about using it? Given some goal $G$, what is the best strategy for finding a proof of $G$? What we would like is a sequence of clauses which are axioms or hypotheses from the given theory and a sequence of ways of resolving them so that we end up with $G$. If we start as this description suggests, choosing any sequence of axioms, then we may end up proving any random theorem. The chance of discovering a proof of the particular goal $G$ is negligible. What is wanted is a goal directed search strategy for the proof of $G$.

The usual approach is a rather contorted method called **proof by failure**. Rather than prove $G$ directly, we show that the negation of our goal, $\neg G$, is not consistent with our theory. The basic idea is that if $G$ itself is an axiom or hypothesis, then the result of resolving

$$G \quad \text{with} \quad \neg G$$

is the 'empty clause', which is *false*. The reasoning behind this may make more sense if you imagine that the two clauses being resolved are not $G$ and $\neg G$ but their equivalents:

$$\textit{false} \vee G \quad \text{and} \quad \neg G.$$

The result of resolving these along $G$ is clearly *false*.

More generally, if ever we can arrive at the 'empty clause' by resolution, then one of the hypotheses used on the way must be false. The usual procedure for proving $G$ runs:

$$C := \neg G$$

WHILE $C \neq \{\ \}$

    Choose any axiom or hypothesis $K$ which can be resolved with $C$.

    $C :=$ the result of resolving $C$ with $K$

If ever this loop ends, then $\neg G$ is inconsistent with the given theory; and then, working backwards up the list of hypotheses $K$, you can construct a (forward) proof of $G$.

This is how Prolog works. Each Prolog clause

$$S :\text{-} P, Q, \ldots R.$$

is a Horn clause,

$$\neg P \vee \neg Q \vee \ldots \neg R \vee S$$

Prolog permits non-Horn clauses too, since one of the subgoals in its body may be negated:

$$S :\text{-} \text{not } P, Q, \ldots .$$

is equivalent to

$$P \vee \neg Q \vee \ldots S$$

except that in Prolog, the resolvent has to be $S$. It cannot be $P$.

The art of resolution theorem proving lies in the sequence of choices of axioms $K$. Prolog makes its choices by simply looking for the first clause in its database which can be resolved with the last disjunct in the current clause.

Resolution theorem proving may not appear to resemble the prototype rule-based searcher, but they are not really so different. In a theorem prover:

- *Start* is $\neg G$;

- $N$ is the current clause, $C$;

- $CS$ is the set of axioms and hypotheses which could be resolved with $C$;

- $R$ is $K$;

- *do-action* is the act of resolving $K$ with $C$;

- the goal of search is the 'empty clause'.


### 7.2.2 CONTRADICTION BACKTRACING

This method was invented by Ehud Shapiro. It is capable of finding factual faults. It works when:

(1)   the flaw is a factual fault;

(2)   the problem concerns theorem proving by resolution;

(3)   the problem solver derives a fallacious clause;

(4)   there is a known false ground instance of this derived clause;

(5)   there is an oracle available.

Remember, a ground clause is a sentence with no variables in it.

An **oracle** is something which will always provide a correct answer to any specific question. You cannot ask it generalities, such as 'If I drop something heavy on my toe, will it hurt?' because the word 'something' is not a specific object. Instead, you have to ask about particular cases: 'If I drop this bag of tomatoes on my toe …?' The oracle can only tell you whether any ground clause is true or false.

We shall demonstrate by an example. Our rules will be written in disjunctive form, with connectives $\lor$ and $\neg$; but if you prefer, you can rewrite them with the connectives $\land$ and $\Rightarrow$. Suppose the rule base of a theorem prover contains the following assumptions:

colour (England, green)
$\neg$part_of $(X,Y) \lor \neg$colour $(Y, C) \lor$ colour $(X, C)$
part_of (field, England).

All variables are taken to be universally quantified. Thus, in the second assumption, we take it that

if $Y$ is *anything*, with *any* colour $C$,
and $X$ is *any* part of $Y$,
then $X$ also has colour $C$.

From these, we can deduce that

colour (field, green).

The proof proceeds as in Table 7.1.

**Table 7.1**

| Hypotheses | Deduced results |
| --- | --- |
| colour (England, green) | |
| $\neg$part_of $(X,Y)$<br>$\lor \neg$colour $(Y,C)$<br>$\lor$ colour $(X,C)$ | |
| | $\neg$part_of $(X,$ England$) \lor$ colour $(X,$ green$)$ |
| part_of (field, England) | |
| | colour (field, green) |

At this point, the oracle reveals that the particular field in question has just been ploughed and is brown (See Fig.7.2). Contradiction backtracing then proceeds.

- Since *colour (field, green)* is false, one of the two assertions from which it was deduced must be false.

- We ask the oracle: is *part_of (field, England)* true? This is a permissible question, because it contains no variables. The oracle replies, yes. Hence,

    $\neg$part_of $(X,$ England$) \lor$ colour $(X,$ green$)$

    must be false.

**Fig. 7.2**  A partly ploughed field—half green, half brown.

Note that all variables are universally quantified, so the assertion which we now know to be false is

if *X* is *any* part of England
then *X* is green.

However, we know more than this. We can conclude that this assertion becomes false in the particular case when *X* is this particular field. One of the two assertions from which this was deduced must be false. The first in the rule base

colour (England, green)

is already a ground clause, so we can ask the oracle directly if it is true. Suppose the oracle says yes; then we can deduce that the fallacious assumption is

$\neg$part_of $(X,Y) \vee \neg$colour $(Y,C) \vee$ colour $(X,C)$.

The contradiction backtracing algorithm works in a cycle. It keeps two working objects. One of them is the 'current clause' *CC*. The other is a substitution, $\vartheta$. At each stage, the clause *CC* is false and $CC\vartheta$ is a false ground instance of it.

To begin with, *CC* is the fallacious result of the proof. In each cycle, *CC* is replaced by one of the two clauses it was derived from and $\vartheta$ gains more values for variables. The algorithm stops when *CC* is a rule in the database.

The algorithm is:

Run the theorem prover and obtain a fallacy.
*CC* := the false conclusion of the proof
Since all variables in *CC* are universally quantified, there is a
false ground instance $CC\vartheta$ of *CC*. Say
$\vartheta$ is any most general such substitution.

WHILE the current clause $CC$ is not a rule
    $CC$ was deduced by resolution from two clauses:
        $K$ containing $S_1$ and $L$ containing $\neg S_2$. Suppose
        $\sigma$ is the most general unifier of $S_1$ and $S_2$.
        Then $CC$ is the result of resolving $K\sigma$ with $L\sigma$
        and $CC\vartheta$ is the result of resolving $K\sigma\vartheta$ with $L\sigma\vartheta$.

*1:  Choose ground values for all variables in $K\sigma\vartheta$ and $L\sigma\vartheta$ in any way you like.
     Say $\tau$ is the substitution which replaces these variables with these values.
     $S' = S_1\sigma\vartheta\tau = S_2\sigma\vartheta\tau$

     Ask the oracle if $S'$ is true.
*2:  IF $S'$ is true
     THEN
        $CC := L$
     ELSE
        $CC := K$

     $\vartheta := \sigma\vartheta\tau$

At this point, $CC$ is a fallacious rule in the database. The substitution $\vartheta$ provides a false ground instance of it.


Notes:

1.    There may be more than one flawed rule in the rule base.
    Different choices of the substitution $\tau$ may cause the algorithm to branch different ways at the next step and so may lead to different flawed rules.
    Since $CC\vartheta$ is ground, any free variables in $K\sigma\vartheta$ and $L\sigma\vartheta$ can only occur in $S_1\,\sigma\vartheta$ and $\neg S_2\,\sigma\vartheta$.

2.    Recall, $CC$ is a false clause and $CC\vartheta$ is a false ground instance of it. $CC\vartheta$ was formed by resolving $K\sigma\vartheta$ with $L\sigma\vartheta$. Since $CC\vartheta$ is ground, it is also the result of resolving $K\sigma\vartheta\tau$ with $L\sigma\vartheta\tau$, so one of these is false.
        If $S'$ is true, then clause $K\sigma\vartheta\tau$ is true, so $L\sigma\vartheta\tau$ must be false. Since all variables in $L$ are universally quantified, $L$ is false.
        If $S'$ is false then clause $L\sigma\vartheta\tau$ is true, so $K\vartheta\sigma\tau$ must be false, so $K$ is false.
    In any event, the new $CC$ is false.


    Brazdil has invented an interesting approach to detecting faults in a learner's knowledge base when the knowledge base itself is not accessible. His method lets a teacher analyse a pupil's misconceptions. It combines features of contradiction backtracing with explanation-based generalization which will be discussed below.

## 7.3    INDUCTIVE METHODS

These methods are good for learning conditions of rules. They are also used for learning 'concepts'. In the world of artificial intelligence, a concept is often treated as if it were a formal definition or predicate. In fact, the human notion of a concept often behaves quite unlike a predicate, but let that pass.

Induction is useful when there is no underlying theory for the actions. That is, the learner learns from similarities, without considering why the similarities occur. Hence, methods in this class are often called **similarity-based**. The algorithm for learning the notion of an arch, sketched in Chapter 1, is based on similarities.

For most of these methods to work, the domain of learning has to satisfy certain quite stringent assumptions. At the outset, it is assumed that there are certain basic predicates which can be used to test a state. These basic conditions are given in advance to the learner.

There is nothing very unusual about this assumption. The predicates are the essential part of the language with which the task is represented. A set of such predicates, and any known relations between them, is sometimes called a **description space**. Differences between similarity-based methods usually arise through different forms of description space.

Similarity-based methods also usually assume that there is some ideal set of rules which would be adequate for the problem solver. There may be more than one. This assumption amounts to saying that there is something which can be learned. If no such set of rules could exist, then the learner would be on a wild goose chase.

The third common assumption is that the training set is clean. If it were noisy, then any similarity between examples might be pure luck or perhaps even misleading. In that case, the learning has to employ some statistical test, so that noisy cases can be overridden. Some sort of correlation method, such as building decision trees, would be more suitable.

### 7.3.1    FOCUSING

This is one such similarity-based method. It is characterized by a detailed description space, which we shall describe below.

Focusing demands some extra assumptions, in addition to the ones above. The first is that the training set contains counter-examples.

The virtue of these is that with them, the learner can often decide when learning is complete. Counter-examples narrow down the field of possible answers. The learner may demonstrate that the right answer must be a generalizaton of some predicate. Then it may also show that any strict generalization cannot be right, from the counter-examples. Hence, the predicate it has found is the answer.

Another restriction required by focusing is that the basic predicates can be partitioned into independent groups. Within each group, the predicates are mutually exclusive and cover all possibilities.

*Example: basic predicates for language acquisition*

Suppose that the objective of the system is to compose English sentences. Its input is some representation of a sentence's meaning and a dictionary. It has to learn rules of syntax. One possible action is to prefix a noun or noun phrase $X$ with the indefinite article 'a' (or 'an'). The possible basic predicates which might apply to $X$ are

*Group 1*

    definite $(X)$: $X$ is a definite description, such as a person's name.

    indefinite $(X)$: $X$ is some other sort of description.

*Group 2*

    singular $(X)$      plural $(X)$

*Group 3*

agent $(X,Y)$:     $X$ is the agent of the clause $Y$.

object $(X,Y)$:     The relation of the clause $Y$ is transitive and $X$ is its object.

*Group 4*

    past $(Y)$      present$(Y)$      future$(Y)$

The third group, *agent* and *object*, take a second argument $Y$ which is the clause containing $X$. The last group, *past/present/future*, tests $Y$ rather than $X$. Thus, in a sentence such as 'A dog runs', '$X$ is 'dog' and $Y$ is the whole sentence. The system's aim is to learn the right conditions in a rule

    IF *Conditions*

    THEN prefix $(X,$ 'a'$)$

Let us consider the group *past/present/future*. These can be combined in disjunctions as in Fig. 7.3. The top disjunction is invariably true. The bottom ones are the mutually exclusive basics. Lower conditions are more restrictive and imply the ones above. The lines drawn between possible conditions depict implication. There are no cycles, so the diagram is a partial order.

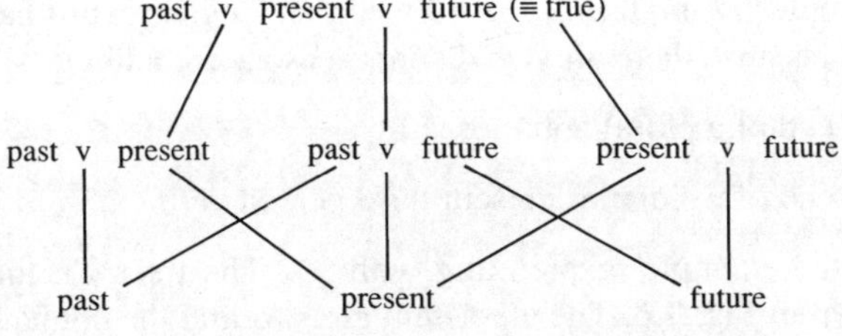


**Fig. 7.3** An upper semi-lattice.

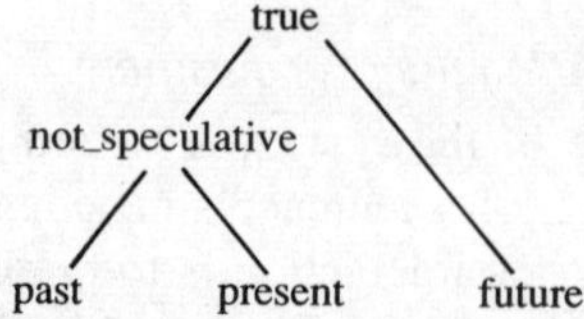

Fig. 7.4   An simpler upper semi-lattice.

There are other similar diagrams which you can construct by omitting some disjunctions from the full version. One such is shown in Fig. 7.4 where *not_speculative* means (*past* ∨ *present*). Another trivial, one is Fig. 7.5.

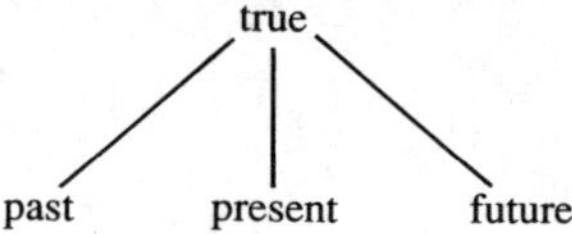

Fig. 7.5   An even simpler upper semi-lattice.

The description space consists of diagrams like these, one for each group of predicates. The essential point about the diagram is that in it, any two conditions $P$ and $Q$ have a **least upper bound**. This is some condition $R$ in the diagram which is above them both and any other condition which is above $P$ and $Q$ is also above $R$. Such a diagram is sometimes called an **upper semi-lattice**. Any rooted tree is automatically an upper semi-lattice and the diagrams of conditions given to a focusing learner are often trees.

Before starting, the learner is given one such upper semi-lattice of conditions for each group. The focusing algorithm also assumes the following:

- Within the ideal rules which it is trying to learn, each action can only be the action of a single rule.

- The condition of a rule is a conjunction, with just one conjunct from each diagram. That is to say, the rule must be of the form

    $$P \wedge Q \wedge \ldots R \Rightarrow \text{Action}$$

where $P$, $Q$, … and $R$ are from different diagrams. (Of course, any conjunct which is *true* can be left out.)

Thus, focusing could *not* be used to learn when an English word has an extra 's' added at the end, because there are *two* distinct reasons for adding 's':

- add 's' at the end of a plural noun; or

- add 's' at the end of a singular present third person verb.

Let us continue the example of prefixing with 'a'. The trees for the four groups might be as shown in Fig. 7.6. The algorithm aims to find the conjuncts in a rule's condition. In each diagram, it keeps two pointers called *lower*, L and *upper*, U. L points to the most restrictive node which could be this diagram's conjunct. The

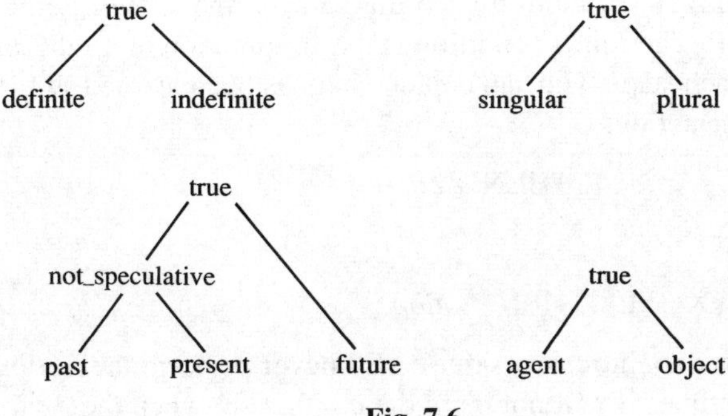

Fig. 7.6

right conjunct is a generalization of all the basic predicates at or below L. U marks the least restrictive node which could be the conjunct. All the basic properties which are not at or below U prohibit the rule's action.

The algorithm is:

Find a positive training instance, *I*, where the rule's action is applied correctly;

FOR each diagram
    find which basic predicate *B* is true of *I*.
    (There will be just one.)
    In the diagram, set *L* to point to *B* (a leaf);

FOR each diagram
    set *U* to the top node (*true*);

WHILE there is a diagram in which $L \neq U$
    *I* := the next training instance;

    IF *I* is a positive instance
    THEN
        FOR each diagram
            *B* := the leaf which is true of *I*;
            *L* := the least upper bound of *B* and the previous value of *L*
    ELSE
        (*I* is a negative instance)
    IF there is a just one diagram in which *B* is below *U* but not below *L*
    THEN in this diagram
        *U* := the least upper bound of all nodes which are
               (1)   at or below the previous value of *U*, and
               (2)   at or above *L*, and
               (3)   not above *B*
    ELSE
        (*I* is a *far miss*: see below).

When the algorithm stops, then in each diagram, $L$ and $U$ will point to the same node, a predicate $P$. The rule's condition is the conjunction of all these $P$s.

If you like, at each stage, you can depict what has been learned so far by a pair of rules with the same action:

IF $U_1 \wedge U_2 \wedge U_3 \wedge \ldots U_k$ THEN *Action*

and

IF $L_1 \wedge L_2 \wedge L_3 \wedge \ldots L_k$ THEN *Action*

The rule with $L$s for conditions is safe. It will never perform the *Action* unless it is justified, but sometimes it will not be in the conflict set when the *Action* is correct. The other version, with $U$s for conditions, may be over-adventurous. It is always in the conflict set when its action is appropriate and maybe at other times too. You can actually run a problem solver with these half-learned rules, so focusing is an incremental learning method.

Let us see how it works. Suppose the training set presented to the focusing learner consists of the sentences (see Fig.7.7)

a **dog** chases balls;

* **dogs** chase balls;

John chased a **dog**;

dogs will chase a **ball**;

* **John** will chase dogs.

In each case, the noun $X$ which might be preceded by 'a' is in bold. The counter-examples have asterisks in front of them.

Suppose that the algorithm examines the examples in this order. The first is a positive example, so after studying it, the algorithm will place lower markers $L$ on the nodes

**Fig. 7.7**   Examples of chasing.

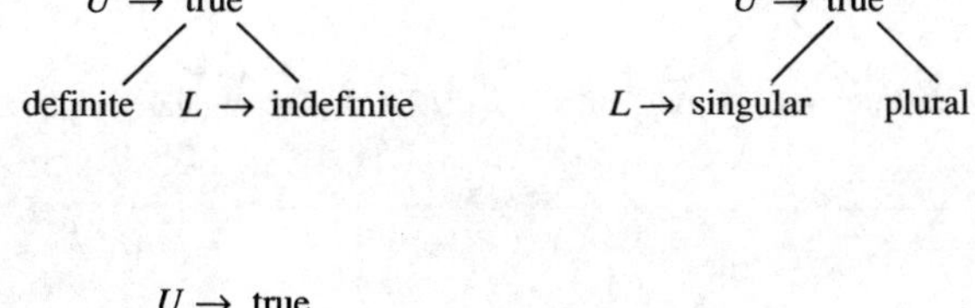

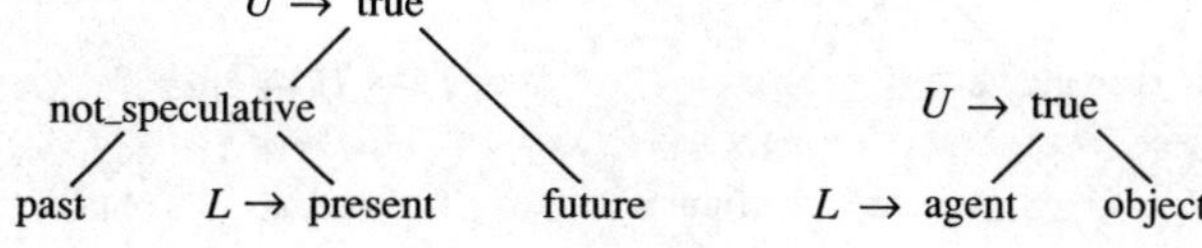

**Fig. 7.8**   The initial positions of pointers while focusing.

indefinite
singular
present
agent.

Its upper markers $U$ will all be at the roots of the trees as in Fig.7.8.
In the second example, the values of the basic predicates B are

indefinite
plural
present
agent.

This is a counter-example. There is only one tree in which B is not below $L$, the *singular/plural* tree, so in this tree we lower $U$. Now $L$ and $U$ coincide in this tree. They both point to *singular*, so the algorithm has learned that *singular*(*X*) is a conjunct in the rule's condition (Fig. 7.9).

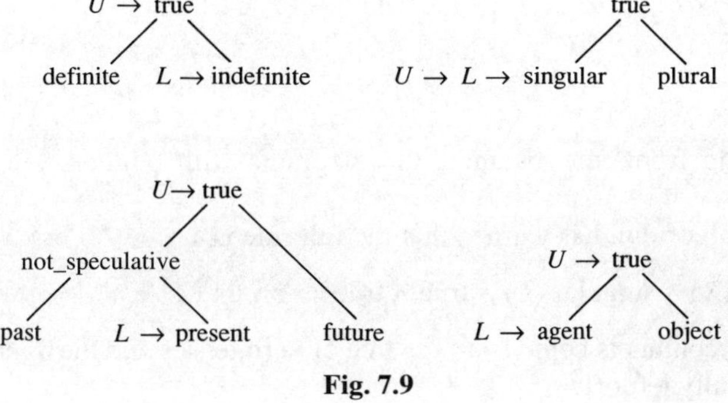

**Fig. 7.9**

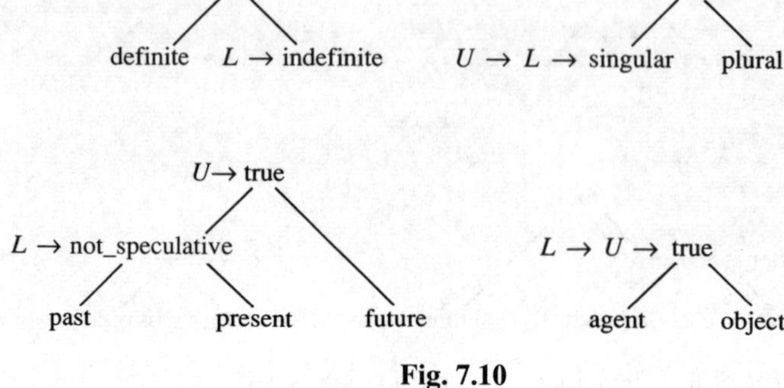

**Fig. 7.10**

The third training case is another positive example, whose basic properties are

indefinite
singular
past
object.

It causes the algorithm to raise its pointers $L$ in the *agent/object* tree to *true*. In the tree of tenses, $L$ is raised to *not_speculative*. The algorithm has then learned that the *agent/object* group has no bearing on the rule (its conjunct in the rule's condition is *true*). Also, it is now known that the conjunct for the tree of tenses is a generalizaton of *past* and of *present*, but the algorithm has not yet found whether the conjunct also covers *future*. See Fig. 7.10.

The fourth sentence shows that the conjunct from the tree of tenses does indeed cover *future*.

The last sentence,

    * **John** will chase dogs,

has properties

definite
singular
future
agent.

After learning from this instance, the diagrams with pointers are as shown in Fig. 7.11.

Thus, the algorithm has learned that the full rule is

indefinite $(X) \wedge$ singular $(X) \wedge$ true $\wedge$ true $\Rightarrow$ prefix $(X,$ 'a')

The two *true* conjuncts come from the tree of verb tenses and the *agent/object* tree. They are usually left off.

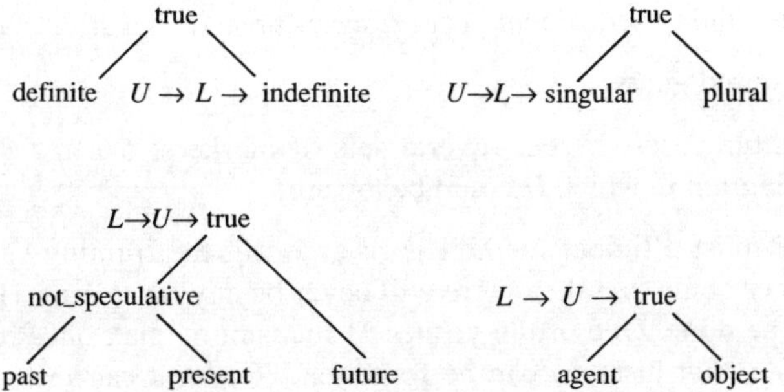

**Fig. 7.11**    The final positions of pointers.

### *Far misses*

In that example, the training set was exceptionally convenient. When the cases in it were examined in the given order, the algorithm was never faced with the last alternative. Let us see what happens if the training set is presented in a different order:

a **dog** chases balls;

* **dogs** chase balls;

John chased a **dog**;

* **John** will chase dogs;

dogs will chase a **ball**.

Learning proceeds just as before for the first three cases (see Fig. 7.10). In what is now the fourth instance, the basic properties are

definite
singular
future
agent.

This is a counter-example, but there are two trees in which $B$ is below $U$ but not below $L$. This $I$ could be excluded by lowering $U$ in either the *definite/indefinite* tree or in the tree of tenses. We do not know which tree to choose.

A training instance of this sort is called a **far miss**. The pleasanter sort of counter-example, which has just one property $B$ below the corresponding $U$ but not below $L$, is called a **near miss**. Focusing can always learn from a near miss.

There are various possible strategies for handling far misses. One can:

- ignore them and learn nothing;

- ask a user at run time which diagram to choose;

- re-order the training set, so that positive cases precede negative ones;

- make an inspired guess;

- try all possible cases. Invent several sets of markers, one for each possible choice of diagram in which $U$ could be lowered.

The safest and most efficient method is re-ordering the training set. If enough positive instances come first then there will never be any far misses. However, this cannot always be done. One of the virtues of focusing is that, once it has learned from an instance, that instance can be forgotten. Hence, it can learn in real time from instances as they occur and the partially learned rule can be used at once. The training set can only be re-ordered if the whole training set is known before learning starts.

The inspired guess is the cleverest solution, if it can be made to work. It has been tried with success. The usual approach is that a property is most likely to be relevant if it applies to a feature near the action's point of application. You can make up your own mind what 'near' means, depending on the situation. In this case, the properties *definite* and *indefinite* apply directly to the noun, $X$. The alternative group of predicates, the verb tenses, are properties of the whole clause $Y$. The action acts on $X$ rather than $Y$, so the *definite/indefinite* group is closer. 'Inspiration' suggests that this is a probably the better group in which to lower U. Indeed, it is.

The last option makes best use of the available information, but is more complex computationally. In the example, it would yield the alternative markers shown in Fig. 7.12, if we assume that the *definite/indefinite* group is significant, and those shown in Fig. 7.13, if the article 'a' depends on the tense of the clause. Given only the first four training cases, the learner cannot choose between these two options. However, if it is prepared to remember both possibilities for a few learning cycles, it will find out which is right. When it comes to the positive case

dogs will chase a **ball**

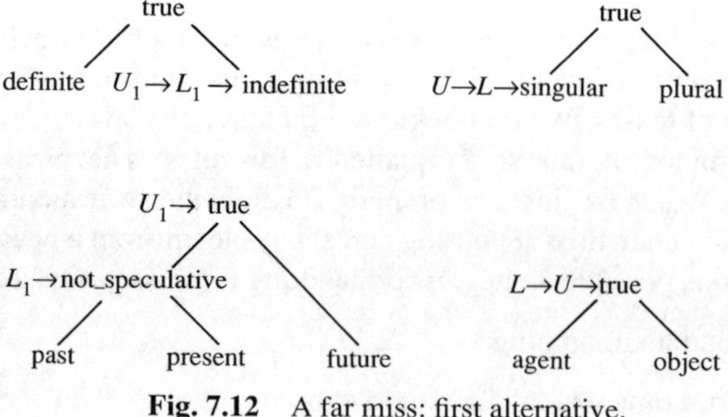


**Fig. 7.12**   A far miss: first alternative.

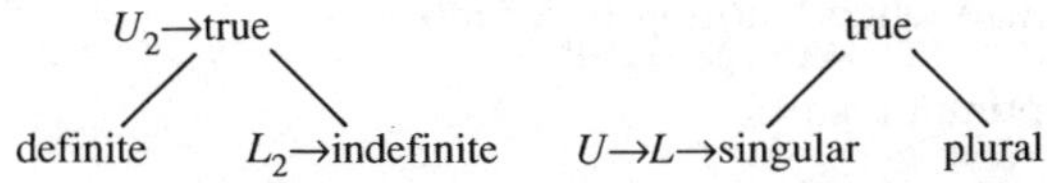

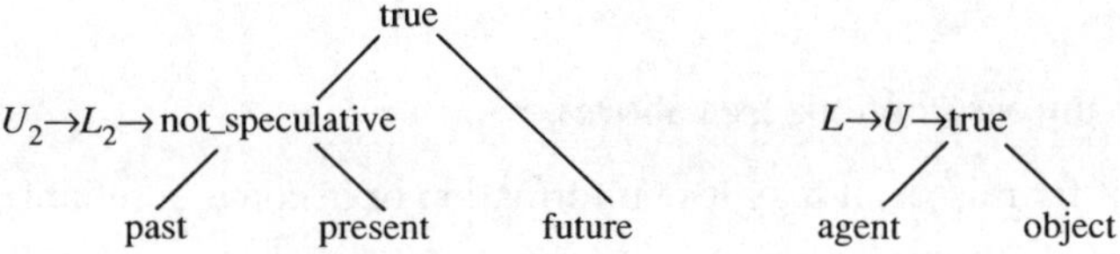

**Fig. 7.13**   A far miss: second alternative.

it will find that the marker $U_2$ is not above $B$ in the tree of tenses. Hence, the choice of markers $(U_2, L_2)$ is not compatible with the training set. At this point, the learner can reject the choice $(U_2, L_2)$ and conclude that the choice $(U_1, L_1)$ is right.

### Cases when focusing breaks down

Recall, the conditions under which focusing works are quite strict. The data must be *clean* and the basic predicates which can occur in the condition must be partitioned into mutually *exclusive exhaustive* groups and the learner must be given a *diagram* (an upper semi-lattice) of predicates for each group and the ideal rule's condition is *conjunctive* with only one conjunct from each diagram. In addition, the desired conjunct must actually occur in the given diagram. If, to start with, you provide too simple a description space which does not contain the right conjunct, then of course the learner cannot find it. Bundy has suggested that a learner might overcome this situation by revising the description space. He calls this 'tree hacking'.

When any one of these preconditions is violated, the same thing happens. In some tree:

- for some positive case, the upper marker is not above a basic property $B$; or

- for a negative case, $B$ is below the lower marker $L$.

Without more information, one cannot tell what has gone wrong. If the training set is noisy, then no form of similarity based learning will ever work. If the data are clean but there is some action which is applicable in two separate sets of cases, then you will have to choose a different algorithm which can learn rules with disjunctive conditions or some equivalent.

### Strengths of focusing

- Learning is incremental, unless the training set has to be re-ordered.

- It does not involve remembering training instances.

- It can recognize when learning is complete.

- It finds an optimal result.

- The result is independent of the order of instances in the training set.

### *Limitations*

- It requires many assumptions (see above).

- In the case of far misses, it may lose information or become unreliable.

### 7.3.2   VERSION SPACES AND CANDIDATE ELIMINATION

There are other similarity-based learning methods. One of the best known is the
**version space** technique, due to Tom Mitchell. The version space is the set, called
$V$, of all possible descriptions of concepts. The method assumes that there is a
partial order on $V$, a bit like the upper semi-lattices of focusing.

   Suppose that we are given a particular training set. Mitchell proposed an
algorithm called **candidate elimination**, very like focusing, which narrows down $V$
to the portion compatible with this training set. The algorithm keeps two sets of
descriptions:

- $G$, the set of *most general* concepts compatible with the training data read so
  far; and

- $S$, the set of *most specific* concepts compatible with the data read so far.

$G$ is analogous to the upper pointers of focusing and $S$ corresponds to the lower
pointers. When the algorithm reads a positive instance, it replaces $S$ with a set of
more general concepts and perhaps removes some concepts from $G$. When it reads
a counter-example, it replaces $G$ with some more specific ones and may remove
some from $S$. This process continues until $G$ and $S$ coincide or until the training set
is exhausted or is found to be inconsistent.

   Mitchell phrased his original account in general terms. He only required that:

- there are two languages, one for describing examples and another for describing
  concepts; and that

- there is a test to see if any given example $I$ is an instance of a given concept $c$.
  This test is called *Match*$(c, I)$.

The languages might both be a single first-order language, but they might be quite
different. The partial order on $V$ can be defined using *Match*:

$$c \le d \iff \text{ for all possible instances } I,$$
$$\text{\textit{Match} } (c, I) \Rightarrow \text{\textit{Match} } (d, I).$$

so $c \le d$ if $d$ is more general than $c$. Also, the candidate elimination algorithm
involves two constructions:

(1)   for each subset $S$ of $V$, and for each possible example $I$, a way to find the set

$$msg(S, I) = \{ v \in V \mid \forall w \, (Match(w, I) \; \wedge \; \exists s \in S \; s \leq w \wedge w \leq v \iff w = v) \};$$

and

(2)   for each subset $G$ of $V$, and for each possible counter-example $N$, a way to find the set

$$mgs(G, N) = \{ v \in V \mid \forall w \, (\neg Match \, (w, N) \; \wedge \; \exists g \in G \; w \leq g \wedge v \leq w \iff w = v) \}.$$

The operations *msg* could be thought of as working in two stages.

First, it finds all concepts in $V$ which match $I$ and which also generalize some $s$ in $S$. Then among these concepts, it finds the least ones.

The operation *mgs* is dual. First, it finds all concepts which exclude $N$ and which are more specific than some $g$ in $G$. Then, from among these concepts, it finds the most general ones.

The candidate elimination algorithm is as follows:

Find a positive training instance, $I$
$S :=$ the set of all least general concepts in $V$ which match $I$
$\qquad \{ v \in V \mid \forall w \, (Match(w, I) \wedge w \leq v \iff w = v) \}$
$G :=$ the set of all most general concepts in $V$ which match $I$
$\qquad \{ v \in V \mid \forall w \, (Match(w, I) \wedge v \leq w \iff w = v) \}$

WHILE $S \neq G$ AND the training set is not exhausted
$\quad$ Choose another instance $I$;
$\quad$ IF $I$ is positive
$\quad$ THEN
$\qquad$ remove from $G$ all concepts $v$ which do not match $I$;
$\qquad S := msg(S, I)$
$\quad$ ELSE
$\qquad$ remove from $S$ all concepts which match $I$;
$\qquad G := mgs(G, I)$
$\quad$ IF $\exists s \in S \; \exists g \in G \; g < s$ OR $S$ is empty OR $G$ is empty THEN *fail*;

$\quad$ RETURN the two sets $S$ and $G$.

The algorithm only fails if the training set is not consistent with any concept in $V$. If it succeeds, then the concepts $v$ which are 'between' $S$ and $G$:

$$\exists s \in S \; s \leq v \qquad \text{and} \qquad \exists g \in G \; v \leq g$$

are just those concepts which are consistent with the data. If $S$ and $G$ are the same set at termination, then you can check that this set only contains one concept:

$$S = G = \{c\}.$$

In that case, the candidate elimination algorithm has learned the concept $c$ from the training set.

In the focusing algorithm, there is just one lower pointer in each tree. One might imagine, by analogy, that the set $S$ can only contain a single concept. This is not so, as Mitchell showed.

*Example*

An instance is an unordered pair of two figures. Each figure has:

- a *size* which may be *Large* or *Small*;

- a *colour* which may be *Red* or *Yellow* or *Blue*;

- a *shape* which may be *Square* or *Circle* or *Triangle*.

Thus, one such instance might be

{(Large   Red   Circle),   (Small   Yellow   Circle)}.

The language of instances consists of all such sets of two triples of feature values.

The language of concepts is the same as that of instances, with one extension. In place of a feature value, a triple may contain the symbol '?'. The *Match* relation is defined thus:

- any feature value matches itself;

- '?' matches any feature value;

- two triples match if their corresponding entries all match;

- a concept $c$ matches an instance $I$ if there is a function $f$ from $c$ onto $I$ so that, for each triple $t$ in $c$, $t$ matches $f(t)$.

Thus,

{(Small   ?   ?),   (Large   ?   Circle)}

matches

{(Large   Red   Circle),   (Small   Yellow   Circle)}.

Now suppose that the training sequence begins with two instances

{(Large   Red   Triangle),   (Small   Blue   Circle)}

and

{(Large   Blue   Circle),   (Small   Red   Triangle)}

which are both positive. After the algorithm has read and taken account of them, the set $S$ will be

$$\left\{ \begin{array}{l} \{(\text{Large} \quad ? \quad ?), \quad (\text{Small} \quad ? \quad ?)\}, \\ \{(? \quad \text{Red} \quad \text{Triangle}), \quad (? \quad \text{Blue} \quad \text{Circle})\} \end{array} \right\}$$

You can check that both concepts in this set $S$ match both instances and that both are as specific as possible (since changing any '?' to a feature value would make some match fail) and that any concept which matches both instances must be at least as general as one of these two. What is more, neither concept in this set $S$ is more general than the other, so $S$ must contain both.

The candidate elimination algorithm is so called because each iteration of its loop may eliminate many candidate concepts. A crude search for the goal concept would examine all possibilities. In the example above, there are 1176 conceivable concepts, so such a search would be barely feasible. In a case like this, the candidate elimination algorithm could find a goal concept after examining perhaps fewer than a dozen instances.

Clearly, this algorithm is easily implemented as long as the tests *Match* and '$\leq$' can be calculated and the operations *msg* and *mgs* can be encoded. Its efficiency depends on the efficiencies of these two operations. When the language of concepts is a simple language of clauses, the relation '$\leq$' has the following pleasant property: any finite subset of $V$ has a least upper bound in $V$ and a greatest lower bound in $V$. In that case, as long as all the sets involved are finite, *msg* and *mgs* can be calculated easily. A partial order with this property is called a **lattice**. Note that the upper semi-lattices used by focusing satisfy half this condition. In an upper semi-lattice, any finite set has a least upper bound but it may not have a greatest lower bound. Unlike focusing, the candidate elimination algorithm is almost completely symmetric between $S$ and $G$, and between positive and negative examples. The only asymmetry lies in the initial steps. The first example examined must be positive.

## 7.4   DISJUNCTIVE CONDITIONS

Recall that focusing is only reliable when you can be sure that each operator is the operator of a single rule and each rule's condition is a conjunction. Often, this is not the case. There are several algorithms which learn disjunctive conditions, but they involve saving at least part of the training set. So far, we have met decision trees and the Scott–Markovitch clustering method. Syntactic generalization can also form several rules for one action when it is applied in different ways.The procedure which invents recursive rules may do this. These methods do not all yield output in the form of rules, but such rules can be found from them easily. For instance, the decision-tree algorithm yields the following rules:

- For each operator, build a decision tree from the training set. The concept which the tree recognizes is 'the operator is applicable'.

- For each tree, form a rule.
  The rule's action is the tree's operator.

  The rule's condition has one disjunct for each leaf labelled 'yes, the action is applicable'.

For any leaf, suppose the nodes on the path from the tree's root to this leaf are labelled with the questions

$$Q_1 \quad Q_2 \quad Q_3 \quad Q_4 \quad \ldots$$

to which the answers are, respectively,

Yes   No   No   Yes   ...

Then the disjunct for this leaf will be

$$(Q_1 \wedge \neg Q_2 \wedge \neg Q_3 \wedge Q_4 \wedge \ldots ).$$

This disjunct is the conjunction of all questions on this path. A question is negated if and only if the answer to it along this path is *No*.

For instance, consider the little tree:
   Is unemployment high?
   *Yes*: The London market will rise today
   *No*:  Is the New York market rising today?
       *Yes*: The London market will rise today
       *No*:  The London market will not rise today

which we learned before. From it, we can invent the rule

IF      unemployment is high
OR     unemployment is not high
        AND the New York market is rising today
THEN  the London market will rise today.

In this case, the negated question in the second disjunct is redundant.

In a sense, the deductive methods which we shall study next could be said to learn alternative rules for a predicate, which are equivalent to disjunctive conditions. However, deduction can never discover essentially new information. The conditions in the new rules are just new formulations of the other conditions in old rules.

There is another way to learn genuine alternative conditions which we shall explore. It is part of a system whose primary purpose is nothing to do with learning conditions of rules. It is designed to learn how to resolve conflicts by a technique called **rule ordering**. Before studying it, though, here is an entirely different approach.


## 7.5   DEDUCTIVE METHODS

The three methods studied under this heading are very similar. They all deduce a conclusion from known rules and then form a new rule with this as its conclusion. The new rule is designed to shorten future search.

**Explanation-based generalization** (EBG) is the classic prototype method. It depends on an explicit form of the deduction which amounts to a proof that the conclusion can be drawn. The other two approaches, **lemma generation** and **chunking**, are elaborations of EBG. Lemma generation includes a feature which lets the learner recognize when learning is complete. Chunking on its own is just a particular implementation of EBG, but it is worth studying because it can be combined with a search technique called **universal subgoaling**. This lets the learner apply EBG to the process of search control, as well as to a user's basic task.

All three are discussed here in some detail. Each reveals a different approach to encoding the deduction. Also, the three have been studied and tested in different ways and the settings of these studies are interesting for their own sakes.

### 7.5.1 EXPLANATION-BASED GENERALIZATION

This is radically different from similarity based learning. It requires only a single positive training instance, but by way of complement, it depends on a substantial body of rules which are already reliable.

As described by Mitchell, Keller and Kedar-Cabelli, the EBG method requires the following inputs:

- a domain theory. This is a set of rules and facts;

- an 'operationality criterion', which is a procedure for deciding whether any property is observable or easily calculated;

- a goal predicate. This may be defined in any way;

- a single instance of the goal, described just by properties which satisfy the operationality criterion.

Given these, the learner attempts to find a reformulation of the goal, describing it just by 'operational' predicates.

The result can be expressed as a rule:

IF *reformulation*
THEN *goal.*

The learner attempts to prove that the training instance is an example of the goal. As axioms, it takes the database of facts and rules. For hypotheses, it can use any operational properties of the instance. Since theorem proving is not decidable, this step of the learner is not an 'algorithm' in the formal sense.

If the learner finds such a proof, then it generalizes constants within the proof to variables. The learner's output is:

- the set of operational predicates which were used as hypotheses in the generalized proof. These make up the reformulation of the goal predicate.

- the generalized proof. This is often referred to as the 'explanation'.

Thus, the learner does not just find a sufficient operational condition for the goal. It also finds a proof that anything which matches this generalization is also an example of the goal.

*Example: stacking*

Suppose that a learner wants to discover when it is safe to stack one thing on top of another. In this case, we consider that stacking is 'safe' if the thing underneath will not get crushed. Maybe it is not safe to stack lots of wooden blocks in a column, because they may fall over, but blocks are hard and cannot be crushed, so for the present purposes, stacking blocks is regarded as 'safe'.

The goal concept is *safe_to_stack*. It is safe to stack if the thing on top is lighter than the thing underneath or if the thing underneath is strong. The learner's rule base already contains:

$$\text{lighter } (x, y) \quad \Rightarrow \quad \text{safe\_to\_stack } (x, y)$$
$$\text{strong } (y) \quad \Rightarrow \quad \text{safe\_to\_stack } (x, y).$$

The domain theory contains the rules

$$\text{volume } (z, v) \wedge \text{density } (z, d) \quad \Rightarrow \quad \text{weight } (z, v \times d)$$
$$\text{is\_a\_table } (x) \quad \Rightarrow \quad \text{weight } (x, 5)$$
$$\text{weight } (x, W_x) \wedge \text{weight } (y, W_y) \wedge W_x < W_y \quad \Rightarrow \quad \text{lighter } (x, y).$$

In this case, the operational predicates are the ones used to describe examples, including *volume* and *density* (but not *weight*), and elementary properties of numbers.

The given training instance consists of one object, $obj_1$, which is safely stacked upon another, $obj_2$. The given facts about this instance are

on $(obj_1, obj_2)$
is_a_box $(obj_1)$
is_a_table $(obj_2)$
colour $(obj_1, \text{red})$
colour $(obj_2, \text{blue})$
volume $(obj_1, 3)$
density $(obj_1, 0.6)$
...

There is a simple proof that
safe_to_stack $(obj_1, obj_2)$
Its steps form a tree as in Fig. 7.14.

Having found a proof for this particular instance, the algorithm generalizes it. To start with, it generalizes the conclusion as far as it can:
safe_to_stack $(obj_1, obj_2)$ becomes safe_to_stack $(x, y)$.

It works from the top downwards. Once it has found a generalization of some node $G$, which was derived by a rule in the database:

$$C_1 \wedge C_2 \wedge \ldots \Rightarrow G$$

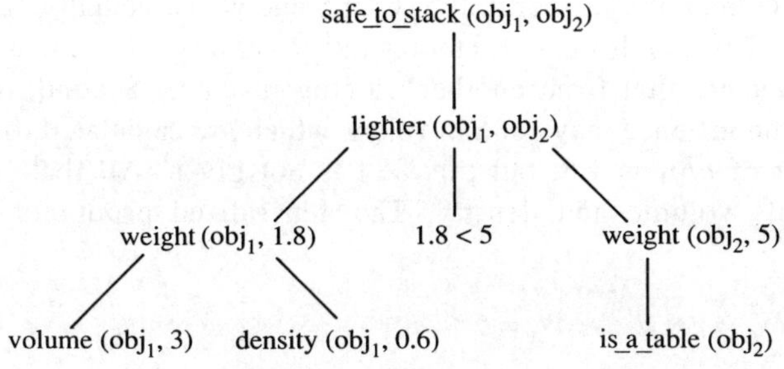

**Fig. 7.14**   A specific proof tree.

it looks for a generalization of the conditions $C_j$. In this case, the resultant generalized tree is shown in Fig. 7.15.

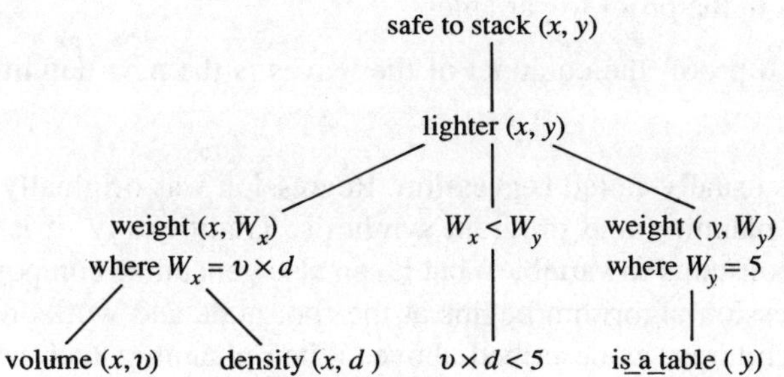

**Fig. 7.15**   The corresponding generalized proof tree.

The ultimate learned rule is

$$volume\ (x,\ v)$$
$$\wedge\ density\ (x,\ d)$$
$$\wedge\ v \times d < 5$$
$$\wedge\ is\_a\_table\ (y) \Rightarrow safe\_to\_stack\ (x,\ y)$$

The left side of this rule is the reformulation of the goal. Observe that the conditions consist entirely of operational properties. If they didn't, this reformulation would not suffice.

One benefit of EBG is that it concentrates attention on relevant properties. In this stacking example, the colours of $obj_1$ and $obj_2$ are given. A similarity-based learner might be led astray by such irrelevant detail. An EBG learner will not be.

Two aspects of generalization are not quite trivial. First, the reformulation which the learner finds is not the most general possible. If there are alternative ways of deriving a node $G$, then the learner only uses the one which actually occurs in the training instance. Thus, in this case it ignores the possibility that $y$ might be *strong*. EBG could only learn that from another training instance. Second, the specific proof for the given instance may contain values which are calculated. In the above case, the weight of $obj_1$ is 1.8, but this fact is not given. All that is given by observation is its volume and density. The generalized proof contains extra expressions

$$v \times d < 5 \qquad W_x = v \times d \qquad W_y = 5$$

which do not appear in the first proof.

This stacking example is the classic case which you will find in most introductory papers on EBG. The algorithm for EBG which is used to construct it runs:

Find a proof that the given instance is an example of the given concept. The proof should be in the form of a tree, as shown.

Generalize terms in the proof to variables.

In the generalized proof, the conjunct of the leaves is the new definition of the goal.

The second stage is usually called **regression**. Regression was originally proposed as a technique for planning and program synthesis. Traditionally, it is supposed only to generalize constants to variables, but it can also generalize composite terms.

The classic regression algorithm begins at the root node and works recursively through it and its children, as described above. When planning, each node of the tree is an operator. Waldinger suggested labelling the tree of operators with conditions, one immediately before and after each operator. The *post*condition after the operator says what should be true when the operator has just been applied. The *pre*condition before it says what has to be true just before applying it, if we want to be sure that the postcondition will hold. The root's postcondition asserts the outcome of the entire plan. Regression is the process of calculating the right precondition for any operator, given its postcondition.

If we want to be sure that the plan will work, then each operator's precondition should follow from the postconditions of its children. Planning is more involved than theorem proving because at some stage the operators have to be placed in a strict sequence. If we are careless, some operator in the sequence may undo a condition which was established by an earlier operator for a later one. Waldinger studied how one can shuffle operators around, so that the plan will work.

The generalizing stage of EBG can be done in two different ways. If the rules' actions contain operators with side effects, then Waldinger's regression method is the right one. However, we are in a different position because our example is an exercise in theorem proving, not planning. This makes the generalizing process simpler, although it is still not trivial. This is how it goes.

Start with the original proof tree, *PT*

$PR$ := a copy of this tree, but with different nodes.
   FOR each node $N$ in *PT*
     IF $N$ is a leaf
     THEN
       $RN$: = a generalization of $N$, replacing all terms with new variables
     ELSE
       say $R$ is the rule in the database which was used to derive $N$.
       $RN$ := a copy of $R$, with new variables.
   The nodes of *PR* are the copied rules *RN*.

Thus, the node

$$\text{lighter } (\text{obj}_1, \text{obj}_2)$$

of the original proof is replaced by

$$\text{weight } (x_1, W_{x_1}) \quad \wedge \quad \text{weight } (y_1, W_{y_1}) \quad \wedge \quad W_{x_1} < W_{y_1} \quad \Rightarrow \quad \text{lighter } (x_1, y_1)$$

and the node

$$\text{weight } (\text{obj}_1, 1.8)$$

will be replaced by

$$\text{volume } (z_2, v_2) \quad \wedge \quad \text{density } (z_2, d_2) \quad \Rightarrow \quad \text{weight } (z_2, v_2 \times d_2)$$

and the leaf

$$\text{density } (\text{obj}_1, 0.6)$$

is replaced by

$$\text{density } (z_3, d_3).$$

Except at the root, the conclusion of each new node will match some condition in its parent. (If it matches more than one, the original proof shows which one is wanted.)

   FOR each node *RN* in *PR* except the root, working down the tree from the root,
     unify *RN*'s conclusion with the corresponding condition in its parent node.
     Apply the unifying substitution throughout *PR*.

The nodes of the copy become linked together by unifying. For example,

$$\text{weight } (z_2, v_2 \times d_2) \quad \text{corresponds with} \quad \text{weight } (x_1, W_{x_1}).$$

The desired unifier is

$$\{z_2/x_1, (v_2 \times d_2)/W_{x_1}\}.$$

The conclusion at the root cannot be unified with anything, because of course there is no higher condition for it to be unified with.

Here comes the twist in the tail. The kind of unification we need is not always Prolog's simple string unification. The proof is in some first-order theory which probably has equality and may have lots of extra laws such as associativity and symmetries of various sorts, so this unification is subject to all these laws. In a bad case, there may not be any most general unifier or there may be scores of distinct most general unifiers. The best that EBG can offer you is general guidance: use the unification method which worked for the specific example.

### *Doing away with the training instance*

EBG uses its training instance in order to discover the proof. It is actually possible to do without any training instance at all. The idea is this: in the EBG algorithm, replace the steps

Find a proof that the given instance is an example of the given concept.

Generalize terms in the proof to variables.

with a single step, which finds a proof of the given concept from operational conditions:

$S$: = {the given concept}
WHILE $S$ contains some predicate $P$ which is not operational
   choose a rule $R$ whose conclusion is $P$;
   $S$: = $S \setminus \{P\} \cup R$'s conditions.

If and when this stops, if none of the rules chosen has a negated condition, the set $S$ will consist of operational predicates which imply the original concept.

Van Harmelen and Bundy give a neat little piece of code for a simple form of the EBG algorithm. In Prolog, it goes:

```
ebg (Leaf, GenLeaf, GenLeaf):-
    operational (Leaf),
    !,
    call (Leaf).

ebg ((Goal1, Goal2), (GenGoal1, GenGoal2), (Leaves1, Leaves2)):-
    ebg (Goal1, GenGoal1, Leaves1),
    ebg (Goal2, GenGoal2, Leaves2).

ebg (Goal, GenGoal, Leaves):-
    clause (GenGoal, GenClause),
    copy ( (GenGoal:- GenClause), (Goal:- Clause) ),
    ebg (Clause, GenClause, Leaves).
```

It can be called thus:

?-ebg (Case, GoalPredicate, Reformulation).

The *Case* is the given goal predicate applied to the instance. The predicate *ebg* calculates *Reformulation*, which is a list of all the operational predicates which are leaves of the generalized proof tree. To generalize the example above, we call

?-ebg (safe_to_stack (obj$_1$, obj$_2$),
     safe_to_stack ($x$, $y$),
     Reformulation).

The Prolog database holds the domain theory and everything known about the instance.

This particular rule, *ebg*, is a bit too naïve for practical application. The proof remains hidden somewhere inside the Prolog interpreter. Also, this theorem prover may loop indefinitely if the goal concept is defined recursively.

Prolog makes the algorithm particularly slick because it discovers the proof automatically and then it also makes copies of rules and performs all unifications automatically too. The first and second arguments correspond to the original proof and to the copy, respectively. A naïve implementation of EBG would fall into two separate parts: one which finds the proof and another which generalizes it. Steven Minton and his collaborators have written another form (in Lisp) which combines the two stages.

What van Harmelen, Bundy, and Prieditis all appreciated is that the first argument may be unnecessary. If it is omitted, we obtain the predicate

```
peval (Leaf, Leaf):-
   operational (Leaf),
   !,
   call (Leaf).

peval ((Goal1, Goal2), (Leaves1, Leaves2)):-
   peval (Goal1, Leaves1),
   peval (Goal2, Leaves2).

peval (Goal, Leaves):-
   clause (Goal, Clause),
   peval (Clause, Leaves).
```

This too is somewhat too simple for applications, but in principle it is almost a *partial evaluator*. It becomes a partial evaluator if, in its first clause, the call to *call(Leaf)* does not instantiate any variables. If one is willing to resort to some messier aspects of Prolog, this can be arranged. Just replace the first clause with

```
peval (Leaf, Leaf):-
   operational (Leaf),
   !,
   not not call (Leaf).
```

*peval* finds conditions which imply *Goal* but which are easier to calculate. A lot of work has gone into the study of partial evaluation, since it makes code more efficient. Van Harmelen and Bundy point out that this little rule can solve the *safe_to_stack* example above and, what is more, it can find *all* operational forms of it, by backtracking. In this case, the particular training instance is quite unnecessary.

Perhaps this study reveals the distinction between artificial intelligence and the rest of computer science. Both EBG and partial evaluation attempt intractable problems for which there are no terminating algorithms. They depend on proof. When partial evaluation works, it is simpler than EBG, but it is less reliable and it will probably take longer, because its theorem prover starts with less information and so has to do more work. In difficult cases, EBG is better because it uses heuristic guidance.

### EBG and partial functions

There is a hidden subtlety in EBG. The proof for the single instance may involve calculations with *partially defined functions*. It is all too easy to do this, particularly during a search with rules which are not totally reliable. For instance, if Johnny is naughty at school, the headmaster may advise his junior member of staff to write a note to Johnny's father. The junior learns this remedy, but later he finds it doesn't work for Matthew because Matthew was begotten during a happy evening his mother spent with a passing soldier. In this case, the function in question is *father*$(x)$. Not every little boy has a father, so this function is partial.

The example *safe_to_stack* does not meet this problem. It uses just one function, multiplication $(\times)$, which is defined for all numbers and, also, the numbers involved above are real. By contrast, division is a partial function, when it is applied to integers and we want an integer answer. The following example is actually quite a tricky case. We shall examine it again in the next chapter.

*Example: trigonometric integrals*

One indefinite integral of $\cos^7 x$ is

$$\sin x - \sin^3 x + 3/5 \, \sin^5 x - 1/7 \, \sin^7 x.$$

The first few steps of its derivation transform the integrand through the stages:

$$\int \cos^7 x \, \mathrm{d}x$$
$$\int \cos^6 x \cos x \, \mathrm{d}x$$
$$\int (\cos^2 x)^3 \cos x \, \mathrm{d}x$$
$$\int (1 - \sin^2 x)^3 \cos x \, \mathrm{d}x$$
$$\int (1 - u^2)^3 \, \mathrm{d}u$$
$$\int (1 - 3 \, u^2 + 3 \, u^4 - u^6) \, \mathrm{d}u$$
$$u - u^3 + 3/5 \, u^5 - 1/7 \, u^7.$$

In this case, the generalization begins by changing the power of *cos x* to an arbitrary whole number:

$$\int \cos^k x \, dx.$$

The next couple of steps become

$$\int \cos^{k-1} x \cos x \, dx$$

and

$$\int (\cos^2 x)^{(k-1)/2} \cos x \, dx.$$

The remainder of the generalized proof will only work if $(k-1)/2$ is a whole number. Otherwise, the integrand cannot be transformed into a simple polynomial in $u$. The EBG algorithm has to introduce the extra conditions: $(k-1)/2$ is a whole number, or alternatively $k$ is odd.

You may be wondering, in what sense is division not defined? The quotient $(k-1)/2$ is perfectly well defined. The only problem is, it is not whole. You are quite right, of course.

There are two different ways of expressing the necessary restriction. We may find that either some function $f$ is not defined, or this function may be defined but its value may have an unsuitable type. The extra condition imposed by the algorithm could be a restriction on one of its arguments, say $C$, or on the value

$$f(A, B, C, D, \ldots)$$

if it exists.

Examples such as integration are a bit unusual. The first numbers which people ever thought of were the natural whole numbers and, for them, division is not always defined. Later, mathematicians *invented* fractions precisely so that division *is* always defined (except by zero). No doubt, the first person who thought of fractions was very clever and some of his friends thought he was talking rubbish, but mathematicians often invent new notions, simply in order to make something possible.

It has been suggested that a learning program might use such an example as this to invent the predicate '*odd*'. To start with, the learner may well not have this predicate in its vocabulary. Since the proof will only work when $k$ is odd, the program might define this condition to be

$$\text{odd } k \Leftrightarrow \exists\, m \, (k = 2m + 1).$$

This is a form of so-called **poof analysis**. A *poof* is an attempt at proof which doesn't quite work. Poof analysis invents a new concept for which it does work. Note though that there are many ways in which a poof may fail and Bundy, who suggested the idea, concentrates on a different one.

### Other examples of EBG

EBG, and extensions of it which are commonly called explanation-based learning (EBL), are popular and widely applicable. A practical case, also suggested by Mitchell, arises in the design of integrated circuits.

*Example: integrated circuit design assistant*

A (human) designer sits at a terminal and lays out his design for a piece of silicon. The design process goes through several stages. Before the designer lays out any patterns for doped layers on the chip, he types in a specification of what the chip should do. If this is simple, then he can describe the chip's physical layout immediately. Otherwise, he divides its operation into simpler steps and:

(1)   specifies each step;

(2)   specifies an area of the chip for each step;

(3)   describes how the various areas are to be connected up electrically;

(4)   designs each area, similarly.

Thus, a big design is divided in a recursive hierarchy of simpler parts.

There is a 'learning assistant' (a program) built into the terminal. Its problem-solving component is intended to accept a specification and then lay out a design on silicon which will satisfy it. Ultimately, we hope, the assistant will be able to design an entire chip from its specification, but to start with, it only knows the basic electrical properties of doped silicon and a bit about geometry. It has almost no knowledge of design.

The learner's training set consists of the human designer's specifications and designs. If the assistant cannot prove that a design fulfils the specification, then it learns nothing. (It may still serve a useful purpose, if it warns the designer that his method is suspect.) When it finds a proof, it generalizes the specification, design, and proof as far as possible. Then it forms a rule, which says that any matching specification can be satisfied by a similar design. The learner may also learn how to optimize chip layout, by rearranging areas.

In this case:

- the domain theory is the assistant's knowledge of electronics and geometry and whatever it may have already learned about layout;

- the operationality criterion is that the resultant design is in terms of layers of doped silicon;

- the goal predicate is the specification;

- the single example is the designer's design.

Nowadays, semi-automatic checking of chips is practical technology, although checking with learning is still considered adventurous. In principle, the same approach should work with software, but in practice, verification of software is a lot harder. This is partly because software systems can be bigger than even the biggest chips and also because software involves much more iteration and recursion than hardware. However, a design assistant for software would be immensely valuable.

*Example: learning to avoid failures*

EBG can also be employed to discover which search paths should be avoided in future. Suppose that a system's objective is to design office floor layouts. These are very like chip layouts. The layout *fails* if two objects, such as the manager's desk and the plumbing, come too close. In this case:

- the domain theory consists of physical shapes and sizes of basic constituents and such notions as adjacency, overlap, and sound transmission;

- a feature is operational if it is describable by simple measurements;

- the 'goal' concept is failure!

- any failed attempt at design will do for a training instance.

If the learner can prove that a certain design fails, then it will learn that any other attempt at a design which puts the plumbing through the wrong corner of the manager's suite will also fail. The system learns where not to put the plumbing.

That is the original form of EBG. It is somewhat analogous to contradiction backtracing. Both methods construct a proof from given rules and then work back through it.

### Strengths of EBG:

It epitomizes all the benefits of rule-based systems.

- It extracts a lot of information from a single instance.

- It bypasses extraneous detail.

- The learned information is accompanied by a proof that it works. This is the 'explanation' which gives the method its name.

### Limitations

- It involves theorem proving, which is hard.

- It may also involve unification subject to laws, which is also hard.

- It depends on a domain theory, which is assumed correct. If the proof depends on a faulty assumption, then the method may produce misleading results.

- The operationality criterion is somewhat arbitrary.

- EBG can only be used to improve on the description of a predicate which is already defined. It is not a way to discover totally new concepts (although it may lead to poof analysis, which *does* invent new concepts).

### 7.5.2  LEMMA GENERATION

This is a method which avoids recalculating. It is rather like a limited form of EBG
which just learns facts (rules with trivial conditions). If a goal is proved true once,
then the learning system makes a note that it is true and does not invent the proof
again. It has been shown to work in a special case, for certain goals. However, it is
only effective if the system is given a lot of specific knowledge about the learning
domain.

Stephen Owen and Richard Hull found experimentally that lemma generation is
the best single technique for improving search in a large Prolog knowledge base.
The Imperial Cancer Research Fund compiled a Prolog database of rules which
predict how a large protein is likely to fold as it is formed from amino acids. This
database worked, but only slowly. Owen and Hull first simplified its logic, so that it
was in pure declarative form. They then wrote a sequence of interpreters for it (in
Prolog). The simplest, called the 'vanilla A' interpreter, runs:

```
solve (Theory, [ ]).

solve (Theory, Goal_List):-
   Goal_List = [_|_],
   select (Theory, Goal_List, G, Rest),
   solve (Theory, G),
   solve (Theory, Rest).

solve (Theory, not (Goal)):-
   not solve (Theory, Goal).

solve (Theory, Goal):-
   system (Goal),
   call (Goal).

solve (Theory, Goal):-
   member ( (Goal:- SubGoals), Theory),
   solve (Theory, SubGoals).
```

(On this occasion, please accept the algorithm in the form of Prolog code. This is
how Owen and Hull present it and their form can hardly be improved on.)

The first two clauses assert that a list of goals can be solved whenever all the
individual goals in the list can be solved separately. The third clause does the
obvious thing with negation. The fourth recognizes system predicates and passes
them straight to the underlying Prolog system. The last clause traces a rule in the
theory of proteins and asserts that the Goal is true if all its subgoals are true.

This interpreter would have exactly the same logic as the underlying Prolog
system if *select* just took the head of *Goal_List* for *G* and let *Rest* be *Goal_List*'s
tail. Owen and Hull experimented with other forms of *select*. They also tried a
technique called 'clause selection'. This avoids certain rules in the theory, in the
last clause of *solve*, when it is known that these rules cannot lead to a solution.

They tested three different ways of saving lemmas. The most elaborate of them:

(1)  avoided recomputation, on backtracking, of results which had already been recorded as lemmas;

(2)  explicitly recorded failure of goals and used positive lemmas to deduce failure;

(3)  controlled the number of lemmas generated. Certain Prolog predicates were labelled as 'lemmable'. Results were only stored as lemmas for these.

In all attempts, the revised interpreters had the same first four clauses for *solve* and the same last clause. New clauses were inserted before the last clause.

The simplest lemma generator detected goals which could have at most one solution and recorded lemmas for them. It did not use positive lemmas to deduce failure. Its additional clauses were:

```
solve (Theory, Goal):-
  unique_soln(Theory, Goal),
  lemma (Goal),
  !.

solve (Theory, Goal):-
  unique_soln (Theory, Goal),
  !,
  member ( (Goal:-SubGoals), Theory),
  solve (Theory, SubGoals),
  add_lemma (Goal),
  !.
```

The rule *add_lemma* has a side effect: It *asserts* a new fact

```
lemma (Goal)
```

into the Prolog database. The rule *unique_soln* was hand coded for specific goals.

The second version had an extended form of *unique_soln* which found a generalization *Goal*1 of *Goal* and looked for a lemma for the generalization.

```
solve (Theory, Goal):-
  unique_soln(Theory, Goal, Goal1),
  lemma (Goal1),
  !,
  Goal1 = Goal.

solve (Theory, Goal):-
  unique_soln(Theory, Goal, _),
  !,
  member ( (Goal:- SubGoals), Theory),
  solve (Theory, SubGoals),
  add_lemma (Goal),
  !.
```

If the generalization is already recorded as a lemma and this lemma matches *Goal*, then of course *Goal* is solvable. If the lemma for *Goal*1 does not match *Goal*, then since *Goal*1 can have at most one solution, *Goal* cannot have any solution.

The third and most general lemma generator also coped with goals which might have several solutions. For them, the interpreter used two extra facts:

    lemmable (Goal)

which is true if *Goal* might have several solutions, but it still merits lemmas, and

    complete (Goal)

which is true when it is known that all possible lemmas for *Goal* have been found. The predicate *lemmable* was hand coded, like *unique_soln*. The predicate *complete* is defined by the system as it runs. For each goal, *complete* records that learning for this goal is complete.

After the two clauses above for uniquely solvable goals, it included:

```
solve (Theory, Goal):-
  lemmable (Goal),
  complete (Goal),
  !,
  lemma (Goal).

solve (Theory, Goal):-
  lemmable (Goal),
  !,
  (lemma (Goal);

  member ( (Goal:- SubGoals), Theory),
  solve (Theory, SubGoals),
  not lemma (Goal),
  add_lemma (Goal);

  add_complete (Goal),
  fail).
```

The line *not lemma* (*Goal*) prevents the interpreter from adding a lemma twice. Note that it may cause the interpreter to miss a solution if *Goal* is recursive, because in that case a lemma for it might be discovered and recorded while proving one of the *SubGoals*. Such a *Goal* may not be lemmable.

After learning for a while, this interpreter made about one reference to the *Theory* for every 300 references by the vanilla A interpreter. When it was combined with clause selection, it ran faster than the underlying Prolog system.

The protein topology problem is comparatively simple, because it is an exercise in pure theorem proving. Once a lemma is proved, it will remain true. Lemmas can also be used in planning systems, but then a lemma may cease to be true if the assumptions underlying its proof change. David Wolstenholme has experimented with a lemma generating expert system shell which lets a user give different answers to a repeated question and which also lets him change the rules in the theory. His system uses built-in knowledge about the theory's predicates, such as which of them are lemmable. Whenever it records a lemma, it notes:

- which rule of the theory was used to prove it; and

- what special case of each subgoal was proved, in this rule.

Thus, suppose the theory contains the rules

    R1 : a (X) :– b (X), c.
    R2 : b (m).
    R3 : c :– h.

and the fact that $h$ is a question which the user will answer. If $a$, $b$, $c$, and $h$ are all lemmable and the user answers *yes* to $h$, then the system will record the lemmas

| | | |
|---|---|---|
| a(m) | because | b(m) $\land$ c $\land$ rule R1 |
| b(m) | because | rule R2 |
| c | because | h $\land$ rule R3 |
| h | because | the user said so. |

Note that the system does *not* record all the reasons together as a single condition for *a(m)*. This means that, if the system ever wants to use the lemma *a(m)* again, then it has to check the immediate reasons for *a(m)* and then check all the reasons for those reasons, recursively. This makes using lemmas harder, but recording and updating their reasons is easier.

The system can do quite a lot more than this example suggests. Not all predicates are lemmable and the reasons skip ones which aren't. Its approach also allows for proof of failure. For this, it has features rather like Owen's *complete*ness test.

The purpose of lemmas is to save searching. This is well worthwhile in a theorem prover. It may not be so valuable in a planner. Every time a planning system quotes a lemma, it has to check that the lemma is still true, by checking its reasons. This checking is itself a searching process. It may be simpler than a full-blown proof, but it may still involve quite a lot of work.

### Strengths of the method

- It works. The final protein topology system runs faster than the original, even though it carries the overhead of an explicit interpreter.

- It incorporates all the benefits of EBG.

- This learner recognizes when learning is complete.

### *Limitations*

- The user has to write the rule *lemmable*.

- Owen and Hull experimented with several interpreters for their knowledge base. They say that choosing the right interpreter for a particular knowledge base is hard.

- The method is designed for pure theorem proving. It is not so suitable for other forms of search.


### 7.5.3   CHUNKING

This approach was incorporated into a program developed by Allen Newell and his students, Paul Rosenbloom and John Laird. They developed a heuristic problem solver, which they call SOAR, to which they added chunking almost as an afterthought.

The idea of chunking developed among psychologists in the 1950s. It is justified in part by the so-called 'power law' of learning, which relates the speed of a human skill to the number of times the skill is practiced. A system which learns by chunking shows the same effect.

The chunking process is not quite like most other learning algorithms because it is more intimately integrated into the problem solver. For that reason, a description of chunking must include a relatively detailed description of the associated solver. The solver will be a heuristic searcher, with one elaboration. Sometimes some stage of the algorithm will fail. For instance, the conflict resolution step may fail to produce a unique preferred rule instance. Such a situation is called an **impasse**. The solver resolves any impasse by further heuristic search. It sets up a subgoal to find a solution of the parent level's impasse.

While the solver solves this recursive subgoal, it has access to all data in the parent search space. For each branch of its search, it keeps records of all entities occurring in the parent search space which are tested by any rule applied down this branch. When it has solved the subgoal, it forms a new rule called a **chunk**. The chunk's conditions are all tests made down the successful branch on facts known to the parent task. Its action is to assert that the solution is the one found by this subgoal.

Note that the chunk's action is an assertion. It adds to the searcher's state of knowledge, but does not delete any previously known fact. Hence, chunking is best suited to monotonic search. It can also be applied to non-monotonic search tasks. The example below is a planning task involving operators which negate some of their conditions, but as we shall see, in such situations, chunking can be unsafe.

Within a chunking program, the accumulated facts can be stored in a data structure resembling a stack. (Laird calls it a 'stock'.) Each recursive call of the searcher will correspond to a level of the stack. The facts below this level are the

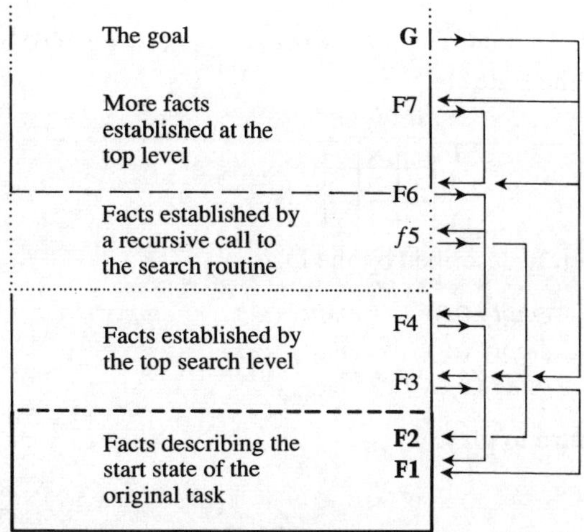


**Fig. 7.16**  A stack of facts derived by recursive search, suitable for chunking.

ones known before the call and the ones above are the ones added during and after the call.

Figure 7.16 is meant to show how facts accumulate on the stack. The arrows down the side show how each fact depends on other facts discovered earlier. The bold dotted line (– – –) shows where the first level of search begins for the goal **G**. The normal dotted line above it (·········) shows where the first level of search created a subgoal to establish F6, which was solved by subsidiary search. The line above it (—— ——) marks the resumption of the first level of search.

The learner forms a new chunk at the end of each level of search. In this example, it will form two. There will be one for the original task:

If **F1** $\wedge$ **F2** then **G**

and one for the subgoal

If **F2** $\wedge$ F3 $\wedge$ F4 then F6.

All the other facts appearing on the stack are either parts of the description of the original task or else are established by previously known rules, so there is no point in creating chunks for them.

The condition for a chunk is found by following the arrows down from its conclusion (**G**). If an arrow points to a fact **F** which was known when the conclusion was posed as a subgoal, then **F** is a condition of the chunk. If it points to a fact (F or *f*) which was established at a more deeply nested level of search, then this fact is not itself a condition, but other facts which it depended on will be conditions. Thus, F6 depends on F4 and *f*5. F4 is a condition in the chunk for F6. *f*5 is not, but its precursors **F2** and F3 are.

*Example (Laird* et al.*): finding a forced win, in noughts and crosses*

Search starts in the state

|   | a | b | c |
|---|---|---|---|
| 1 |   | × |   |
| 2 |   | ○ |   |
| 3 |   |   |   |

which is completely described by the facts

$empty\ (a_1)$     $empty\ (a_2)$     $empty\ (a_3)$     $empty\ (b_3)$
$empty\ (c_1)$     $empty\ (c_2)$     $empty\ (c_3)$
$filled\ (b_1, ×)$     $filled\ (b_2, ○)$

and, if it is ○'s turn to play,

$to\text{-}move\ (○).$

The available rules include

IF *to-move* $(○) \wedge linear\ (x, y, z) \wedge$
   $filled\ (x, ○) \wedge empty\ (y) \wedge empty\ (z)$
THEN
   $play\ (y, ○); play\ (z, ×)$

IF *to*-move $(○) \wedge linear\ (x, y, z) \wedge linear\ (x, y', z') \wedge$
   $\{y, z\} \neq \{y', z'\} \wedge$
   $empty\ (x) \wedge$
   $filled\ (y, ○) \wedge empty\ (z) \wedge$
   $filled\ (y', ○) \wedge empty\ (z')$
THEN
   any move beginning *play* $(x, ○)$ will lead to *win* $(○).$

This leads to a forced win, through the position

|   | a | b | c |
|---|---|---|---|
| 1 | × | × | ○ |
| 2 |   | ○ | * |
| 3 | * |   | ○ |

○ plays at $c_3$, forcing × to play at $a_1$ and then ○ plays at $c_1$. However × replies, ○ wins at one of the squares marked *.

The first move is found by an instance of the first rule, when

$x = b_2$     $y = c_3$     $z = a_1$

and then the win is recognized by an instance of the second rule, in which

$x = c_1$     $y = b_2$     $z = a_3$
$y' = c_3$     $z' = c_2.$

In this example, the facts which were used directly to establish the goal are

> *to-move* ($\circ$)     *linear* ($c_1, b_2, a_3$)     *linear* ($c_1, c_3, c_2$)
> $\{b_2, a_3\} \neq \{c_3, c_2\}$
> *empty* ($c_1$)
> *filled* ($b_2, \circ$)     *empty* ($a_3$)
> *filled* ($c_3, \circ$)     *empty* ($c_2$).

Of these, *filled* ($c_3$) was not true when the search started. It was established by application of the first rule, whose conditions are

> *to-move* ($\circ$)     *linear* ($b_2, c_3, a_1$)
> *filled* ($b_2, \circ$)     *empty* ($c_3$)   *empty* ($a_1$).

All these are true in the start state, so a searcher can form the chunk

> IF *to-move* ($\circ$) $\wedge$ *linear* ($b_2, c_3, a_1$)
>   *filled* ($b_2, \circ$) $\wedge$ *empty* ($c_3$) $\wedge$ *empty* ($a_1$) $\wedge$
>   *linear* ($c_1, b_2, a_3$) $\wedge$ *linear* ($c_1, c_3, c_2$) $\wedge$
>   $\{b_2, a_3\} \neq \{c_3, c_2\}$ $\wedge$
>   *empty* ($c_1$) $\wedge$
>   *filled* ($b_2, \circ$) $\wedge$ *empty* ($a_3$) $\wedge$
>   *empty* ($c_2$)
> THEN
>   *win* ($\circ$).

We shall study two implementations of chunking. Newell's searcher is very well adapted for chunking because it can detect many kinds of impasse, but it is unconventional and so the chunking process is obscured by the details of searching. Therefore, we begin with a more conventional implementation which only forms subgoals to resolve conflicts. This particular algorithm is very simple, with nothing much to recommend it except its simplicity and the fact that it shows what chunking is all about.

### First implementation of chunking

This simple algorithm takes as inputs:

- goal *G*. This is a conjecture which the searcher should prove;

- a database *db* of rules, including any learned chunks;

- a set *Facts* of facts.

It may succeed or fail. In either event, it may augment the data base *db* with additional learned chunks.

This searcher has just one kind of impasse. If ever it has a conflict set *CS* containing more than one rule then it calls itself recursively to choose a best rule out of *CS*. Thus, all its conflict resolution is done by subsidiary search. The process of calling a search algorithm recursively within itself, to solve a subsidiary task, is called **subgoaling**. Most search algorithms have some built-in conflict resolution method. This one does not. In simple tasks, conflict resolution by subgoaling may be unduly cumbersome and slow, but for hard problems it can be very effective.

Note that the subsidiary search is for a *rule*, whereas the original is for a *goal* in the original search space. Thus, these two searches are at different levels of abstraction. The database of rules, *db*, will have to include rules about the choice of rules as well as rules about solving the user's tasks.

```
IF db contains a rule R of the form If Conds then G
    where Conds ⊆ Facts
THEN
    succeed
ELSE
  Fs := Facts;
  WHILE db does not contain a rule of the form
       If Conds then G where Conds ⊆ Fs
    CS := { R ∈ db | R = (If Conds then A ) ∧ Conds ⊆ Fs ∧ A ∉ Fs };
    IF CS is empty
    THEN
       fail
    ELSE
    IF CS is a singleton set
    THEN
       R := the unique rule in CS
    ELSE
       search (a best rule R in CS, db, Fs);
    Fs := Fs ∪ {A}
       where A is the conclusion of R;

    Save a note that the precursors of A are Conds;

  Cs := Conds;
  WHILE Cs contains a fact C which is not in Facts
    Cs := Cs less C ∪ the precursors of C;
  db := db ∪ { If Cs then G }
```

Observe that the last few lines of the algorithm, from

    Save a note that the precursors of *A* are *Conds*;

onwards, could be omitted. The searcher would still work without them. This last section does all the chunking. When chunking is included, then when the algorithm is called again later in similar circumstances, its very first test will succeed and it will not enter its main loop.

### Chunking in SOAR

The searcher which Newell *et al.* invented is quite different.

- SOAR applies all possible rules, so its set of facts grows much faster.

- SOAR has a finer classification of impasses.

- Conflict resolution in SOAR is by a novel scheme of *preferences*. A preference is a fact which can be asserted by a rule, like any other fact.

- SOAR will always try to solve the highest-level goal. If it finds a solution to a high-level goal by chance, before solving all intermediate subgoals, it will forget the subgoals at once.

SOAR is now an old program and it has been through several versions. Hence there is no unique program called SOAR and its description is a bit arbitrary. The following outline is only a rough approximation to Newell's conception. You should study the original references if you want a definitive description.

When SOAR runs, there are actually two levels of search in progress. The upper level is a search in an application domain. The lower search is for a single step in the upper search space. The upper search space is not part of SOAR. It is defined by facts and rules, which are encoded in a form which SOAR can apply. The lower search is performed by a machine which is the core of SOAR. In order to describe SOAR thoroughly, one should specify:

- the allowed structure of an application domain;

- the language of facts and rules in SOAR;

- the machine.

This account will concentrate on the machine.

An application consists of

- a goal, and

- a problem space, which consists of

- states and

- operators.

While the SOAR machine runs, it can keep many goals and many problem spaces and, perhaps, several states in each space. The machine tries to form quadruples of the form

$$(g, p, s, o)$$

where $g$ is a goal, $p$ is a space, $s$ is a state in $p$, and $o$ is an (instance of an) operator of $p$ which might be applied to $s$. Such a quadruple is called a **context**. The last few entries in a context may be unknown, in which case they appear as hyphens: '-'.

The machine starts from the user's goal $g$ which it records in an initial context:

$(g, -, -, -).$

It then searches for a suitable problem space $p$, with which it forms a new context

$(g, p, -, -)$

and then successively a state and then an operator. Whenever it forms a complete context

$(g, p, s, o)$

it tries to apply the operator $o$ to the state $s$, thus creating another state $o(s)$ in $p$. However, it does not then immediately choose $o(s)$ as its next state. This new state is just added to the machine's collection of available states which it might choose from. In fact, all the intermediate choices may be abandoned and it may start afresh in a new problem space or even on a new goal.

The machine's memory consists of

- a *database* of *rules*, and

- a *stack* of accumulated *facts*. The searcher has access to all entries on it, not just to the one most recently pushed on to it.

A condition of a rule is a pattern which can be matched against any fact on the stack or can test for the absence of any fact. The rules' actions are all *monotonic*: they add facts to the stack, but an action can never delete a fact from the stack. The database usually remains unchanged. The only way to change it is by adding a new chunk.

Facts are of three kinds:

(1)   incomplete contexts $(g, \dots )$;

(2)   relations between goals, problem spaces, states, operators, and any other features in them such as blocks and tables. According to Newell, an attribute can also occur as a feature in a relation, so SOAR's language is higher order;

(3)   *preferences*. A preference is a particular kind of fact which is used while making a choice.

The machine's algorithm will be presented twice, first without chunking and then again with chunking. It contains a loop. In each cycle, it adds as many facts as possible to the stack with its rules; then it chooses one entry, a space or state or operator, to extend a context; and then, if the context is completed, it tries to apply the operator and create a new state.

WHILE there is still an incomplete context on the stack
*elaborate*:
   WHILE any instance of any rule can be performed
      do that rule's action;
*decide*:
   try to choose $E$ and $C$, the best entry and context, according to all preferences;
   IF there is an impasse
   THEN
      create a new goal $g$ to solve it;
      push a new context $(g, -, -, -)$ on to the stack
   ELSE
      $C' := $ a new context formed from $E$ and $C$;
      IF $C'$ is incomplete
      THEN
         push $C'$ onto the stack
      ELSE
         $\{C' = (g, p, s, o)\}$
         try to perform the operator $o$ in $C'$;
         IF there is an impasse
         THEN
            create a new goal $g'$ to solve it;
            push a new context $(g', -, -, -)$ onto the stack
         ELSE
            $\{$Say $C''$ is the parent context $(g, -, -, -)$ consisting of just the goal in $C$
              and $Fg$ is the essential feature of the new state $o(s)$ which solves $g\}$
            prune the stack back up to above $C''$;
            push $Fg$ onto the stack.

A **preference** is a structured object consisting of

- a candidate entry $E$;

- the type of $E$: *goal, problem space, state* or *operator*;

- a candidate context $C$;

- a vote; and perhaps

- another candidate entry $E'$.

The possible values of the vote are

*acceptable    reject    better    best    indifferent*

where

- *reject* means that $E$ cannot be added to $C$;

- *better* means that $E$ is preferable to $E'$ for $C$;

- *best* means that $E$ is preferable to any other entry which might be added into $C$;

- *acceptable* means that $E$ is worth considering in $C$;

- *indifferent* means that, if $E$ and $E'$ have the same number of acceptable votes in $C$, then either will do. It is not worth creating a *tie* subgoal to choose between them (see below).

Preferences are added to the stack by the actions of rules, just like any other facts. The choice of $E$ and $C$ proceeds as follows:

```
CS := the set of all pairs (E, C) which have at least one acceptable preference,
          less those with a reject preference;
IF CS is empty
THEN
    impasse: all vetoed
ELSE
    BS := the subset of pairs (E, C) in CS for which
            E has not already been chosen for C, and
            for all E", (E", C) is not better than (E, C);
    IF BS is empty
    THEN
        impasse: no change
    ELSE
        HS := the subset of pairs in BS with earliest context C;
        TS := the subset of best pairs in HS, or if none are best,
                the subset of pairs in HS with most acceptable votes.
        IF TS contains more than one pair and not all
           pairs in TS are indifferent,
        THEN
            impasse: tie
        ELSE
            return E and C.
```

Impasses can arise at two stages: while choosing $E$ and $C$, as above, and also while trying to apply an operator $o$ to a state $s$ in a complete context. The kinds of impasse which can occur at this second stage are:

- *incomplete operator*: not all of $o$'s parameters are specified;

- *inapplicable*: $s$ does not satisfy $o$'s preconditions;

- *not applied*: $o$ was just too complex. Perhaps the computer ran out of memory.

Notes

1. SOAR is provided with an initial set of rules which are designed to solve the subgoals $g$ arising through impasses. These rules are independent of the upper-level task.

2. The particular rules in SOAR for solving a *tie* impasse work by exploring all alternatives exhaustively. Hence, SOAR with these rules is no use for search in infinite spaces.

3. Different accounts of SOAR mention different kinds of impasse.

4. The references do not always keep a clear distinction between SOAR's two levels of search.

5. Laird wrote in his thesis that the chosen entry $E$ overwrites an entry in $C$ and all later entries in $C$ are removed; that is, they become '-'. Thus, if $E$ is a problem space $p$ and $C$ already contains a problem space and a state, then the original $C$: $(g, p', s' \text{ -})$ will become $(g, p, \text{-}, \text{-})$.

6. According to Laird, the machine keeps only one context at any instant. This is called the **current context**. Newell's account (1990), page 172, also refers to 'the existing decision context', but then on the next page he writes of 'the total goal context, which includes the current context, the supercontext, the supersupercontext, and so on'.

7. On page 162 of the same reference, Newell writes

   *'only a single state exists ... at any one time'*

   but then on page 170 he states that

   *'the decisions that must be made are ... what state should be used ...'*.

   If there were only one state then there would be no need for a decision about it.

The major novel feature of SOAR is that every impasse gives rise to a subgoal and all subgoals are created automatically by the impasse recovery scheme at the lower search level. This feature is called **universal subgoaling**. The upper level contains no instructions about forming subgoals, although of course the database may contain rules about solving them.

That completes the outline of the problem solver without chunking. The modifications for chunking are quite simple.

1. During the *elaboration* stage, each application of a rule adds some fact to the stack. For chunking, the fact is tagged with pointers to the rule's conditions,

and for each condition, with pointers to the objects earlier on the stack which that condition tested.

2.  During the *decision* stage, whenever a context $(g, p, s, o)$ is completed and its operator is applied successfully, the machine creates a chunk. The chunk's action is to add the feature of $o(s)$ which is required by the goal $g$. Its conditions are found by tracing back along all tags from this context, up to just above the incomplete parent context

$$(g, -, -, -).$$

Thus, the complete form of SOAR with chunking goes:

WHILE there is still an incomplete context on the stack
*elaborate*:
   WHILE any instance of any rule can be performed
      do some such rule's action:
         Push a new fact $F$ on to the stack;
      Tag $F$ with a pointer to this rule;
      FOR each of the rule's formal parameters $P$
         tag $P$ with a pointer to its actual value
*decide*:
   try to choose $E$ and $C$, the best entry and context, according to all preferences;
   IF there is an impasse
   THEN
      create a new goal $g$ to solve it;
      push a new context $(g, -, -, -)$ onto the stack
   ELSE
      $C' := $ a new context formed from $E$ and $C$;
      IF $C'$ is incomplete
      THEN
         push $C'$ onto the stack
      ELSE
         try to perform the operator $o$ in $C'$;
         IF there is an impasse
         THEN
            create a new goal $g'$ to solve it;
            push a new context $(g', -, -, -)$ onto the stack
         ELSE
            {Say $C''$ is the parent context $(g, -, -, -)$ consisting of just the goal in $C$
              and $Fg$ is the essential feature of the new state $o(s)$ which solves $g$}
            form a new chunk;
            prune the stack back up to above $C''$;
            push $Fg$ onto the stack.

The procedure *form a new chunk* has to collect the chunk's conditions:

$Facts := \{Fg\}$;
$Conds := \{\ \}$
WHILE *Facts* contains any $F'$ later than $C''$ on the stack
   $Conds := Conds$ less any conditions which tested $F'$
               $\cup$ all conditions of the rule which established $F'$;
   $Facts := Facts$ less $F'$
         $\cup$ all facts on the stack which were tested
           by the rule which established $F'$
The new *Chunk* has
  conditions: *Conds*
  action: *assert Fg*

### 7.5.4 SOME PROPERTIES OF DEDUCTIVE METHODS

All the features mentioned below can appear in any deductive method. They have been studied particularly carefully in the case of chunking, so we shall present them just for chunking.

Clearly, if a single task is presented twice then a chunking system can solve it the second time in a single search step. However, there is a penalty. As the learner finds more and more chunks, the branching ratio of its search space grows. At each step of the search, the conflict set of applicable rules becomes larger, so conflict resolution becomes more tricky.

This problem of excess chunks can be overcome to some extent by slowing down formation of new chunks. There is a modified form of the algorithm which only forms a new chunk if the subgoal was solved without further subgoals. The effect is that a big chunk is only formed from a little chunk when the goal solved by the little chunk has been posed at least twice.

Chunking can also increase speed of solution in a single complex problem with several similar component steps. This phenomenon was exhibited by Laird *et al.*, who studied how quickly a chunking system solved the 'Eight puzzle'. The system had to move tiles around a $3\times3$ array, starting at the position

```
1  6  3
2  7  4
8     5
```

until it reached the position

```
   2  3
8  1  4
7  6  5
```

The solution consists of nine basic moves. When the system was allowed a simple heuristic, but no learning, it made ten false steps. When it also chunked all its intermediate steps, it made only five false steps. Thus, knowledge learned early in the task was used later during the same task.

A chunk cannot be guaranteed always to work. Its conditions may be too lax, so it may guide search down a false path. The situation can occur thus: during a first search, some condition *Cond* is true and is necessary for the solution, but is not tested by the rules which solve the task. Hence, *Cond* is not incorporated as a condition of the chunk. Then during a second search, *Cond* is false. The chunk's condition will be satisfied and so its action is applied, but it does not lead to a goal. The chunk's condition should include a conjunct of the form

$$... \wedge Cond$$

This phenomenon was shown very simply by Laird, Rosenbloom and Newell. Their system was presented with the position in noughts and crosses which we studied earlier, in which $\circ$ is to play and force a win.

|   | a | b | c |
|---|---|---|---|
| 1 |   | × |   |
| 2 |   | ○ |   |
| 3 |   |   |   |

Recall, this search succeeds through the position

|   | a | b | c |
|---|---|---|---|
| 1 | × | × | ○ |
| 2 |   | ○ | * |
| 3 | * |   | ○ |

The rules which led to this solution have conditions which only test the occupied squares and the squares marked '*'. They do not test that square $a_2$ was empty, so no such test was incorporated into the chunk's condition. If × had already played in $a_2$, so that the start state is

|   | a | b | c |
|---|---|---|---|
| 1 |   | × |   |
| 2 | × | ○ |   |
| 3 |   |   |   |

then this sequence loses the game for ○.

If search is performed with the original rule base, without chunking, this fallacy may not arise. Other rules can test for the missing conditions and then conflict resolution can make these other rules override the incomplete ones. The rule base for this example might contain rules which we have not used yet, such as

    IF to-move (○) ∧ linear $(x, y, z)$ ∧
        filled $(x, ×)$ ∧ filled $(y, ×)$ ∧ empty $(z)$
    THEN
        play $(z, ○)$

    IF to-move (○) ∧ linear $(x, y, z)$ ∧
        filled $(x, ○)$ ∧ filled $(y, ○)$ ∧ empty $(z)$
    THEN
        play $(z, ○)$; win (○)

The weakness of the new chunk arises in part because we do not know how to resolve conflicts between it and other rules.

### Strengths

- Chunks are intended to remove the need for repeated deep search. This they do well. If the same impasse occurs in two different searches, then on the second occasion the searcher can solve it by a single step.

- Some impasses arise through lack of knowledge about the user's domain. Others can arise through lack of information about search control. Chunking can learn both sorts of knowledge.

- Chunking is an incremental learning algorithm. A newly learned chunk can be applied later in the solution of the task in which it was learned.

- Learning continues as long as the system is presented with new kinds of tasks which it can solve. After that, it stops.

- Chunking in a computer program has quantitative similarities with some kinds of human learning.

- Chunking can be applied in any rule-based system that has subgoaling.

### Limitations

- There is a trade-off between deep search and broad search. Chunking may generate many irrelevant rules which hamper search. After several cycles, conflict sets are large.

- The learned chunk's condition may not be restrictive enough.

Both these limitations can be overcome by good conflict resolution, but the chunking procedure does not suggest how to resolve conflicts involving new chunks. This limitation can only be overcome when the conflict resolution method is specified in detail.


## 7.6   CONFLICT RESOLUTION: RULE ORDERING

This method is the subject of Pavel Brazdil's thesis. Bundy, Silver, and Plummer (1985) give a summary of it. As with so many algorithms, the description here will not do Brazdil's work full justice. His program actually has three separate ways to learn preferences between rules and it has four ways to invent conditions for new rules. The original was written in Prolog and it appears quite unlike the algorithm described here because it depends subtly on Prolog's unification mechanism.

Brazdil's program, called ELM for 'Experimental Learning Model', constructs a partial order '<' among its rules. If two rules $P$ and $Q$ are both in a conflict set and the program has already learned that $P < Q$, then the searcher will prefer $P$ to $Q$. At the same time, ELM can also discover disjunctive conditions. As seems to be common in such methods, the algorithm has to remember some essential details from past examples in the training set.

The algorithm's inputs fall into three sections:

1.  The *context for learning*:

    - a set of basic predicates which the learner may incorporate into conditions;

    - implications between these basic predicates;

2.  Information *already learned*:

    - a set of rules with partly learned conditions;

    - a partial order '<' on the rules, which may be empty;

    - for each learned case $P < Q$ of '<', a note of the example from which it was learned. This example is called the learned case's 'context';

3.  A *training instance*:

    - a sample task;

    - an ideal trace for this task or some equivalent.

Its output consists of:

- a revised set of rules; and

- a revised partial order; and

- if the new partial order is extended, then notes that the extensions were learned from this training instance.

Say that some training instance is called $Ex$ and the conflict set of rules which could be applied to $Ex$ is $CS$. In the ideal trace, the rule which should be chosen from $CS$ is $P$, say. ELM's principle is that $P$ should be preferred to every other rule in $CS$:

$P < Q$ for all other $Q$ in $CS$, if this makes sense.

The algorithm is complicated because it may not make sense. The relation '<' is a partial order. Hence, we cannot require that $P < Q$ if the program has already learned that $Q < P$. This situation arises in practice. In such cases, one clever part of the algorithm invents a new rule $P'$ which is a restricted form of $P$ and it adds the orders

$$P' < P \qquad \text{and} \qquad P' < Q.$$

$P'$ has the same action as $P$, but its condition is more specific. In a simple case, the algorithm invents the condition for $P'$ from $P$ and $Q$ and the current context $Ex$ and from the context $ExQP$ in which it learned to prefer $Q$ over $P$. However, the general form of the information already learned can be quite elaborate and the full algorithm has extra features to handle all possible situations. Before examining the algorithm itself, consider a couple of the complications it has to cope with.

Since '<' is a transitive relation, the algorithm may not learn $Q < P$ directly from an example. Perhaps it learned that

$$Q < R \qquad \text{and} \qquad R < S \qquad \text{and} \qquad S < T \qquad \text{and} \qquad \ldots \qquad U < P.$$

It will keep notes of the contexts for each of these separate learned preferences, but there will be no one context $ExQP$.

The new rule $P'$ should be fitted into the order as thoroughly as possible. The algorithm may have already learned several preferences

$$\ldots < R_3 < R_2 < R_1 < P < S_1 < S_2 < S_3 < \ldots$$

Since $P' < P$, $P'$ will inherit preferences over the $S$s automatically:

$$P' < S_1 < S_2 < S_3 < \ldots$$

but without further action, there is no way to prefer each $R$ over $P'$. This is significant, because the action of $P'$ is the same as that of $P$ and so any other rule $R_j$ not in $CS$ and with a different action should still be preferred to $P'$.

To remedy this, the algorithm adds preferences

$$R < P'$$

if it could have learned them earlier, had $P'$ existed. One has to be careful because any preference $R < P$ may not have been learned directly. If the program has already learned that $T \leq P$ where $T$ is in $CS$ and that $R < T$ in context $ExRT$ and if also the conditions of $P'$ are satisfied by $ExRT$ and $R$ is not in $CS$, then it will learn to prefer $R$ to $P'$ in context $ExRT$ too.

Figure 7.17 shows the situation when $T$ is actually $P$.

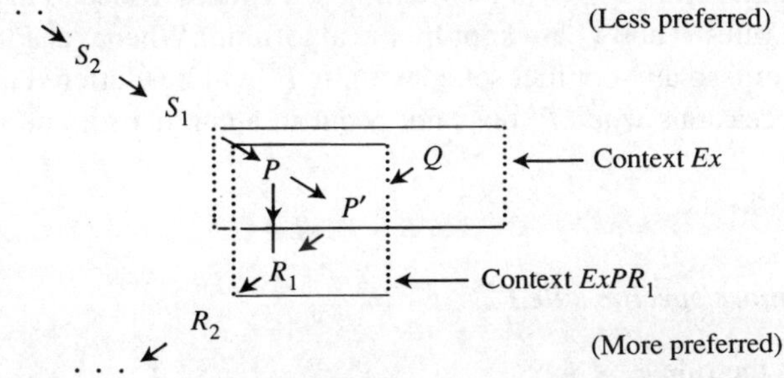


**Fig. 7.17** $R_1$ is preferred to the new rule $P'$, in context $ExPR_1$.

The algorithm runs:

```
IF there is no Q in CS for which Q < P
THEN
    FOR all Q in CS except P
        IF P is not already preferred over Q
        THEN
            learn that P < Q in context Ex
ELSE
    invent a new rule P';
        (The details will be explained below)
    FOR all rules Q in CS, including P,
        learn that P' < Q in context Ex;
    FOR all rules T in CS, including P,
        IF T ≤ P THEN
            FOR all rules R which the algorithm has learned to prefer over T,
                R < T in a context ExRT,
            IF R is not in CS
            AND the condition for P' is true in the context ExRT
            THEN
                learn that R < P' in context ExRT
```

Brazdil experimented with modifications of this algorithm. One of them went further still. Whenever it found some $Q$ in $CS$ which was already preferred to $P$, it invented a new rule $Q'$ too, with a more specific condition than the condition for $Q$ and preferences

$$Q' < Q \quad \text{and} \quad Q' < P.$$

The logic of this is that it makes the preferences on rules rather less dependent on the order of examples in the training set. If $Q < P$ were learned in context $ExQ$, then the algorithm shown above would invent $Q'$ if $ExQ$ follows $Ex$ in the training set, but invent $P'$ if $Ex$ follows $ExQ$ in the training set. Brazdil chose to invent both.

The two old rules $P$ and $Q$ are kept by the algorithm. Whenever $P'$ is applicable then $P$ will be too, so any conflict set containing $P'$ will be cluttered up with $P$, but there may be occasions when $P'$ does not occur in a conflict set and then $P$ is still useful.


### Inventing the more specific rule P'

Let us say, $P$ is the rule

IF $C_P$ THEN $A_P$

$P'$ will be a rule with the same action $A_P$ but a stronger condition

IF $C'$ THEN $A_P$.

The condition $C'$ for $P'$ is designed so that

- $P'$ can be applied to the example $Ex$; and

- if $Q$ is any rule in $CS$ which is preferred to $P$, $Q < P$, then the sequence of examples which led to the preference $Q < P$ would not lead to a preference for $Q$ over $P'$.

ELM has two ways to construct $C'$. It first tries to *remove disjuncts* from $C_P$, so that $C'$ is simpler than $C_P$. If that does not work, then it *adds extra conjunct*s to $C_P$.

Say $RExs$ is the set of contexts

$$\{(Ex' \mid \text{There is some } Q \text{ in } CS \text{ for which the algorithm has}$$
learned preferences

$$Q < R < \dots S < P$$

and $Ex'$ is the context in which it learned that $S < P\}$

Note that in Fig. 7.18 $S$ could be $Q$; so if the algorithm learned $Q < P$ directly in context $ExQP$, then $ExQP$ will be in $RExs$. The training instance $Ex$ is called a **selection context** for $P'$ and each context $Ex'$ in $RExs$ is a **rejection context** for $P'$.

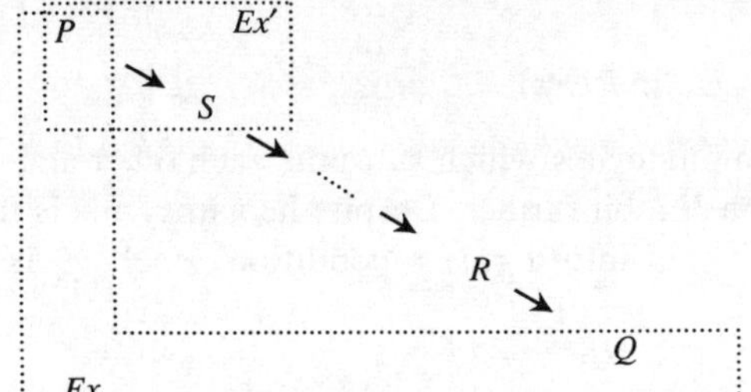


**Fig. 7.18**   The selection context $Ex$ and a rejection context $Ex'$.

### *Removing disjuncts*

Express $C_P$ in disjunctive normal form:

$$C_P = C_1 \vee C_2 \vee \dots C_k$$

where each $C_i$ is a conjunction of basic predicates or negated basic predicates. Since $P$ is in $CS$, at least one $C_i$ is true of $Ex$. Say
$SC := \{C_i \mid C_i$ is true in context $Ex$ and
$\qquad\qquad C_i$ is false in every context $Ex'$ in $RExs\}$
IF $SC$ is empty
THEN
    the method of removing disjuncts *fails*
ELSE
    $C' :=$ the disjunction of all $C_i$s in $SC$.

**Adding conjuncts**

If removing disjuncts fails then $P'$ will be of the form

    IF $C_P \wedge H$ THEN $A_P$.

The new condition $H$ should be true of the context $Ex$ and false in enough contexts so that the algorithm won't learn to prefer any $Q$ over $P'$. The chosen $H$ is

- true in $Ex$; and

- false in every context $Ex'$ in the set $RExs$.

Suppose the basic predicates available to the algorithm are

    $B_1$   $B_2$   $B_3$   $B_4$   ...


    FOR each rejection context $Ex_i$ in $RExs$
      $BS_i$   $:=$ the set of all the basic predicates $B_j$ which
                are true in context $Ex$ and false in $Ex_i$;
      $BS2_i := \{B \mid B \in BS_i \wedge$ for all other predicates $B'$ in $BS_i$,
                    $B'$ does not imply $B\}$
      $H_i$   $:=$ the disjunction of all basic predicates in $BS2_i$
      $H$    $:= H_1 \wedge H_2 \wedge H_3 \ldots$
         (one conjunct for each context $Ex_i$ in $RExs$)

The various conjuncts $H_i$ may contain predicates which subsume each other and so ELM may simplify the extra condition $H$ a bit further. Despite its name, this is the step which may introduce new disjuncts into a rule's condition. Each $H_i$ is a disjunction.


*Example: solving simple equations*

This is the sort of task on which Brazdil originally tested his methods. It is so basic that you may wonder what artificial intelligence is coming to. The objective is to transform some equation such as

    $3 + 2 = x_1$

into another in which one side is the unknown $x_1$ and the other is a number:

    $5 = x_1$     or     $x_1 = 5$.

It is not trivial, however, because the available rules are only the most basic laws of arithmetic. Thus, for instance, the associative law

    $a + (b + c) = (a + b) + c$

has to be invoked explicitly. The equation

    $3 + (1 + 1) = x_1$

is not the same as

$$(3 + 1) + 1 = x_1.$$

A state of this task is a set of equations. The start state is the set consisting of just the given equation:

$$\{3 + 2 = x_1\}.$$

A goal is any state containing one of the desired form:

$$\{\ldots n = x_1 \ldots\}$$

Each rule has:

- a *name*;

- a *pattern*;

- a set of *new patterns*;

- an *additional condition*.

Such a rule will be written thus:

$$name:\ pattern \rightarrow \quad new\ patterns$$
$$\text{IF } additional\ condition$$

The patterns consist of equations whose terms are variables written with capital letters, $X_i$, to distinguish them from the unknowns $x_i$ in the equations to be solved. The additional condition is some heuristic test of the variables in the pattern, composed from basic predicates. One such rule is

$$\text{rule}_1:\ X_1 + X_2 = X_3 \rightarrow X_1 = X_3 - X_2$$
$$\text{IF } X_2 := 4$$

The basic predicates are tests on the variables in a pattern. In this example, they are

| | | |
|---|---|---|
| var $(X)$ | : | true if $X$'s value is unknown (some $x_i$). |
| int $(X)$ | : | true if $X$ is an integer. |
| $X := Y + Z$ | : | true if $X$ is a sum of two subterms. |
| $X := n$ | : | true if $X$ is a constant, $n$. |

The learner is also told in advance that any condition of the form $X := n$ implies *int* $(X)$.

A rule will occur in the conflict set if its pattern matches some equation in the current state and the additional condition holds. Each rule may match several equations and it may match one equation in more than one way. Each match gives rise to an instance of the rule. The conflict set is a set of such instances. When a pattern matches an equation, its variables gain values. The value of a variable is a term in the equation. Thus, the equation

$$3 + (1 + 1) = x_1$$

is matched by the pattern

$$X_1 + X_2 = X_3$$

and the values for variables following the match are

| variable: | $X_1$ | $X_2$ | $X_3$ |
|---|---|---|---|
| matched term: | 3 | $(1 + 1)$ | $x_1$ |

After this match, the basic predicates

$int\ (X_1)$
$X_1 := 3$
$X_2 := X_4 + X_5$
$X_4 := 1$
$X_5 := 1$
$var\ (X_3)$

are all true.

The rule

$$\text{rule}_1:\ X_1 + X_2 = X_3 \ \rightarrow\ X_1 = X_3 - X_2$$
$$\text{IF } X_2 := 4$$

will not occur in the conflict set for the start state

$$\{3 + 2 = x_1\}.$$

Its pattern matches the equation, but its additional condition is false.

The action of a rule consists of forming a new set of equations, by:

(1)    removing the matched equation;

(2)    adding a new equation for each new pattern.

The new patterns can contain the unknowns in the matched pattern and other unknowns too. In new equations, any old variable keeps its matched value. The value of any new variable is a new unknown. For instance, the rule

$$\text{subz:}\ X_1 + X_2 = X_3 \ \rightarrow\ X_2 = X_4 \wedge X_1 + X_4 = X_3$$
$$\text{IF } true$$

matches the start state

$$\{3 + 2 = x_1\}$$

(in just one way). The result of applying it will be the state

$$\{2 = x_4,\ 3 + x_4 = x_1\}.$$

Here, $x_4$ is a new unknown.

The rules which the program has to begin with include

asoc:  $X_1 + (X_2 + X_3) = X_4 \;\rightarrow\; (X_1 + X_2) + X_3 = X_4$
      IF *true*

subs:  $X_1 + X_2 = X_3 \qquad\rightarrow\; X_1 = X_4 \wedge X_4 + X_2 = X_3$
      IF *true*

subz:  $X_1 + X_2 = X_3 \qquad\rightarrow\; X_2 = X_4 \wedge X_1 + X_4 = X_3$
      IF *true*

pred:  $1 = X_1 \qquad\qquad\rightarrow\; 1 + 0 = X_1$
      IF *true*

      $2 = X_1 \qquad\qquad\rightarrow\; 1 + 1 = X_1$
      IF *true*

      $3 = X_1 \qquad\qquad\rightarrow\; 1 + 2 = X_1$
      IF *true*

and so on.

suc:   $0 + 1 = X_1 \qquad\quad\rightarrow\; 1 = X_1$
      IF *true*

      $1 + 1 = X_1 \qquad\quad\rightarrow\; 2 = X_1$
      IF *true*

      $2 + 1 = X_1 \qquad\quad\rightarrow\; 3 = X_1$
      IF *true*

and so on.

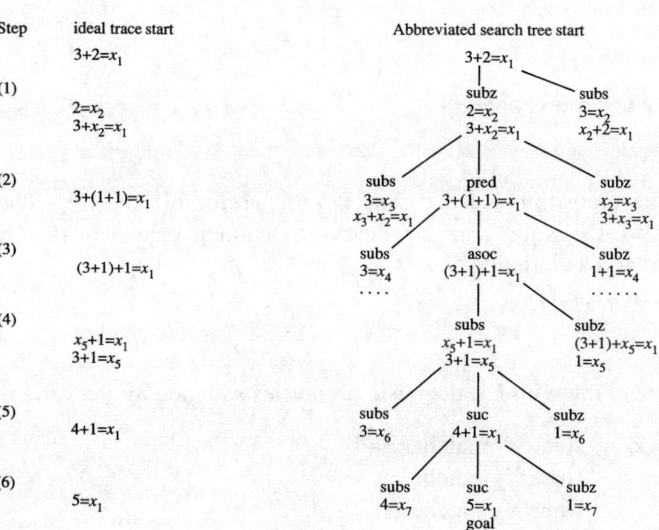



**Fig. 7.19**

As yet the partial order '<' for conflict resolution is empty. The solution of this example can be depicted by a search tree. The one shown in Fig. 7.19 is simplified because it leaves out large parts of the search tree and it also leaves out most equations from each state. Except after steps 2 and 4, only the matched equation is shown.

The learner has to discover that

- *subz* is preferable to *subs* at step 1;

- *pred* is preferable to both *subs* and *subz* at step 2;

- *asoc* is preferable to both *subz* and *subs* at step 3;

- *subs* is preferable to *subz* at step 4;

- *suc* is preferable to *subs* and *subz* at steps 5 and 6.

The results of the first few stages are:

(1)    $subz < subs$ in context $\{3 + 2 = x_1\}$

(2)    $pred < subs$ in context $\{3 + x_2 = x_1, 2 = x_2\}$

       $pred < subz$ in context $\{3 + x_2 = x_1, 2 = x_2\}$

(3)    $asoc < subs$ in context $\{3 + (1 + 1) = x_1\}$

       $asoc < subz$ in context $\{3 + (1 + 1) = x_1\}$

At stage (4) we meet a case where the algorithm must invent a new rule, $P'$. In this case, $P$ is *subs* and $Q$ is *subz*. We cannot impose the condition '*subs* < *subz*' because that would contradict the preference learned at stage (1). The selection context *Ex* is

$$\{(3 + 1) + 1 = x_1\}$$

and the only rejection context is

$$\{3 + 2 = x_1\}$$

in which the algorithm learned that *subz* is preferable to *subs*. The additional condition $C_P$ for $P$ is just *true*, so removing disjuncts cannot work. Therefore, the learner must add a conjunct.

The pattern for $P$ is

$$X_1 + X_2 = X_3$$

and, after matching with *Ex*, the basic properties satisfied by the variables in it are

$$
\begin{aligned}
&X_1 := Y + Z  &&\text{since } X_1 \text{ matches } 3{+}1 \\
&\text{int } (X_2)  &&\text{since } X_2 \text{ matches } 1 \\
&X_2 := 1  &&\text{since } X_2 \text{ matches } 1 \\
&\text{var } (X_3)  &&\text{since } X_3 \text{ matches } x_1
\end{aligned}
$$

When the same pattern is matched with the rejection context, the satisfied basic
properties are

| | |
|---|---|
| int $(X_1)$ | since $X_1$ matches 3 |
| $X_1 := 3$ | since $X_1$ matches 3 |
| int $(X_2)$ | since $X_2$ matches 2 |
| $X_2 := 2$ | since $X_2$ matches 2 |
| var $(X_3)$ | since $X_3$ matches $x_1$ |

The set *BS* of basics which are true in *Ex* but false in *ExQP* is

$$\{X_1 := Y + Z, X_2 := 1\}.$$

Neither of these is more general than the other. Since there is only one rejection
context, the new conjunct $H$ is formed from these two and so the new rule $P'$ is

$$\text{subs2:}\quad X_1 + X_2 = X_3 \;\rightarrow\; X_1 = X_4 \wedge X_4 + X_2 = X_3$$
$$\text{IF } X_1 := Y + Z \vee X_2 := 1$$

In this case, the new rule's additional condition is a disjunction.

This new rule fits into the preferences:

$$subs2 < subs \text{ in context } \{(3 + 1) + 1 = x_1\}$$
$$subs2 < subz \text{ in context } \{(3 + 1) + 1 = x_1\}$$

The program has already learned that *asoc* and *pred* are preferred to *subs*, but the
contexts in which these preferences were learned are not satisfied by *subs2*'s
additional condition. Hence, the algorithm does not add any more preferences for
*subs2*. This completes the algorithm for stage (4).

### Avoiding repeated errors

Besides its primary learning method, ELM also incorporates focusing and
clustering for other purposes. The heuristic constraint on a new rule may be too
specific or simply wrong. For example, ELM was set the exercise

$$x_1 + 4 = 7.$$

It chose the rule

$$\text{subz:}\quad X_1 + X_2 = X_3 \rightarrow X_2 = X_4 \wedge X_1 + X_4 = X_3$$
$$\text{IF } true$$

which leads to the subgoal

$$\{4 = x_4, x_1 + x_4 = 7\}.$$

Better than this, the ideal trace uses a rule

$$\text{over:}\quad X_1 + X_2 = X_3 \rightarrow X_1 = X_3 - X_2$$
$$\text{IF } true$$

so

$$x_1 = 7 - 4.$$

With various other contexts, ELM invented the new rule

over1:  $X_1 + X_2 = X_3 \rightarrow X_1 = X_3 - X_2$
  IF $X_2 : = 4$

and the preference $over1 < subz$. Later, in the solution of

$$x_1 + 5 = 8$$

ELM made the same mistake. Really, the condition of $over1$ is too constrained. It should have been

$$var(X_1) \wedge int(X_2)$$

but the methods described so far could never find this condition.

The solution of this impasse is a form of focusing. Suppose that an ideal trace selects rule $P$ in context $Ex$, but the problem solver opts for $Q$ because $Q < P$; and furthermore, the system has already learned a special case $P''$ and $P'' < Q$, but $P''$ is not in the conflict set for $Ex$. In this case, ELM invents $P'$ by adding conjuncts, as before, but it uses two selection contexts. One of them is $Ex$. The other is the context $Ex''$ in which it learned that $P'' < Q$. The condition for $P'$ is the conjunct of all basic predicates

$$B_1, \quad B_2, \quad B_3, \quad \ldots$$

which are true in both $Ex$ and $Ex''$ and false in any related rejection context.

    FOR each rejection context $Ex_i$ in $RExs$
        $BS_i$  : =  the set of all the basic predicates $B_j$ which
                are true in all selection contexts $Ex, Ex'', \ldots$
                and false in $Ex_i$;
        $BS2_i$ : =  $\{B \mid B \in BS_i \wedge$
                    for all other predicates $B'$ in $BS_i$, $B'$ does not imply $B\}$
        $H_i$   : =  the disjunction of all basic predicates in $BS2_i$
    $H$  : = $H_1 \wedge H_2 \wedge H_3 \ldots$ simplified
            (one conjunct for each context $Ex_i$ in $RExs$)

In this example,

$P$ is *over*
$Q$ is *subz*
$P''$ is *over1*
$Ex''$ is $x_1 + 4 = 7$
$Ex$ is $x_1 + 5 = 8$

The pattern of $P$ is

$$X_1 + X_2 = X_3$$

and so the possible basic predicates which are true of both $Ex$ and $Ex''$ are

$\mathrm{var}(X_1)$
$\mathrm{int}(X_2)$
$\mathrm{int}(X_3)$

The rejection contexts and corresponding disjunctions $H_i$ were

| | |
|---|---|
| $\{3 + 2 = x_1\}$ | $\mathrm{var}(X_1) \vee \mathrm{int}(X_3)$ |
| $\{(3 + 1) + 1 = x_1\}$ | $\mathrm{var}(X_1) \vee \mathrm{int}(X_3)$ |
| $\{3 + (1 + 1) = x_1\}$ | $\mathrm{var}(X_1) \vee \mathrm{int}(X_2) \vee \mathrm{int}(X_3)$ |
| $\{(x_1 + 3) + 1 = 7\}$ | $\mathrm{var}(X_1)$ |
| $\{x_1 + (3 + 1) = 7\}$ | $\mathrm{int}(X_2)$ |

so the extra condition $H$ learned for rule $P'$ was

$\mathrm{var}(X_1) \wedge \mathrm{int}(X_2)$

as desired.


### Learning from positive instances

In practice, unwanted disjuncts can be eliminated by another way which is a form of clustering. Suppose that a rule $P$ is chosen several times in ideal traces, in selection contexts $Ex_1$, $Ex_2$, ... and that the condition for $P'$ is

$$C_1 \vee C_2 \vee C_3 \vee \ldots$$

If at least one $C_j$ is true in all selection contexts $Ex_n$, then it makes sense to assume that this predicate matters. Any other $C_k$ which is false in some $Ex_n$ is probably irrelevant.

The last of ELM's learning strategies works on this principle. After a decent series of tests, if at least one basic predicate is true in all selection contexts, then ELM simply deletes all the disjuncts which are sometimes false.

Like EBG, ELM's learning mechanism has

- a goal predicate (the heuristic condition of a rule);

- a domain theory (its known rules);

- an operationality criterion (all basic predicates are operational);

- a single instance of the goal (from the ideal trace).

It also resembles EBG in that each learned preference between rules has an explanation, the context in which it was learned.

### Strengths of rule ordering

- It learns a lot from a single example.

- It appears to be the best method for conflict resolution. It is faster than the bucket brigade algorithm and its output is accessible.

- It also appears to be a particularly effective method for learning disjunctive conditions. All such algorithms seem to involve saving parts of the training set. Brazdil's algorithm seems to save the bare minimum of detail.

- Brazdil has included a technique which eliminates some surplus disjuncts.

- It applies in any rule-based system.

- It can be used incrementally.

### Limitations

- It requires an ideal trace.

- As with many methods for learning rules, it is susceptible to noisy data.

- Flawed rules remain in the knowledge base. They eventually become irrelevant, but they still cause extra work for the problem solver.

- It involves remembering some training instances.

- What is learned depends on the order of training instances.

## 7.7   INVENTING NEW ACTIONS

There is only one way to invent a new operator. Any new operator is a sequence of old simpler operators. From the point of view of learning, the interesting questions about a new operator are

- what are its characteristics?

- does it make search easier?

The two approaches presented below are designed to provide satisfactory answers for these two questions.

The first approach can be used in a naïve searcher. When such a system first attempts some task, it solves it by heuristic search. When a similar problem appears, rather than repeat the search, the system will:

(1)   check that the two situations really are similar and then, if they are,

(2)   apply all the actions which solved the first task, in the same order.

The action which the system learns is the composite of the simpler ones which it applied as steps in the first search.

The second approach involves a more clever kind of search associated with planning. Suppose that the goal is described by several properties and a planner has invented a partial solution. This is a sequence of operators which achieves a state with some of the goal's desired properties, but not all of them. Also, suppose that every known operator would upset at least one desired feature. In that case, the searcher needs a new operator. This new operator should move the search on to another state which resembles the goal even more, without upsetting any of the established features.

### 7.7.1 COMPOSING CONDITIONS AND EFFECTS OF ACTIONS

This method was developed carefully by Fikes, Hart and Nilsson. They envisaged a robot which is given tasks such as: 'fetch a box into this room'. Their system, called STRIPS, plans the robot's sequence of actions and then generalizes the plan. The generalized plan could also apply to the later tasks: 'fetch a balloon into this room' and 'move a balloon from the next room into the room beyond'. Furthermore, STRIPS stores its generalized plans in such a way that any part of the plan can be retrieved and used separately; or if the whole plan breaks down half way through, the system has enough information to analyze why.

STRIPS has sets of facts which describe possible states of the robot and its environs. A fact is equivalent to a ground literal in predicate calculus. Each set of facts is called a 'model'. One such model records the current state. This is the start state of the planning process. Others derived from it describe other states which the robot might achieve. For instance, if there is a door called $D_1$ between rooms $R_1$ and $R_2$ then there is a fact

$door\_between(D_1, R_1, R_2)$

which is true in the current state. (This particular fact will probably also be true in all other models, since the robot should not knock down doors.) Another fact, which may well vary between models, is that there is a box $B_3$ in room $R_2$:

$in\_room(B_3, R_2)$

Along with these facts, which vary from model to model, STRIPS has extra laws between the predicates which occur in facts. One such law is

$\forall d \, \forall r_1 \, \forall r_2 \, (door\_between(d, r_1, r_2) \Rightarrow door\_between(d, r_2, r_1))$

which says that a door between any two rooms is still a door if viewed from the other side.

The tasks which are set to STRIPS are not really imperatives like 'fetch a box into this room.' Any task is presented to STRIPS in the form of an assertion, such as:

There is a box $b$ in this room, room $R_1$

or, in predicate calculus notation,

$\exists b \; in\_room(b, R_1).$

If there is already a box in room $R_1$, in the current model, then STRIPS constructs a plan which does nothing. Otherwise STRIPS tries to find a plan for changing the current state. The planner searches for a sequence of actions which transform the present state into another in which this assertion is true.

STRIPS has a set of rules which can produce new models from old ones. Each rule consists of:

- a *name* with *parameters*;

- a *condition*;

- a *delete list*;

- an *add list*.

An 'instance' is formed by giving the parameters values. The instance applies to a model if, with these values, the condition can be proved from facts in the model and the laws. In that case, the rule's action forms a new model consisting of:

- all the facts in the original model

- less all facts in the delete list

- plus all facts in the add list.

The precise algorithm for constructing plans does not concern us. (The one used in STRIPS is actually derived from another program called GPS. It is a heuristic searcher, which uses a measure of the difference between the searcher's current state and a necessary set of conditions to prove the goal task.) What is of interest is the form of the resulting plan. It is stored as a so-called 'triangle table' shown in Fig. 7.20. It contains one row for each application of a rule in the plan, and one more row to hold the plan's outcome.

- $Op_j$ is the action of $j$th rule.

- $Adds_{j,j}$ is the add list of operator $Op_j$; and when $k > j$

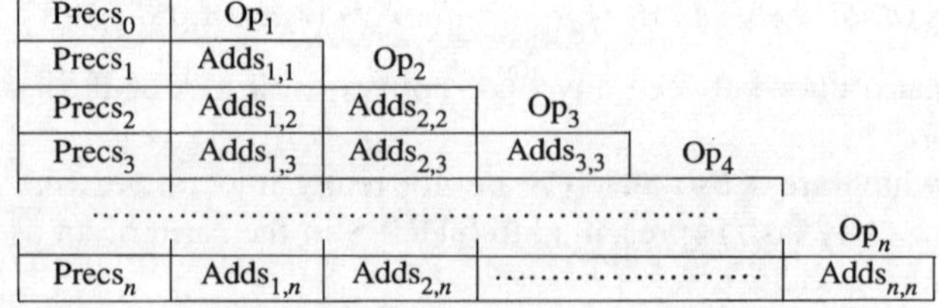

Fig. 7.20   The form of a triangle table.

- *Adds* $_{j,\,k}$ is the subset of all facts in $Adds_{j,\,k-1}$ not in the delete list of $Op_k$. Thus, it consists of all facts added by $Op_j$ and then not deleted by any later operator up through $Op_k$.

- $Precs_0$ consists of all the facts in the start state which are necessary if the plan is to work; and when $k > 0$

- $Precs_k$ is the subset of all facts in $Precs_{k-1}$ not in the delete list of $Op_k$. Thus, it consists of all facts in the start state which are not deleted by any of the actions $Op_1, Op_2, \dots Op_k$.

From this table, one can see how the models evolve. The model produced from the start state by the sequence of operators

$$Op_1, Op_2, \dots, Op_k$$

consists of precisely the facts occurring in

$$Precs_k \cup Adds_{1,\,k} \cup Adds_{2,\,k} \cup \dots Adds_{k,\,k}$$

so the plan's final state is described by all facts in the bottom row. The operator $Op_{k+1}$ can be applied if all its conditions are provable from

$$Precs_k \cup Adds_{1,\,k} \cup Adds_{2,\,k} \cup \dots Adds_{k,\,k}.$$

Assuming that all STRIPS's rules describe the robot's behaviour accurately, the plan is workable if each operator can be applied and if this final state implies the goal.

*Example*: *the blocks world (Chapter 2)*

Recall that this involves four constants

**A   B   C   table**

and two operators, each with its one rule:

1.
*If* on$(x, y) \wedge y \neq$ **table** $\wedge$ clear$(x)$
*then*   *unstack*$(x)$:
       delete the fact   on$(x, y)$
       add the facts   on$(x,$ **table**$)$   clear$(y)$

2.
*If* on $(x,$ **table**$) \wedge$ clear$(x) \wedge y \neq$ **table** $\wedge$ clear$(y) \wedge x \neq y$
*then*   *stack*$(x, y)$:
       delete the facts   on$(x,$ **table**$)$   clear$(y)$
       add the fact   on$(x, y)$.

This example will be more interesting if we omit the condition $x \neq y$ for stacking, so please imagine that a block could in principle be stacked on itself. Thus, the version of the rule for stacking that we shall actually use is

*If* on $(x, \textbf{table}) \wedge \text{clear}(x) \wedge y \neq \textbf{table} \wedge \text{clear}(y)$
*then*     $stack(x, y)$:
        delete the facts    on$(x, \textbf{table})$    clear$(y)$
        add the fact    on$(x, y)$.

Suppose that the initial state is described by the facts

on$(\textbf{A}, \textbf{B}) \wedge$ on$(\textbf{B}, \textbf{table}) \wedge$ on$(\textbf{C}, \textbf{table}) \wedge$ clear$(\textbf{A}) \wedge$ clear$(\textbf{C})$

and in this case the goal is any state in which

on$(\textbf{B}, \textbf{A})$.

This can be achieved by the sequence (see Fig. 7.21):

   $Op_1 = unstack(\textbf{A})$    $Op_2 = stack(\textbf{B}, \textbf{A})$.

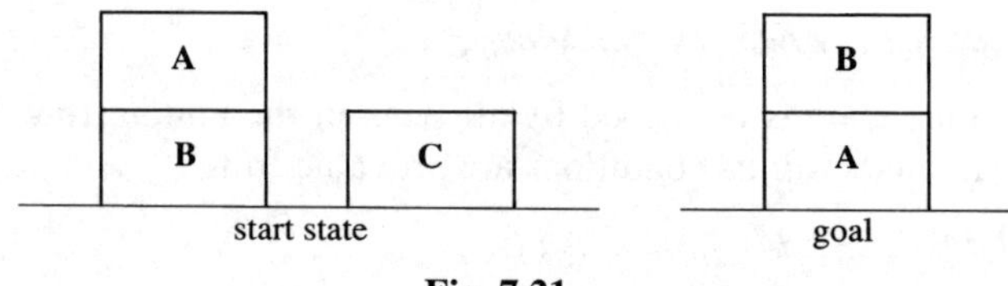


**Fig. 7.21**

The triangle table for this plan is shown in Fig. 7.22.

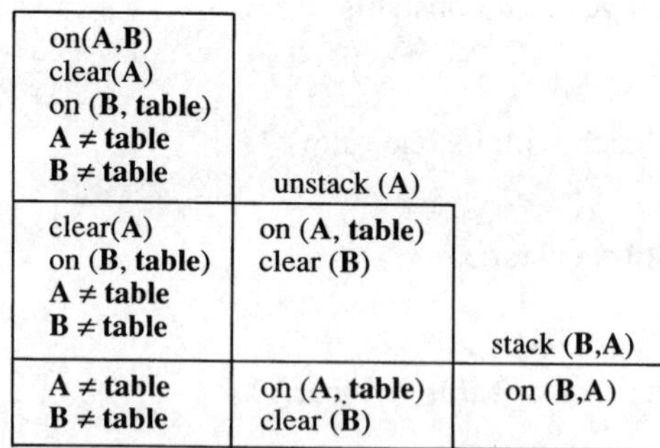


**Fig. 7.22**    The triangle table for a plan for the task in Fig. 7.21.

STRIPS learns generalizations of these triangle tables. Generalizing is done by regression, almost as for explanation-based learning. However, the situation in this learner is a bit more subtle than for EBG because the operators in STRIPS delete facts. They are not monotonic. In order to demonstrate the subtlety, we shall see how generalization would proceed if we ignore it.

### Unsubtle generalization

All terms in $Precs_0$ and in each $Adds_{j,j}$ are replaced with distinct new variables. The entry in any lower box $Precs_k$ or $Adds_{j,k}$ is replaced with the corresponding subset of $Precs_0$ or $Adds_{j,j}$. Then the variables in each row, say row $k$, are unified so that the proof of $Op_k$'s condition will still work.

| | | |
|---|---|---|
| on($a$,$b$)<br>clear($c$)<br>on ($d$,$t$)<br>$e \neq v$<br>$f \neq w$ | unstack($x$) | |
| clear($c$)<br>on ($d$,$t$)<br>$e \neq v$<br>$f \neq w$ | on ($g$,$u$)<br>clear ($h$) | stack ($y$,$z$) |
| $e \neq v$<br>$f \neq w$ | on ($g$,$u$)<br>clear ($h$) | on ($i$, $j$) |

**Fig. 7.23**

In the example from the blocks world, the result of changing terms to variables is shown in Fig. 7.23. After unifying enough to guarantee the correct application of *unstack*, it becomes as shown in Fig. 7.24. In order to apply *stack*, the variable $y$ must be unified with $d$ and $i$, and $z$ with $a$ and $j$ and $e$. In addition, $t$ and $v$ must be **table**. The result is shown in Fig. 7.25.

| | | |
|---|---|---|
| on ($a$,$b$)<br>clear ($a$)<br>on ($d$,$t$)<br>$e \neq v$<br>$b \neq$ **table** | unstack ($a$) | |
| clear ($a$)<br>on ($d$,$t$)<br>$e \neq v$<br>$b \neq$ **table** | on ($a$, **table**)<br>clear ($b$) | stack ($y$, $z$) |
| $e \neq v$<br>$b \neq$ **table** | on ($a$, **table**)<br>clear ($b$) | on ($i$, $j$) |

**Fig. 7.24**

This is still not quite enough to prove that *stack* will work, because it does not guarantee that $d$ is clear. $d$ must also be unified with $b$. The final generalized plan is therefore given in Fig. 7.26.

The penultimate state of the table is interesting because it very nearly portrays a plan which could move $a$ off $b$ and then move a *different* block $d$ on to $a$. There are cases in which generalization in STRIPS can produce new prototype plans which are a lot more versatile than the original. For instance, a robot can move out of one room, $R_1$, to another, $R_2$, and then push a box back from $R_2$ into $R_1$. When this plan

|  |  |  |
|---|---|---|
| on $(a,b)$<br>clear $(a)$<br>on $(d,$ **table**$)$<br>$a \neq$ **table**<br>$b \neq$ **table** | unstack $(a)$ |  |
| clear $(a)$<br>on $(d,$ **table**$)$<br>$a \neq$ **table**<br>$b \neq$ **table** | on $(a,$ **table**$)$<br>clear $(b)$ | stack $(d, a)$ |
| $a \neq$ **table**<br>$b \neq$ **table** | on $(a,$ **table**$)$<br>clear $(b)$ | on $(d,a)$ |

**Fig. 7.25**

|  |  |  |
|---|---|---|
| on $(a, b)$<br>clear $(a)$<br>on $(b,$ **table**$)$<br>$a \neq$ **table**<br>$b \neq$ **table** | unstack $(a)$ |  |
| clear $(a)$<br>on$(b,$table$)$<br>$a \neq$ **table**<br>$b \neq$ **table** | on $(a,$ **table**$)$<br>clear $(b)$ | stack $(b,a)$ |
| $a \neq$ **table**<br>$b \neq$ **table** | on $(a,$ **table**$)$<br>clear $(b)$ | on $(b,a)$ |

**Fig. 7.26**   An incorrect generalized plan.

is generalized, there is no reason why the room that it pushes the box into should be the one it first came out of. Hence, the generalized plan will suggest how the robot could leave room $r$ for room $s$ and thence push a box into a third room $t$.

### The correct form of generalization

The flaw in the unsubtle approach arises because some row in the table contains two facts which can be unified and the operator in this box's row deletes one of them. In this case, the second row contains

    clear$(a)$        and       clear$(b)$

and the operation *stack* deletes *clear*$(a)$. If this generalized plan is ever applied to an instance in which *clear*$(a)$ and *clear*$(b)$ are unified, then they should both be deleted.

The solution is to add to the other facts, in all lower rows, a condition which states that they cannot be unified. In this example, the correct result of applying *stack* is given in Fig. 7.27. (This phenomenon is only apparent in this example because the condition $x \neq y$ was omitted from the rule for *stack*.)

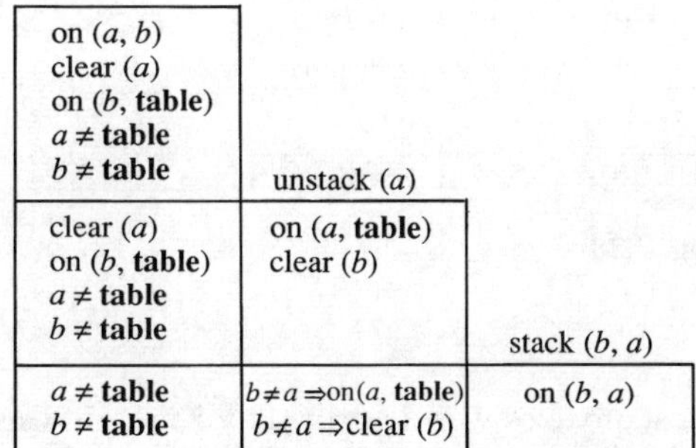

**Fig. 7.27**   The correct generalized plan.

### *Another improvement*

Fikes, Hart, and Nilsson suggest a second change to the unsubtle method. It actually has no effect on the meaning of the generalized table, but it can speed up its use. Suppose that some constant occurs twice in $Precs_0$. It will be generalized to two distinct variables, say $x$ and $y$. At the end of all the unifications, $x$ and $y$ may still be distinct. If one of them (say $y$) does not occur in the bottom row of the table and if the constraints on them are identical, then there is no point in keeping them both. The effect of the two of them is that the table is equivalent to some formula

$$\ldots \exists x \, \exists y \ldots x \ldots y \ldots$$

whereas it could be rewritten as

$$\ldots \exists x \ldots \quad x \ldots x \ldots$$

Every time the generalized table is applied, the extra quantifier just obliges the system to make an extra search for a value for $y$. The value for $x$ will always suffice. STRIPS traces any such pair of variables and unifies them.

Triangle tables are particularly convenient because from them we can easily extract sub-plans. The sub-plan consisting of the first $k$ operators is described precisely by rows 0, …, $k$ of the table. Still more is possible if we add a bit more information to the table: say a fact in any box of the table is 'starred' if that fact is used as an assumption, when proving that the condition for the operator in the same row can be applied. Thus for row 3, $Op_4$ can be applied to any model containing all the starred facts in

$$Precs_3 \cup Adds_{1,\,3} \cup Adds_{2,3} \cup Adds_{3,3}.$$

Now it is clear how to form intermediate sub-plans. The sequence of operators

$$Op_{j+1}, \quad Op_{j+2}, \quad \ldots \ Op_{k+1}$$

can be applied to any model containing all the starred facts in the rectangle of boxes shown in Fig. 7.28.

| $Precs_j$ | $Adds_{1,j}$ | $\cdots\cdots\cdots\cdots\cdots$ | $Adds_{j,j}$ |
|---|---|---|---|
| $Precs_{j+1}$ | $Adds_{1,\,j+1}$ | $\cdots\cdots\cdots\cdots\cdots$ | $Adds_{j,\,j+1}$ |
| $\cdots\cdots\cdots\cdots\cdots\cdots\cdots\cdots\cdots\cdots\cdots\cdots\cdots\cdots$ | | | |
| $Precs_k$ | $Adds_{1,k}$ | $\cdots\cdots\cdots\cdots\cdots$ | $Adds_{j,k}$ |

**Fig. 7.28**

### 7.7.2 MACROS

An alternative way to construct operator sequences was described by Korf. He was interested in search tasks in which:

- each state is described by the values of just a finite set of attributes:

  $\{A : v, A' : v', \dots\};$

- each attribute $A$ has just a finite range of possible values $v$.

In addition, the method works much more simply if:

- each operator is applicable to all states;

- all operators are reversible.

Such tasks can be solved relatively easily whenever the attributes can be ranged in some particular order

$$A_1, \quad A_2, \quad A_3, \dots \quad A_n$$

so that, when any first few of them all have their goal values

$$A_1 : g_1, \quad A_2 : g_2, \quad A_k : g_k$$

then the next attribute can be changed to its goal value

$$A_{k+1} : g_{k+1}$$

without changing any of the previous values $g_j$ $(j \leq k)$. When this last property holds, the task is said to have **serializable subgoals**. Clearly, the property of having serializable subgoals depends on what operators are available.

*Example*: *the blocks world, with four blocks*

For instance, suppose that there is a task in the blocks world whose goal is

$$on(\mathbf{A}, \mathbf{B}) \quad on(\mathbf{B}, \mathbf{C}) \quad on(\mathbf{C}, \mathbf{D}) \quad on(\mathbf{D}, \mathbf{table}) \quad clear(\mathbf{A})$$

starting from the state

$$on(\mathbf{C}, \mathbf{D}) \quad on(\mathbf{D}, \mathbf{A}) \quad on(\mathbf{A}, \mathbf{B}) \quad on(\mathbf{B}, \mathbf{table}) \quad clear(\mathbf{C}).$$

In this case, the property on($\mathbf{A}$, $\mathbf{B}$) is true in both the start state and the goal, so there is some sense in trying to solve the task without ever moving $\mathbf{A}$ off $\mathbf{B}$.

Similarly, one might try to avoid moving **C** off **D**. If the only known operators are *stack* and *unstack* then this is impossible. The task is not serializable.

However, if we invent new composite operators

$$\textit{unstack-two}(x, y) = \textit{unstack}(x); \textit{unstack}(y); \textit{stack}(x, y)$$

and

$$\textit{stack-two}(x, y, z) = \textit{unstack}(x); \textit{stack}(y, z); \textit{stack}(x, y)$$

then it can be done. The solution is the sequence

$$\textit{unstack-two}(\mathbf{C}, \mathbf{D}); \textit{stack-two}(\mathbf{A}, \mathbf{B}, \mathbf{C}).$$

With these two additional operators, the task is serializable.

The search for this solution is quite easy. There are four possible operators and the goal is at depth 2, so crude breadth-first search will find it after examining at most 21 states. (In fact, it will examine at most eight.) If the search strategy excludes operators which move **A** off **B** or **C** off **D**, then there is only one possible operator sequence and it leads straight to the goal!

Of course, this solution has to be expanded into basic operators before we can actually perform it. When it is expanded, the result consists of six basic operators. This is actually a simplest solution; the search with just the basic operators is for a goal at depth 6. At this depth, the search space includes all 73 possible configurations of the four blocks. Thus, search can be made a lot easier by including suitable operators, whether it becomes serializable or not; and if it is serializable and this feature is exploited, then a hard search may become trivial.

(If you want, you can translate this example into the $A_j : v_j$ form. For each block $x$, say the value of attribute $A_x$ is the thing that $x$ is on.)

Korf's objective is to discover operators which make a task serializable. A sequence of basic operators is called a **macro**. His aim is to construct suitable macros.

His method builds up a table of sequences of actions. The table has one row for each attribute, and the $j$th row has one entry for each possible value $v_j$ of $A_j$ except $g_j$. This entry is a macro which:

- can be applied to any state in which $A_i$ has value $g_i$ for $i = 1, \ldots, j - 1$ and $A_j$ has value $v_j$;

- when applied to any such state, produces another state in which $A_i$ has value $g_i$ for $i = 1, \ldots, j$.

Given such a table and any start state, it is easy to find a sequence of actions leading to the goal without further search. Korf gives several examples and he describes three alternative procedures for constructing such a table. All these approaches are based on the assumption that there is a given order

$$A_1, A_2, A_3, \ldots$$

of the attributes.

### Brute force search

> Start with an empty macro table.
> WHILE the table is not full
>   AND resources are sufficient
>     $S := $   a sequence of basic operators.
>          The simplest way is to generate all those of length 1 (the basic operators themselves), then all those of length 2, then all those of progressively greater lengths. Korf suggested doing this by a search method called 'iterative deepening'.
>     Apply $S$ to the goal state $G$
>     IF in $S(G)$ $A_i : g_i$ for $i < j$
>     AND $A_j : v_j$ where $v_j \neq g_j$
>     AND this cell of the table is as yet empty
>     THEN
>       $M :=$ the reverse of $S$
>         (The reverse of a sequence consists of the reverses of the operators in the sequence, in reverse order.)
>       IF for all states $s$ in which $A_i : g_i$ for $i < j$ and $A_j : v_j$
>         $M(s)$ has $A_i : g_i$ for $i \leq j$
>       THEN
>         put $M$ into this cell.

Korf remarks that it is sometimes hard to know when to stop this search. There are tasks whose macro tables can never be filled. For instance, in the eight tiles problem, half the conceivable arrangements of tiles can never be produced from the goal. The graph of all conceivable states is not connected.

This algorithm is practical for construction of macro tables when the search space is not too big, but for larger problems, we need a more efficient approach.

### Composing one sequence with the reverse of another

Brute force search will generate many macros with similar effects. Suppose that $S_1$ and $S_2$ are two sequences of operators, and $k$ is the largest number for which, for all $j \leq k$,

the value of $A_j$ in $S_1(G) = $ the value of $A_j$ in $S_2(G)$.

In that case, there are two macros

$S_1$; reverse($S_2$)     and     $S_2$; reverse($S_1$)

which both leave the first $k$ attributes of $G$ unchanged. What is more, if the problem has a table containing macros whose lengths are at most $2N$, then this method can find them all by constructing sequences $S$ of length at most $N$. When $N$ and the number of basic operations are large, this method provides an immense saving of effort.

It involves remembering all the sequences $S$ as they are formed and storing them so that any pair $S_1$ and $S_2$ can be compared easily. This may be done by recording each sequence at a leaf of a tree whose nodes are indexed by sequences of attribute values. The root of the tree has one child for each possible value of $A_1$. Each child of the root has one child for each possible value of $A_2$, so the root has one grandchild for each pair $(A_1 : v_1, A_2 : v_2)$. Similarly, there is one node at depth $k$ for each possible sequence of attribute values

$$(A_1 : v_1, A_2 : v_2, ..., A_k : v_k).$$

Any operator sequence $S$ is stored at the leaf indexed by the attribute values of $S(G)$.

There is never need to store more than one sequence at any one leaf. When some sequence $S$ has been stored, any other sequence with the same effect on $G$ is better ignored. Later duplicates will be longer and less efficient.

Both this method and the brute force approach will find macros of shortest possible lengths. Korf's third method does not, but it can still prove useful.

### Composing macros

Suppose that the second method, composing $S_1$ with the reverse of $S_2$, is applied whenever possible to all sequences of length at most $l$. This will form all possible macros of length $\leq 2l$. Let all such macros

$$S_1; \text{reverse}(S_2)$$

be stored in the tree too, along with the sequences $S$. These macros can themselves be composed, just like the sequences $S_1$ and $S_2$, to form yet more macros with lengths as large as $4l$.

This approach will probably not find all macros. In fact, it may not even yield macros of minimal length. The reason is that each macro $M$ invented by the first two methods is minimal among the set of possible entries in its cell in the table, but is *not* necessarily the shortest sequence which could be stored at its leaf of the tree.

Korf found that brute force can construct complete macro tables for the eight tiles problem and for the Towers of Hanoi problem with three disks. It requires too much space and time to be practical for Rubik's cube or for the fifteen tiles problem (the problem analogous to the eight tiles but with 15 tiles on a $4 \times 4$ array). Composing sequences can form a complete table for the fifteen tiles problem and it finds all macros bar seven for Rubik's cube. He found these last seven by composing macros.

### Serially decomposable problems

As described so far, there are two conspicuous snags with the method. The simpler is that the last test in the brute force algorithm is at best computationally expensive and at worst not performable if it involves testing on an infinite set of states. The method is only practical when it can be omitted. The second snag is that, in general, there may not be any macros of the type we seek. The brute force algorithm's last test may exclude them all.

The paper by Korf describes a condition on search spaces which guarantees that the last test can be omitted. Let us fix on any suitable order of the attributes

$$A_1, \quad A_2, \quad A_3, \quad \ldots \quad A_n.$$

With respect to this order, we say that an operator $f$ is *serially decomposable* if there are functions

$$f_1, \quad f_2, \quad f_3, \quad \ldots \quad f_n$$

so that, for any state $s$ described by attribute values

$$A_1: v_1, \quad A_2: v_2, \quad A_3: v_3, \quad \ldots \quad A_n: v_n$$

the attribute values for the state $f(s)$ are

$$A_1: f_1(v_1), \quad A_2: f_2(v_1, v_2), \quad A_3: f_3(v_1, v_2, v_3), \quad \ldots$$

Thus, $f$ is serially decomposable iff the $j$th attribute of $f(s)$ only depends on the first $j$ attributes of $s$, for each $j$.

A search space is serially decomposable (with respect to a particular order of attributes) iff all its operators are. This property has the following pleasant consequences:

- All macros are serially decomposable too.

- The inverse of a serially decomposable reversible operator is also serially decomposable.

- It may not be possible to fill the whole macro table, but enough of it can be filled to solve any soluble task.

This last needs explanation. There may be states $s$ from which the goal state $G$ is inaccessible. However, suppose that $s$ is a state from which one can reach the goal by a sequence of operators, say $M$, so $M(s) = G$. If $s$ is not $G$ itself, then for some $k$, the attributes of $s$ are

$$A_i: g_i \quad \text{for } i < k \text{ and } A_k: v_k \text{ where } v_k \neq g_k.$$

Since $M$ is serially decomposable, if $s'$ is any other state with these same first $k$ attributes, then the first $k$ attributes of $M(s')$ are the same as those of $M(s)$, so they are

$$g_1, \quad g_2, \quad \ldots \quad g_k.$$

Thus, $M$ satisfies the brute force algorithm's last condition and its cell in the macro table will be filled.

For a serially decomposable search space, the last test is redundant. The brute force algorithm can be simplified to the following, which is quite practical:

Start with an empty macro table;
WHILE the table is not full
AND resources are sufficient
  $S :=$ a sequence of basic operators;
  IF in $S(G)$ $A_i : g_i$ for $i < j$
  AND $A_j : v_j$ where $v_j \neq g_j$
  AND this cell of the table is as yet empty
  THEN
    $M :=$ the reverse of $S$;
    Put $M$ into this cell.

Rubik's cube is serially decomposable. In fact, it satisfies a stronger criterion: say we fix one corner of the cube and measure all positions relative to it. The attributes of a state are then the positions of the other 25 little cubes. If $f$ is any operator (a twist of the whole cube which leaves the chosen corner fixed), its effect on any one cube will depend *only* on the previous position of that one cube. Thus, this space is serially decomposable for any order of the attributes. Korf calls such an operator *totally decomposable*.

The eight tiles problem is also serially decomposable, when described suitably. Say the attributes are

$A_e$: the coordinates of the empty square

and

$A_i$: the coordinates of the tile marked $i$ ($i = 1, \ldots 8$).

The order must put $A_e$ first, but otherwise the attributes can be in any order. The operations are

$U$: move a tile up
$D$: move tile down
$L$: move a tile left
$R$: move a tile right.

These are not defined for all states. They are partial functions. Korf suggests that, in such a situation, any partial operator can be extended to a total one by defining

$U(s) = s$

if, in state $s$, the empty square is on the bottom row; and similarly, for each other operator $O$,

$O(s) = s$      if not otherwise defined.

The problem with this extension is that the resulting operators are not reversible, but it does not matter because the brute force algorithm always generates shortest possible macros. If a macro contained an operator which could not be applied, then it would not be shortest. Every operator actually used in the table has an inverse on the states to which it is ever applied.

The eight tiles puzzle reveals one other aspect of the method. Instead of the nine attributes $A_e, A_1, \ldots A_8$, it could equally well be described with attributes

$B_{i,j}$ :  the number on the tile in position $(i, j)$
       or '$e$' if this position is empty

where $i$ and $j$ range from 1 to 3. These two representations are dual to each other. The '$A$' representation is a function from tiles to positions and the '$B$' representation is the inverse function from positions to tiles. Although the puzzle is serially decomposable with respect to the first set of attributes, in a suitable order, there is no order on the '$B$' attributes which makes the puzzle serially decomposable.

### Strengths of macros

- They can greatly simplify problem solving.

- The effort needed to compute a table of macros is of the same order as the effort to solve one instance of the task.

### Limitations

- Korf's methods for constructing macros can only obviously be guaranteed to work when:

  (1)  each attribute has only a finite set of possible values;

  (2)  every operator can act on every state;

  (3)  all operators are invertible or at least the search space can be embedded in one in which all operators are invertible.

  However, as in the eight tiles problem, they can also work in some other cases.

- Sometimes there may be no order of attributes which makes the task serially decomposable. There is no simple way to tell if there is such an order.

- The method is sensitive to bias.

- As described, the method does not allow any generalization. However, perhaps it could be combined with the triangle tables of STRIPS, in which case generalized macros might be possible.

## 7.8   POSTSCRIPT

That completes our survey of particular algorithms. Bear in mind that no one on its own leads very far. Most practical learning systems combine two or more methods. One such is Brazdil's ELM. Another has been described by McCluskey. His program finds macro operators and then it uses Vere's criteria to choose relevant conditions for them. A third is MOBAL, which is incorporated into the ESPRIT Project 2154. Recently, Baker and Smaill have written a theorem prover which incorporates program synthesis and EBG.

In broad outline, the subject of practical learning has by now gained a recognizable form. Given any training set which embodies any information worth learning from, there is usually a family of algorithms which can probably learn from it. In many cases, the output of one such family is suitable as input to another family, so, for example, the features discovered by a Kohonen net may be suitable objects for conceptual clustering and the structure discovered by a conceptual clusterer may be a suitable domain for inductive inference and a set of rules formed by inductive inference may be extendable by deductive inference.

There are some obvious gaps still outstanding.

1. *Classification* The scheme used here is not properly defined. Perhaps the perfect scheme would classify algorithms according to properties of the spaces they search in and the quantity and quality of heuristic guidance.

2. *Clustering* The selection of algorithms in Chapter 5 omits a lot from the published literature, but even if many more were included, I do not think the subject would be much clearer. Classification of clustering algorithms is a subject in its own right. I have not seen a classification which attempts to cover all possible forms of correlation and clustering, and there probably will not be one until the literature contains more precise descriptions of more such algorithms. In varying degrees, the same could be said of most classes of algorithms, but clustering is the most obscure case.

3. *Representation and bias* This problem is at least appreciated, if not solved. Bias is principally concerned with languages for representing data and search spaces.

4. *Recognizing what might be learned* Given any training set, how might one guess what could be learned from it? There is a hierarchy of learnable features: functions and predicates, logical interrelationships, facts, rules, forms of heuristic information, the arts of inventing these features, the art of learning these arts, and so on. The hierarchy of learnable features is perhaps best described with meta-languages.

Bias and meta-languages are two of the topics which will be touched on in the last chapter.

# FURTHER READING

The method of proof by resolution was first proposed by Alan Robinson (1965). It is also described by David Luckham (1967) and Alan Bundy (1983).

Contradiction backtracing was described by Shapiro (1983). The account here follows Bundy, Silver and Plummer (1985), which is the source of much of the material in this chapter. Pavel Brazdil's extension for diagnosing a pupil's misconceptions appears in Brazdil (1986, 1989).

The arch-learner mentioned in Chapter 1 learns from similarities. It is described in more detail by Winston (1992).

The focusing algorithm was first reported by Young, Plotkin, and Linz (1977), but their account is very brief. The description here follows Bundy, Silver, and Plummer. The example used above is due to them. Maarten van Someren (1984) describes the 'inspirational' method, and similar approaches, for handling far misses. Mitchell (1977, 1982) invented version spaces and the candidate elimination algorithm.

As well as the candidate elimination algorithm, Tom Mitchell invented explanation-based generalization. The original source is Mitchell, Keller, and Kedar-Cabelli (1986). Keller (1988) has studied the operationality criterion. The standard references on regression are R. Waldinger (1977) and also pages 288–92 of N. J. Nilsson (1980). There is an extension of string unification called paramodulation, which covers cases involving equality. It is described in Chapter 5 of the book by Alan Bundy (1983). For a survey of unification subject to laws, see Siekmann's paper, referred to in the last chapter. The trick of doing without training instances was discovered by van Harmelen and Bundy (1988), and independently by A. E. Prieditis. Minton's implementation appears in Minton, Carbonell, Knoblock, Kuokka, Etzioni, and Gil (1989) and also S. Minton (1988). They provide a thorough discussion which relates EBG to other algorithms and further examples. The idea of using EBG to avoid future failures has been tried by Mostow and Bhatnagar (1987). For a study of EBG with faulty knowledge bases, see R. Doyle (1986) and S. Rajamoney and G. DeJong (1987).

The problem of solving trigonometric integrals was investigated by Utgoff and Mitchell (1982). Boswell (1989) has also studied it, though he concentrates on another aspect. The notion of a poof was described by Alan Bundy (1985).

The presentation given here of lemma generation is reproduced from the accounts by Owen and Hull (1988) and Wolstenholme (1989).

There is a large literature on SOAR, including Rosenbloom and Newell (1983), Laird (1984), Laird, Rosenbloom, and Newell (1984, 1986), and Newell (1990). The standard reference for the original psychological theory of chunking is Miller (1956). The account on pages 181–90 of Laird *et al.* (1986) incorporates the modification which slows down the chunking process. The statement that an attribute can be a feature in a relation occurs in Newell (1990), page 169. The different papers on SOAR use non-standard terminology and they are sometimes somewhat obscure and inconsistent. For instance, Newell (1990, page 169) says that there is no conflict resolution in SOAR and Laird *et al.* (1984) state that in a chunk, constants are generalized to variables, but they do not say how.

Rule ordering was investigated by Brazdil (1978, 1981). Bundy, Silver and Plummer (1985) give a summary of his ideas.

The paper by Fikes, Hart and Nilsson (1972) is a classic, and the one by Korf (1987) is also very readable. McCluskey (1987) presents a composite system incorporating Korf's method. The theory behind MOBAL is presented by Morik (1991). The system of Baker (1992) and Baker and Smaill (1993) finds proofs of universally quantified formulas in arithmetic by generalizing from proofs of instances.

# EXERCISES

1.  A rule-based problem solver discovers the following 'proof' that the ship *Queen Elizabeth II* will sink.

| *database clauses* | *inferred clauses* |
| --- | --- |
| dense(steel) | |
| made_of$(x, y) \wedge$ dense$(y) \Rightarrow$ dense$(x)$ | made_of$(x,$ steel$) \Rightarrow$ dense$(x)$ |
| made_of(qe2, steel) | dense(qe2) |
| dense$(z) \Rightarrow$ sinks$(z)$ | sinks (qe2) |

An oracle asserts the truth of the following ground clauses:

    ¬sinks(qe2)
    dense(qe2) $\Rightarrow$ sinks(qe2)
    dense(steel)
    made_of(qe2, steel)

Show which database clause is wrong.

2.  A rule for providing correct change at a shopping till has two clauses. In Prolog notation, they appear as

```
correct_change(Sum, Change) :-
    coin(Sum),
    Change = [Sum].

correct_change(Sum, Change) :-
    coin(S),
    difference(Sum, S, S2),
    correct_change(S2, Other_change),
    Change = [S | Other_change].
```

There are seven facts defining *coin*:

```
coin(100).
coin(50).
coin(20).
coin(10).
coin(5)
coin(2).
coin(1).
```

The predicate *difference* $(A, B, C)$ is true when $A$ is the sum of $B$ and $C$. This predicate and *coin* are operational.

How could one solve the query

```
correct_change(3, Change)
```

for the unknown, *Change*, in terms of the operational predicates? What rule can be learned by the EBG algorithm from this solution? How might this rule be used in the solution of a second query

    correct_change(17, Change)?

3.   Consider the example of focusing which involves learning when to prefix a noun with the indefinite article 'a'. How does the algorithm behave when it is given the same description space but the training set is presented in the order

      John chased a **dog**
*   **John** will chase dogs
*   Dogs chase **balls**
      A **dog** chases balls
      Dogs will chase a **ball**

At the first far miss, consider all possibilities. At the second far miss, make an inspired guess and justify it.

4.   Consider the following assertions

   (a)   If the maintenance team is competent then every engine is in working order.

   (b)   If a bus's engine is in working order then the bus runs on time.

   (c)   If a bus runs on time then it leaves each stop as predicted by the timetable.

   (d)   The maintenance team is competent.

   (e)   There is a timetable entry for the bus to Southampton leaving Winchester at 8.03 a.m.

   (f)   There is a timetable entry for the bus to Portsmouth leaving Winchester at 9.15 a.m.

Translate these assertions into disjunctive clauses.
Assertion (c) is a definition, so it can be assumed correct. If a reliable oracle confirms assertion (d), (e), and (f), and in addition it says that

- the bus for Southampton does not leave Winchester at 8.03;

- the bus for Portsmouth does not leave Winchester at 9.15;

- the Southampton bus's engine is in working order;

- the Portsmouth bus's engine is not in working order

then what can you conclude about the assertions?

5.   A very important person gains a new robot personal assistant. One of the new assistant's jobs is to make appointments for the VIP. The first such appointment is with another VIP and all the arrangements are made with the second VIP's secretary. By EBG, the robot learns to expect that all arrangements for its master's appointments are made through other VIPs' secretaries.

    Some days later, the VIP wants an appointment with someone not quite so important who has no secretary. Suggest in outline how the robot might correct its erroneous learned information.

6.  The *FunG/PredG* method for inventing recursive definitions, explained in Chapter 6, can be guided by heuristic rules. Some such can be found by rule ordering. Here are five basic predicates which may be relevant.

   1.  *condition* is true of any atomic formula in the condition of $R$.

   2.  *symb* describes the leading symbol of a term or atomic formula:

       *symb(t,s)*

       is true when $t$ is a term and $s$ is its leading function or predicate symbol or when $t$ is a constant or variable and $s$ is the same as $t$.

   The easy way to describe other properties is by means of places, as in Chapter 6.

   3.  *place* is a predicate with three arguments:

       *place(t, p, v)*

       which is true when $t$ is a term, $p$ is a place, and $v$ is the subterm of $t$ at place $p$.

   4.  '$\leq$' relates a place to its sub-places.

       $p \leq q$

       if $q$ is an extension of $p$, so for any $t$ which has a place $q$, the subterm at place $q$ in $t$ is a subterm of the subterm at place $p$ in $t$:

       $$place(t, p, u) \wedge place(t, q, v) \Rightarrow \exists r\, place\,(u, r, v).$$

       The last of the five is

   5.  identity of symbols, '$\equiv$'. This may apply to rules, literals, terms, predicate and function symbols, constants and variables, and places.

   (There are other basic predicates which could be included. For instance, one could invent a predicate which counts the occurrences of one symbol in a rule or term. Another predicate might test to see whether two terms can be unified.)

   (a)   Write down a selection of properties of the rules

   $$\begin{array}{ll}
   on(a, \textbf{table}) & \Rightarrow column([a]) \\
   on(b, \textbf{table}) \wedge on(c, b) & \Rightarrow column([c, b]) \\
   on(d, \textbf{table}) \wedge on(e, d) \wedge on(f, e) & \Rightarrow column([f, e, d])
   \end{array}$$

   which can be described with these predicates.

   (b)   Starting from a state in which these three are the only known rules, write down an ideal trace leading to a state containing the definition of *column*. At each point along this trace, find an alternative search step which would lead off the ideal trace.

   (c)   Write minimal rules for the operators *FunG* and *PredG*. Each operator will have just one such rule. This rule's condition is the minimal necessary condition for applying the operator.

   (d)   With the ideal trace and false search steps which you chose in part (b), use the rule-ordering technique to develop more sensitive rules for *FunG* and *PredG*.

7.  Prove the following properties of the candidate elimination algorithm.

    (a)  All the time the algorithm runs, if $s \in S$ and $I$ is a positive instance which the algorithm has already chosen, then
         $s$ matches $I$.

    (b)  All the time the algorithm runs, if $s \in S$ and $g \in G$ then
         $s \le g$.

    (c)  If the algorithm does not fail then, when it finishes, either
         $$S \ne G \quad \text{or} \quad S = G = \{c\}$$
         for some unique concept $c$.

8.  Show that the blocks world is serially decomposable and construct a macro table for it.

9.  Construct a few macros for part of a table for the eight tiles problem. (The whole table is rather big.)

# 8 Further developments

## SUMMARY

The three major theoretical approaches to learnability are centred on:

- Gold's theorems;
- PAC theory;
- complexity and what is computable.

Ease of learning can be changed by:

- choice of representation;
- meta-level views of a task.

The subject offers many topics for further research.

## INTRODUCTION

So far, apart from occasional sections on classification, search, and related methods, we have treated learning as an experimental subject. You dream up an algorithm, encode it, test it, and if it works, you describe it and publish it. This introduction to algorithmic learning will be rounded off with brief notes on some theoretical topics.

## 8.1   THEORIES OF LEARNABILITY

Up to this point, we have been proceeding in a mood of blind optimism. We have taken for granted that all the algorithms presented will work efficiently on a selection of realistic training sets and that for any realistic training set there will be a suitable algorithm.

Theoreticians take a different approach. The subject has been revolutionized in the past few years by notions of **learnability**: in what situations is learning possible and feasible? Whereas experimenters are generally optimistic, theoreticians tend to be pessimists. They keep telling us that learning is sometimes impossible and often impractical. Gold's theorem, described below, is a classic case. It asserts that there is *no* algorithm capable of learning context-free grammars. In a sense, he is undoubtedly right, but in many cases, his theory is best ignored. There are useful learners, such as Wolff's, which do *often* provide grammars which *approximately* describe the target language.

Valiant studied the notions 'often' and 'approximately'. His theory predicts that learning is sometimes possible but not feasible. Again, he is undoubtedly right, but still I urge you, be not dismayed. If you think that you can devise a practical learner for a specific situation, then you are probably right. These theories of learnability depend on strict assumptions about what is to be learned and what information is available in a training set. Some theories show that learning becomes impractical in complex situations. They also tend to assume that training data are random.

In the situation which you face, any or all of these assumptions may be wrong. A realistic training set may contain much more information than is used by the first algorithm you consider. A different algorithm, or the same algorithm applied to a different description of the task, may work better. Do not be put off by talk of complexity. Your task has a fixed size. Perhaps your algorithm would be impractical on much larger tasks, but if so, that is only a matter of theoretical interest. Furthermore, your training set may well not be random. Recall, when a mother talks to her young child, her sentences are highly redundant with lots of repetition. You can probably arrange your training set so that it is repetitious and redundant too, and then all theoretical results will be void. We live in a world of low entropy and learners take advantage of it.

Despite their gloomy attitude, theorists still have useful ideas. They often provide positive results which state that learning *is* possible and even practical, and these results are sometimes proved by providing explicit algorithms with demonstrable properties. In addition, when a theory presents a negative result, it at least warns us that the approach of that theory will not work under the theory's assumptions, so if those assumptions apply, it would be wise to concentrate on other methods.

There are several theories of this sort. One, due to Lakshmivarahan, studies how estimates of probabilities of events evolve. Say some experiment $X$ can have outcomes $x_1$, $x_2$, ... $x_M$. The learner observes outcomes of the experiment many times. After the $k$th observation, he estimates that outcome $x_j$ may occur with probability $p_j(k)$. The $M$-tuples

$$(p_1(k), p_2(k), ... p_M(k))$$

form a Markov sequence and Lakshmivarahan studies its properties. Perhaps the most profound theory of learning to date is so-called PAC learnability, invented by Valiant. We shall study it, but we begin with a much earlier theory due to Gold.


### 8.1.1 LEARNABILITY OF GRAMMARS: GOLD'S THEOREM

Gold asked: Can there exist an algorithm $A$ whose input is a sample of a *language* and whose output is a *grammar* for that language? Say the language is called $L$. It is a set of finite strings ('words') of symbols out of some given set called the *alphabet. L* could be any one of a large class of languages, such as the class of all context-free languages over the alphabet. The input will be an infinite sequence

$$w_1 \quad w_2 \quad w_3 \quad w_4 \quad \ldots \quad w_{622} \quad w_{623} \quad w_{624} \quad \ldots$$

of words from *L*. Clearly, this sequence must contain enough words. If it consisted of just one word *w* repeated over and over again, then no algorithm could possibly distinguish between *L* and the language $\{w\}$ consisting solely of *w*, so we must assume at least that *L* has *no* strictly smaller context-free sub-language containing every word in the sequence.

Gold proved that there is no such algorithm.

Suppose that there were, so *A* exists. Suppose also that *L* is infinite. The essence of an algorithm is that it produces its answer in a finite number of steps. It takes at least one step to read a word from the sequence, so while it runs, *A* can examine only a finite initial sub-sequence of the sequence:

$$w_1 \quad w_2 \quad w_3 \quad w_4 \quad \ldots \quad w_N$$

for some number *N*. This number *N* may be very large. What is more, *N* may depend on *L*, so the time that *A* runs for could be made as long as you wish by giving it difficult training sets; but still, for each training set, there is some such *N*.

Let *L'* be the finite sub-language of *L* consisting of just the words occurring among $w_1, w_2, \ldots w_N$. Note that *L'* is another context-free language, so *A* should be capable of learning a grammar for *L'* too. Consider a new infinite sequence formed by repeating this finite one over and over again:

$$w_1 \quad w_2 \quad w_3 \quad w_4 \quad \ldots \quad w_N \quad w_1 \quad w_2 \quad w_3 \quad w_4 \quad \ldots \quad w_N \quad w_1 \quad w_2 \ldots$$

This contains every word of *L'* and no words not in *L'*, so *A* with this input should produce a grammar for *L'*. However, by definition of *N*, we know that *A* will only examine the first *N* words in the sequence and so *A* cannot distinguish between this sequence and the original one. Hence, its output with this sequence will be a grammar for *L*. Therein lies a paradox, so *A* cannot exist.

Despite this fact, many people have experimented with programs which invent grammars from samples of languages. Gold's theorem is limited because it assumes that the conjectured algorithm *A* is infallible. All such algorithms will produce wrong grammars in many cases, but they are still valuable, even though they are unreliable.


### 8.1.2 PAC LEARNABILITY

The letters PAC stand for 'Probably Approximately Correct'. Valiant's paper revolutionized the subject. At least one world-class researcher has given up since.

Valiant's approach is the opposite of Gold's. He considers algorithms which may sometimes go wildly astray, producing output which bears negligible resemblance to what we hope to learn. Also, even when the algorithm is working well, it may not produce an exact description of the target concept. It may just produce a good guess at it. In PAC learnability, these notions are made precise.

Say $X$ is any set. We take it as the set of all conceivable examples and counter-examples. There is a subset $C \subseteq X$, the *concept* of examples. If $X$ is finite then we might reasonably imagine that $C$ could be any subset of $X$. However, if $X$ is infinite, there are just too many possible subsets, so we restrict attention to those in some class $\mathscr{C}$ of concepts. $\mathscr{C}$ is a set of subsets of $X$.

According to PAC theory, the objective of learning is to discover a good approximation to $C$. That is to say, a learning function $f$ takes as input a finite training set $T$ from $X$ and produces a description of another subset $f(T) \subseteq X$. We hope that $f(T)$ will approximate $C$. In general, $f(T)$ might not be in $\mathscr{C}$ but in some other class $\mathscr{H}$ of *hypotheses*. For convenience here, we shall take it that the possible hypotheses are just the potential concepts, so $\mathscr{H} = \mathscr{C}$. The training 'set' $T$ is actually a sequence of the form

$$((x_1, s_1) \quad (x_2, s_2) \quad (x_3, s_3) \quad \ldots \quad (x_m, s_m))$$

where each $x_j$ is in $X$ and

$$s_j \;\; = \;\; 1 \text{ if } x_j \text{ is in } C$$
$$\phantom{s_j} \;\; = \;\; 0 \text{ otherwise.}$$

The only restrictions on $f$ are that for each training set $T$

$$f(T) \in \mathscr{C}$$

and $f$ is *consistent* with $T$: for each $j$,

$$x_j \in f(T) \Leftrightarrow s_j = 1.$$

If learning is to be useful, then the training set $T$ should not have to be too big. If $T$ contains only a very few instances then of course the learning process will inevitably be unreliable, but there should be some number $m$ so that, for most training sets $T$ containing $m$ samples, the difference

$$C \,\Delta f(T) = \{x \in X \mid x \in C \wedge x \notin f(T) \vee$$
$$x \notin C \wedge x \in f(T)\}$$

between $C$ and $f(T)$ is small.

The notions of 'most training sets' and 'small' only make sense when there is a *probability* measure $P$ on $X$, so we assume there is one. Suppose that $\varepsilon$ is some small positive number. When we say that $f(T)$ is a good approximation to $C$, we mean that

$$P(C \,\Delta f(T)) < \varepsilon.$$

There is a product measure $P^m$ on the set of $m$-tuples $X^m$. Suppose that $\delta$ is another small positive number. The phrase 'for most training sets' means that in $X^m$

$$P^m \{(x_1, x_2, \ldots x_m) \mid P(C \,\Delta f(T) \geq \varepsilon\} < \delta.$$

It is a strange and beautiful property of the theory that often the choice of $P$ *does not matter*.

*Example: approximately learning the notion of 'medium build'*

Suppose that we want to characterize the class of people of 'medium build'. Each person has a height and a weight, two real numbers. Thus $X$, the set of all possible examples, consists of pairs of real numbers: $X = \mathbb{R}^2$. The class $\mathscr{C}$ of possible concepts could be quite complicated, but for simplicity we take $\mathscr{C}$ to be the class of closed rectangles

$$\{(h, w) \mid a \le h \le b \wedge c \le w \le d\}.$$

The target concept $C$ is one such rectangle:

$$\{(h, w) \mid s \le h \le t \wedge l \le w \le o\}.$$

Somebody is of medium build if his height is in the range $s \dots t$ and his weight is in the range $l \dots o$. The distribution of peoples' heights and weights, $P$, is not known. Despite that, we shall see that there is a learning function $f$ and a bound $m$ on the necessary size of the training set. The learned hypothesis, $f(T)$, will be another such rectangle (see Fig. 8.1).

Figure 8.1 shows the target concept $C$, the solid rectangle. It also shows four strips around the inside edges of $C$. The one on the left is shaded. These strips are chosen so that the measure of each is small—less than $\varepsilon/4$. In this particular example, there is a suitable function $f$ which ignores all counter-examples. $f(T)$ is the smallest rectangle which just contains all the positive examples (see Fig. 8.2).

If every strip round the edge contains an example then $f(T)$ will encompass the middle section of $C$ inside the strips, so $C \mathbin{\Delta} f(T)$ lies within the strips. Therefore

$$P(C \mathbin{\Delta} f(T)) < \varepsilon/4 + \varepsilon/4 + \varepsilon/4 + \varepsilon/4$$

as desired. If $T$ contains enough examples (positive and negative) then, in most cases, all four strips will contain an example. Blumer *et al.* show that the chance of $T$ missing an edge strip is less than $\delta$ whenever $m$ is large enough. In this case, it suffices if

$$m \ge \max\left\{\frac{4}{\varepsilon} \log_2 \frac{2}{\delta}, \quad \frac{32}{\varepsilon} \log_2 \frac{13}{\varepsilon}\right\}$$

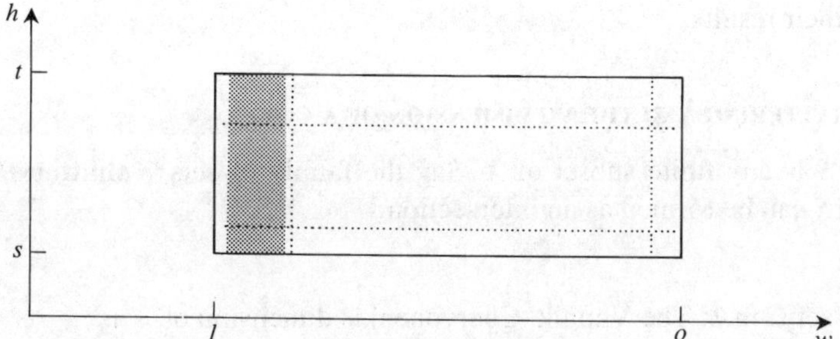

**Fig. 8.1** The zones of error, in an approximation to a concept.

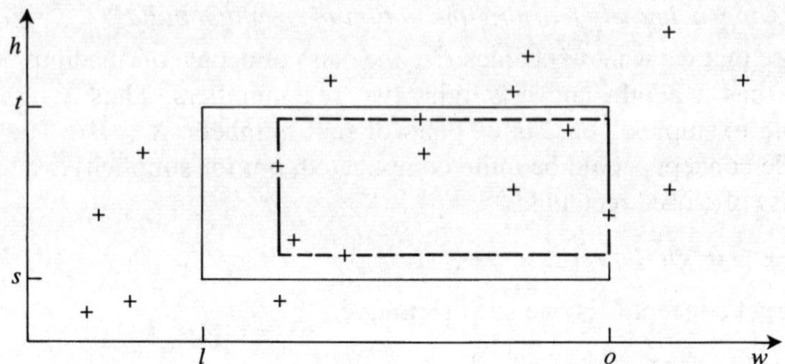


**Fig. 8.2**   The approximation found from a training set.

where logarithms are taken to base 2. Notice that the probability distribution $P$ does not affect this result. We had to assume that some $P$ exists, so that we could describe the edge strips. Different distributions will give rise to different widths of strip, but $P$ does not affect the necessary value of $m$.

The natural questions of PAC theory are:

- Does there exist such a function $f$, independent of $P$? If there is such an $f$, we say that the class $\mathscr{C}$ is *uniformly learnable*. The learning function $f$ should work for any distribution $P$, as long as $P$ can assign probabilities to sets such as rectangles. In that case:

- Is there an algorithm $A$ which computes $f$?

One can also ask how fast $m$ rises with the required precision and reliability of $A$, $1/\varepsilon$, and $1/\delta$.

There are cases where $\mathscr{C}$ is the union of a sequence of simpler classes $\mathscr{C}_n$ and each $\mathscr{C}_n$ is learnable. The next questions that arise are:

- How fast does the complexity of $A$ rise with increasing $n$? and,

- How fast does $m$ rise with $n$?

Blumer *et al.* investigate all these issues in depth. This account will only outline some of their results.


### 8.1.3   SHATTERING AND THE VC DIMENSION OF A COVER

Suppose $S$ is any finite subset of $X$. Say the family of sets $\mathscr{C}$ **shatters** $S$ if every subset of $S$ can be formed as an intersection

$$S \cap C$$

for some set $C$ in $\mathscr{C}$. The Vapnik–Chervonenkis dimension of $\mathscr{C}$ is

$$\dim \mathscr{C} = \max \{\, |S| : \mathscr{C} \text{ shatters } S \,\}.$$

It is the number of points in a largest set which $\mathscr{C}$ shatters. If for any number $n$ there is a set of $n$ points shattered by $\mathscr{C}$, then the dimension of $\mathscr{C}$ is infinite.

*Examples of classes and their dimensions*

1.  If $\mathscr{C}$ is finite then dim $\mathscr{C} \leq \log_2 |\mathscr{C}|$. (If $S$ is any subset of $X$ containing more than $\log |\mathscr{C}|$ points then $S$ has more than $|\mathscr{C}|$ subsets, but there at most $|\mathscr{C}|$ subsets of the form $S \cap C$ so some subset of $S$ cannot be of that form.)

2.  $X$ is infinite and $\mathscr{C}$ is the set of finite subsets of $X$. Then dim $\mathscr{C}$ is infinite.

3.  $X$ is infinite and $\mathscr{C}_n$ is the set of subsets $C$ with at most $n$ points. Then dim $\mathscr{C}_n$ is $n$. (If $|S| < n$ then every subset of $S$ is in $\mathscr{C}$. If $|S| > n$ then $S$ itself is not of the form $S \cap C$.)

4.  $X$ is the real line, $\mathbb{R}$, and $\mathscr{C}$ is the family of closed intervals

    $$[a, b] = \{\, x \in \mathbb{R} : a \leq x \leq b \,\}.$$

    Then dim $\mathscr{C}$ is 2. (If $S = \{3, 6\}$ then the subsets of $S$ are

    $$
    \begin{aligned}
    \{\ \} \quad &= S \cap [0,1] \\
    \{3\} \quad &= S \cap [2, 4] \\
    \{6\} \quad &= S \cap [5, 7] \\
    \{3, 6\} &= S \cap [2, 7]
    \end{aligned}
    $$

    so dim $\mathscr{C} \geq 2$. If $S$ is any set of three distinct points in $\mathbb{R}$ then any interval containing the greatest and the smallest must also contain the one in between, so dim $\mathscr{C} < 3$.)

5.  $\mathscr{C}$ is the class of rectangles

    $$\{(h, w) \mid a \leq h \leq b \wedge c \leq w \leq d\}$$

    of the example above. This class has dimension 4. It is easy to see that it shatters the set shown in Fig. 8.3 so dim $\mathscr{C} \geq 4$. If $S$ is any subset of $\mathbb{R}^2$ containing five points then the smallest rectangle containing them all must either have two points of $S$ on one face or have a point of $S$ in its interior. In either event, no rectangle can exclude the odd point and include the rest, so dim $\mathscr{C} < 5$.

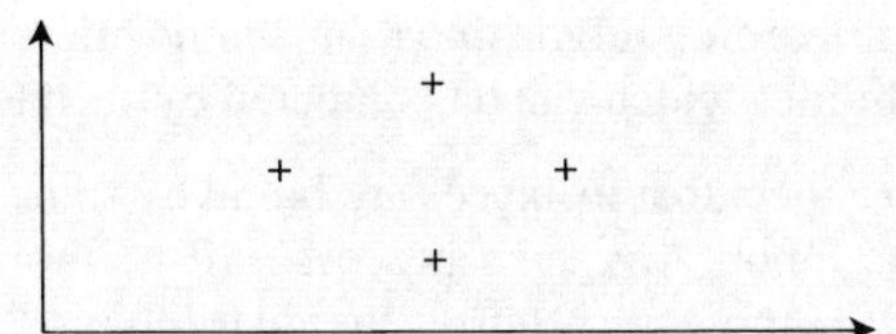

**Fig. 8.3**  The class of rectangles has dimension $\geq 4$.

The main theorem of PAC learning theory is the following:

*If $\mathscr{C}$ is a well-behaved family of subsets of $\mathbb{R}^n$, then it is uniformly learnable if and only if it has finite dimension.*

*When* dim $\mathscr{C}$ *is d, finite, and* $0 < \varepsilon < 1$ *and*

$$m \geq \max\left\{\frac{4}{\varepsilon}\log_2\frac{2}{\delta}, \quad \frac{8d}{\varepsilon}\log_2\frac{13}{\varepsilon}\right\}$$

*then any consistent function f is a learning function for $\mathscr{C}$.*
*If* $0 < \varepsilon < 1/2$ *and*

$$m < \max\left\{\frac{1-\varepsilon}{\varepsilon}\log_e\frac{1}{\delta}, \quad d\,(1-2(\varepsilon+\delta-\varepsilon\delta))\right\}$$

*then no function f is a learning function for $\mathscr{C}$.*

This result does not just apply to subsets of $\mathbb{R}^n$. It can also be used in discrete learning situations. Say *true* is represented by the number 1 and *false* by 0. Any situation described by $n$ predicates corresponds to a vertex of a unit cube in $\mathbb{R}^n$. (Recall the discussion of how to build a decision tree by partitioning its training set, in Chapter 5.) Thus, any of the discrete situations we have considered before can be embedded in $\mathbb{R}^n$, for some $n$, and any concept described in terms of a finite family of predicates corresponds to a set of corners of such a cube. The class of possible concepts corresponds to a class $\mathscr{C}$ of such sets of corners. Either of them is learnable if and only if the other is.

The method can be extended to cover learning from noisy data and it has obvious applications to conceptual clustering.


## 8.2   COMPUTABILITY AND LEARNING

The approach adopted by PAC theory is unsatisfactory from three points of view.

1.   The possible training instances are assumed to form a set such as Euclidean space. The set of computable instances is countable and Euclidean space is not countable, so most points in Euclidean space cannot be computed. It would be a strange world which might ever generate an uncomputable training instance.

2.   PAC theorists assume that the samples in the training set may be distributed randomly according to (almost) *any* probability distribution $P$. When learning, particularly when learning algorithms, it seems more natural to restrict attention to probabilities which can be computed by such algorithms.

3.   It seems natural to expect that instances are found by some computable process with random finite input. Any such process will be more likely to generate instances with low complexity. Hence, the distribution $P$ should assign high probabilities to simple instances.

### 8.2.1 COMPUTABLE EXTRAPOLATIONS

Solomonoff has developed a theory of complexity which can be regarded as a universal way of learning. He and Kolmogorov both discovered essentially the same notion of complexity. The following account is adapted from his paper, which contains proofs of the theorems quoted below and is perhaps still the best introduction.

For any organism, be it natural or artificial, learning is a method for predicting likely future events from past experience. Solomonoff's learner is a passive creature which accepts its lifetime's experience as a stream of bits:

00011011111000101101001111001010100111100011...

The creature's objective is to predict, at each stage, what the next bit of its input will be. Our real lives appear a good deal more interesting than this. For a start, we are not passive. In addition, our experiences appear to have more features than a bit stream. Solomonoff's idea appears reasonable, though, if one allows that the creature might be a young Victorian lady who is allowed to do nothing but read romantic novels all day, every day. In that case, her experience will just consist of the stream of characters printed in the novels. This stream of characters can be translated into such a bit stream. After a while, she will probably become quite skilled at predicting the plots of novels.

The creature has to make some assumption about the world it finds itself in. In Solomonoff's approach, it assumes that there is a 'probability' $P$ on finite bit strings. $P(x)$ is the probability that the finite string $x$ will be the start of its life experience. When it has actually lived for a while and it has seen the initial segment $y$ of its experience, then by Bayes' rule, the chance that the next bit will be 0 is the conditional probability

$$P(y0 \mid y) = \frac{P(y \mid y0)\,P(y0)}{P(y)}$$

$$= \frac{P(y0)}{P(y)} \quad \text{since } P(y \mid y0) = 1.$$

The creature does not know what $P$ is. The learning process consists of discovering a good approximation to $P$.

$P$ should satisfy three conditions:

1.  For any initial string $x$,

    $$P(x) = P(x0) + P(x1).$$

This makes sense, since $x$ must be followed by either 0 or 1 and cannot be followed by both.

2.  For any length $n \geq 0$, the sum of probabilities of all strings of length $n$ is 1:

$$\sum_{|x|=n} P(x) = 1.$$

('$|x|$' is the length, the number of bits, in $x$. The sum is over all strings of length $n$.) This is the condition which justifies calling $P$ a probability. It is actually an easy consequence of condition 1 and the postulate that $P(\varepsilon) = 1$: the creature certainly begins its life with no input.

3.  The probability $P$ is *computable by a UIO machine.*

The definition of this term will be explained below. In outline, it means that, although the world may not be deterministic, the probabilities of various possible experiences are determined by some mechanism equivalent to a computer program. The world may be random, but there is order in its randomness.

Any non-negative real-valued function on finite bit strings which fulfils 1, 2, and 3 is called a **computable** probability. The key to this approach is condition 3. There are uncountably many functions satisfying conditions 1 and 2, but there are only countably many possible computer programs. Condition 3 narrows down the search for $P$ into a tractable space of probabilities.

This is in contrast to other theories such as Valiant's, in which the probability distributions under consideration form a continuum. Li and Vitányi show that there are situations in which learning is deemed impossible according to PAC theory, but in which one can learn if search is restricted to what is computable.

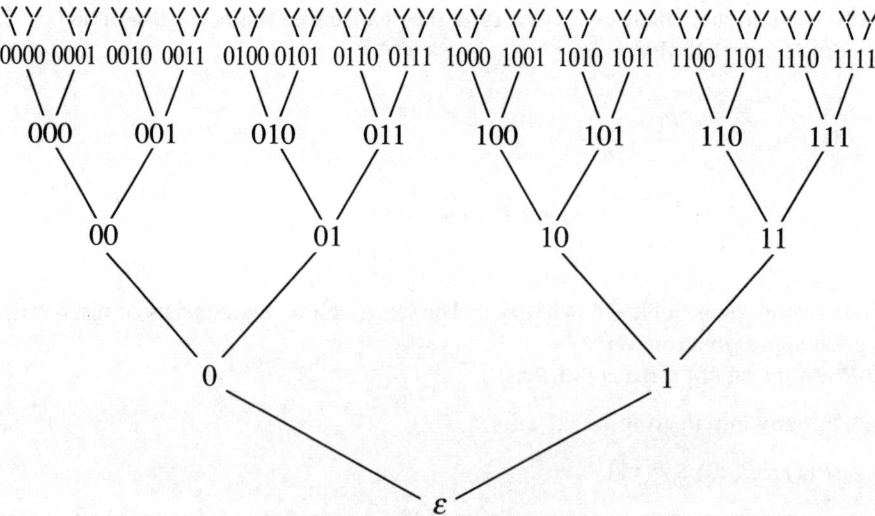


**Fig. 8.4**   The first few layers of bit strings approximating $\Omega$.

It is not strictly accurate to call $P$ a probability. $P$ is not a probability measure on the set of finite strings. However, $P$ is equivalent to a true probability measure on the set $\Omega$ of all *infinite* binary strings (Fig. 8.4). $\Omega$ has a natural topology. Its basic open sets are the sets

$$\Omega_x = \{ \omega \mid x \text{ is an initial segment of } \omega \}$$

where $x$ is the sort of finite string we have discussed before. With this topology, $\Omega$ is homeomorphic to the Cantor set. By a simple adaptation of a theorem of Kolmogorov, any probability $P$ satisfying conditions 1 and 2 defines a true probability measure, also called $P$, on the Borel sets of $\Omega$. On the basic open sets, $P$ is given by

$$P(\Omega_x) = P(x).$$

### Computable probabilities on binary sequences

The letters UIO stand for 'Unidirectional Input and Output'. A **UIO machine** is a Turing machine with three tapes.

1. An *input* tape. The machine's head on this tape is initially at the left end and can only move to the right.

2. An *output* tape. This is initially blank. The head on this tape can only move right.

3. A *work* tape. The head on this tape may move in either direction.

All the machines considered here will have alphabet $\{0, 1, b\}$ where $b$ is the blank symbol. If a machine reads a blank symbol on the input tape, it always stops.

With respect to a UIO machine called $M$, we say that a string $s$ is a **code** for another string $x$ if, when the input tape begins with $s$, its output $M(s)$ always begins with $x$. The string $s$ is a **minimal code** for $x$ if $s$ is a code for $x$ and any shorter initial segment of $s$ is not a code for $x$.

A UIO machine is said to **compute** $P$ if, when its input tape initially contains $x$ followed by $b$, its output consists of the successive bits of the binary expansion of $P(x)$. If $P(x) = 0$ then it should write 0 and then halt. Otherwise it need not halt. For instance, if $P(x)$ does not have a finite binary expansion, the machine will not halt. A probability $P$ is **computable** if some UIO machine computes it.

### A universal probability

The computable probabilities are the ones which we can construct. There are no others which any algorithm could ever learn, but in order to study them, it is helpful if we conceive of others. We shall study a different way to define a probability called $P'$. The virtues of $P'$ are that it satisfies conditions 1 and 2 and if $P$ is any computable probability then $P'$ is a good approximation to $P$ for long strings $x$.

Theorems 2 and 3 below show how. Although $P'$ is not computable, it is a limit of computable probabilities.

A Turing machine $M$ is called **universal** if, for any other Turing machine $N$, there is a finite string $n$ so that, for every string $x$, $M$ with input $nx$ behaves like $N$ with input $x$:

$$M(nx) = N(x)$$

and if one halts then so does the other.

*Theorem (Turing). There is a universal Turing machine.*

By analogy, a universal UIO machine $M$ is one which can simulate any other UIO machine. Choose any universal UIO machine, say $U$. All the constructions which follow depend on the choice of $U$, so perhaps they should all have a subscript $U$ written after them, but we shall not bother to mention $U$ often. The choice of $U$ does not matter very much.

Any set $K$ of strings is called a **prefix set** if no one string in $K$ is a proper initial segment of any other. Any prefix set satisfied Kraft's inequality:

$$\sum_{s \in K} 2^{-|s|} \le 1.$$

Clearly, the set of all minimal codes is a prefix set. Kraft's inequality is what is needed to ensure that the quantity $P'$ we describe satisfies condition 2, so when we describe it, we shall only use minimal codes.

The idea underlying the construction is that $P'(x)$ should be large if calculating $x$ is easy, so $P'(x)$ is large if $x$ has many short minimal codes. A first guess at $P'(x)$ might be

$$\sum 2^{-|s|}$$

where the sum is over all minimal codes for $x$. By Kraft's inequality, this sum is $\le 1$. Unfortunately, that is where its virtues end. It does not obviously satisfy conditions 1 and 2. Furthermore, it is not clear how to set about computing it. Since $U$ is universal, there are inputs $t$ which will make $U$ loop, running forever without writing any output. Even worse, the 'halting problem' says that it is impossible to decide which inputs will cause $U$ to loop. This means that it is impossible to write a program which will list all $x$'s minimal codes. We cannot decide what codes $s$ to sum over.

The solution is as follows. Let $U_T$ be a UIO machine which behaves like $U$ for its first $T$ steps and then halts. ($U_T$ halts sooner if $U$ halts within $T$ steps.) One *can* compute all the minimal codes for $x$ with respect to $U_T$. Define

$$\tilde{P}_T(x) = \sum_s 2^{-|s|}$$

where the sum is over minimal codes $s$ for $x$ with respect to $U_T$. $\tilde{P}_T$ is computable. Observe that any minimal code for $x$ with respect to $U_T$ is also one with respect to $U$, and any minimal code with respect to $U$ is one for $U_T$ as long as $T$ is big enough. Therefore, we define

$$\tilde{P}(x) = \lim_{T \to \infty} \tilde{P}_T(x)$$

This is at least well defined and it is **semi-computable**: one can design a machine, or write a program, which will compute better and better approximations to it if left to run indefinitely. It is still not computable because we cannot tell how good any particular approximation may be, but let that pass. The probability we require is a normalized form of it.

$$P'(x) = \tilde{P}(x)\, C(x)$$

where

$$C(x) = \prod_{i=0}^{|x|-1} \frac{\tilde{P}(x_{1:i})}{\tilde{P}(x_{1:i}0) + \tilde{P}(x_{1:i}1)}.$$

Here, $x_{1:i}$ is the initial substring of $x$ consisting of its first $i$ bits. When $i$ is 0, it is the null string, $\varepsilon$.

$$x_{1:i}0 \qquad \text{and} \qquad x_{1:i}1$$

are the two possible extensions of $x_{1:i}$ by one bit. The normalizing constant $C(x)$ is the product of $|x|$ factors, one for each initial substring of $x$ except $x$ itself.

*Theorem 1. $P'$ fulfils conditions 1 and 2*

*Theorem 2. If P is any computable probability, there is a constant $k$ depending only on P and U so that, for any x,*

$$P'(x) \geq 2^{-k} P(x).$$

The constant $k$ measures the complexity of a machine which computes $P$. Say $M$ is some such machine. Since $U$ is universal, there is a bit string $m$ which encodes $M$ for $U$. If $x$ is any string,

$$U(mx) = M(x).$$

The constant $k$ can be the length of $m$.

This theorem suggests that $P'$ can detect any feature which might occur in any possible life history and which could ever be detected by any computable probability. The existence of any such $P'$ is a remarkable result. However, Solomonoff proved more. He showed that, for any particular $P$ and for strings which are long enough, $P'$ usually makes predictions close to those of $P$. The conditional probabilities $P'(y0 \mid y)$ and $P'(y1 \mid y)$ tend to converge in the mean square sense to those of $P$.

*Theorem 3. For each possible length n of strings, if $n \geq 1$ then*

$$\sum_{x:|x|=n} P(x) \sum_{i=0}^{n-1} (P(x_{1:i}0 \mid x_{1:i}) - P'(x_{1:i}0 \mid x_{1:i}))^2 \; \leq \; \frac{k}{2} \log_e 2$$

The outer sum is over all strings of length $n$. The inner sum, for one string $x$, is the sum of squares of the differences of the conditional probabilities predicted by $P$ and by $P'$ over the initial segments of $x$. The constant $k$ is the same as in Theorem 2. By rearranging terms in the sum and observing that the 0 in Theorem 3 could equally well have been 1, one can derive the following corollary.

*Corollary*

$$\sum_{x} \sum_{i=0,1} P(x) \, (P(xi \mid x) - P'(xi \mid x))^2 \; \leq \; k \log_e 2$$

*where the outer sum is over all finite strings x.*

Gács discovered a related result:

*With probability 1, measured by P, if y is a fixed finite string and $\omega$ is any infinite string,*

$$\frac{P'(\omega_{1:i}y) / P'(\omega_{1:i})}{P(\omega_{1:i}y) / P(\omega_{1:i})} \to 1 \quad \text{as} \quad i \to \infty.$$

*Here, $\omega_{1:i}y$ is the string formed from the first i bits of $\omega$ followed by y.*

These results imply that, for almost all possible life histories (as measured by $P$), after a while the predictions made by $P'$ will converge to those of $P$. Thus, it makes sense to call $P'$ a **universal** probability.

Theorem 3 is remarkable, but it is worth remembering its limitations. It implies that if a creature employs a universal probability then it will almost certainly learn *eventually* how to predict its future experiences, but there is no knowing how long the learning will take. Furthermore, in practice, no real creature ever could employ a universal probability, because universal probabilities are not computable.

### 8.2.2 COMPLEXITY AND OCCAM'S RAZOR

There are actually two distinct varieties of complexity. The kind mentioned in Chapter 2 can be described more precisely as follows. Choose any Turing machine $M$. The complexity of $x$ relative to $M$ is

$$C_M(x) = \min \{|y| \mid \text{When the input to } M \text{ is } y$$
$$\text{then } M \text{ terminates and } M(y) = x\}.$$

(If the choice of $M$ is unfortunate, then there may be no such strings $y$. In that case, $C_M(x)$ is not defined.) The choice of $M$ is equivalent to the choice of a programming language in Chapter 2.

The second variety is called **prefix complexity**, $K_M(x)$. The only difference between it and $C_M(x)$ is that the machine $M$ is a UIO machine, rather than any general Turing machine. $K_M$ is preferable to $C_M$ because, by Kraft's inequality,

$$\sum_x 2^{-K_M(x)} \leq 1$$

but no such inequality holds for $C_M$.

The best choice for $M$ is a universal UIO machine, $U$. With this choice, the prefix complexity of $x$ is called $K(x)$ without a subscript. If $M$ is any UIO machine, there is a constant $k$ for which

$$K(x) \leq K_M(x) + k \qquad \text{for all finite strings } x.$$

($k$ is as in Solomonoff's Theorem 2.) Since two universal machines can simulate each other, their two forms of $K$ will vary by at most a constant. This is our excuse for claiming that the choice of $U$ does not matter much.

Another reason for preferring $K$ to $C$ is that there is a simple relationship between $K$ and the universal probability $P'$. There are positive constants $c_1$ and $c_2$ so that, for any $x$,

$$c_1 \leq P'(x)\, 2^{K(x)} \leq c_2.$$

### The virtue of simplicity

In the early fourteenth century, William of Occam wrote that '*Entities should not be multiplied beyond necessity.*' This assertion is called **Occam's razor**. Nowadays, it is understood to mean that simple solutions should be adopted in preference to complex ones. Suppose that we are faced with a choice between possible objects, typically called $f$. For instance, each $f$ might be a child node of some current node in a search task. Then Occam's razor says that the best choice will be the $f$ for which $K(f)$ is least.

### Approximations: the MDL principle

$K$ is not computable. This is a serious blemish on the theory. In order to use it for practical purposes, one must find an approximation to it. No computable approximation will suffice in all situations, but there may be one which works well in a limited set of cases.

Suppose that the objects of study form a countable set and that we have actually enumerated them:

$$f_1, \quad f_2, \quad f_3, \quad f_4, \quad \cdots \quad f_n, \quad f_{n+1}, \quad \cdots$$

What exactly is an enumeration? It is a function whose argument is a positive number $n$ and whose value is $f_n$. When discussing complexity, the function we can handle are the ones computed by Turing machines, so for current purposes, an enumeration is (equivalent to) a machine $E$ whose input is a string describing a number, $n$, and whose output is a string describing $f_n$. Thus, if we forget the distinction between things and their descriptions, $n$ is a code for $f_n$.

The set of finite binary strings has a natural **lexical** order:

$$x \leq y \quad \text{iff} \quad \text{either } x = y$$
$$\text{or } x \text{ is shorter than } y\colon |x| < |y|$$
$$\text{or their lengths are equal}\colon |x| = |y|$$
$$\text{and for some string } z$$
$$z0 \text{ is an initial segment of } x$$
$$z1 \text{ is an initial segment of } y.$$

Hence, there is a unique order-preserving correspondence between the set of numbers, $\mathbb{N}$, and the set of finite strings:

| 0 | 1 | 2 | 3 | 4 | 5 | 6 | 7 | 8 | 9 | 10 | 11 | 12 | ... |
|---|---|---|---|---|---|---|---|---|---|----|----|----|-----|
| $\varepsilon$ | 0 | 1 | 00 | 01 | 10 | 11 | 000 | 001 | 010 | 011 | 100 | 101 | ... |

This correspondence gives us a convenient description of numbers in terms of bit strings. We choose to use it, so, with respect to the enumeration $E$, the complexity of an object $f_n$ is the length of the bit string corresponding to $n$:

$$K_E(f_n) \approx \log_2 n.$$

Some subscripts will be for objects $f$ which are no use. (If each $f$ is a rule, then some rules' conditions will not be fulfilled.) Among the remaining subscripts $n$ whose objects $f_n$ are possible solutions, we select the first subscript, say $j$. The simplest object is then taken to be $f_j$.

All this is a rather roundabout way of arguing that the best solution is likely to be the first one you can think of. This is only true if your enumeration is good. Really, we want the $f$ with a simplest code with respect to the universal machine $U$. In general, there is no way to determine minimal codes with respect to $U$, but when the $f$s are enumerated, we can find *some* codes for them. These codes may not be minimal, but they may be good approximations to minimal codes.

Since $U$ is universal, there is a string $e$ which encodes $E$ for $U$:

$$\text{For any string } x, U(ex) = E(x).$$

Thus, $en$ is a code for $f_n$ with respect to $U$. If $e$ is short then it seems likely that $en$ will not be much longer than a minimal code for $f_n$ with respect to $U$. Hence, when $e$ is short, the first $f_n$ really is likely to be the simplest solution.

The **minimum description length** (MDL) principle applies when one has to choose a best solution and there is a variety of models available by which one can describe possible solutions. A model is equivalent to a machine $E$ which enumerates the objects $f$, and a description in this model is equivalent to a code $n$ for $f$

with respect to $E$. The MDL principle can only be applied when we are given a restricted class of models $E$, each with a code $e$. Then the MDL principle asserts that the best solution is likely to be $f_n$ calculated by $E$, where $|n| + |e|$ is minimal. This sum is called the combined description length of $f_n$.

When Rissanen enunciated the MDL principle, he assumed that the model is statistical and that its parameters are real numbers. However, he did not restrict the number $k$ of parameters, so the parameters can be a tuple $\vartheta$ of arbitrary length. In this case, he showed that the combined description length is approximately

$$-\log_2 P(x \mid \vartheta) + \log^*(C_k \, (\| \, \vartheta \, \|_{M(\vartheta)})^k).$$

Here:

$P(x \mid \vartheta)$      is the likelihood of the datum described by $x$, according to the statistical model described by $\vartheta$;

$C_k$      is the volume of the $k$-dimensional unit ball;

$\| \, \vartheta \, \|_{M(\vartheta)}$      is the square root of an inner product
$$\vartheta' \, M(\vartheta) \, \vartheta$$
where $M(\vartheta)$ is the $k \times k$ matrix of second partial derivatives of $-\log_2 P(x \mid \vartheta)$;

$\log^* a$      is the sum
$$\log_2 a + \log_2 \log_2 a + \log_2 \log_2 \log_2 a + \ldots$$
taken as far as all terms are positive.

Other approximations can be justified as special cases of the MDL principle. Among them are the *maximum likelihood* principle and the principle of *maximum entropy*.

### 8.2.3   FURTHER ASPECTS OF COMPLEXITY

The presentation here is only a short summary. Much more good work has been done, extending the theory in several directions.

1. The subject can be presented using Turing machines, as above, or using recursive functions. The references tend to use a mixture of both formalisms. Recursive functions are attractive, but perhaps the Turing machine approach is more natural. Complexity is defined in terms of the lengths of bit strings and bit strings are most easily manipulated with machines.

2. There is a notion of **relative** complexity:

$$K(x \mid y)$$

is the length of a shortest string $z$ for which $U(yz) = x$. The relative complexity is designed to measure how much information is needed to construct $x$ when $y$ is already given. Simple complexity is a special case:

$$K(x) = K(x \mid \varepsilon).$$

3.  There are connections between complexity, entropy, and information theory.

4.  Gold's theory of inference can be justified by Occam's razor, as presented in terms of complexity.

5.  Bit strings can be used to describe anything which can be described, but they are primitive and working with them is tedious. Flajolet has developed a subtle theory of complexity which captures much of the structure in both data and algorithms.

6.  Li and Vitányi define a class of measures which they call *simple*. Computable probabilities can be used to define simple measures on the unit interval in $\mathbb{R}$ and so also can the universal probability $P'$. They show that there are concepts which are not PAC learnable, but which are learnable if the probability measures considered are all simple.


## 8.3   THE DESCRIPTION LANGUAGE AND BIAS

Learning in any situation depends on how the situation is described. The choice of a language for descriptions is one example of **bias.** This has been discussed already, briefly, in the section on search in Chapter 2 and in the introduction to clustering and for inference of recursive rules in Chapter 6. When focusing, the partial order among possible conditions is also a form of bias. At the end of Chapter 7, we saw how the eight tiles puzzle can be solved easily with macros if it is described in one representation, but not in the dual representation.

Pitt and Valiant have discovered another case. It involves Valiant's notion of PAC learnability and two families of languages called *k-term-DNF* and *kCNF*. The precise definitions of these languages can be found in the references. Their results are:

1.  for all $k$, *k-term-DNF* is learnable by *kCNF*; and

2.  if RP $\neq$ NP then, for all $k \geq 2$,

    *k-term-DNF* is not learnable by *k-term-DNF*.

When $A$ and $B$ are languages, the notion '*A is learnable by B*' is a variant of PAC learnability involving complexity of algorithms. The classes RP and NP are also aspects of complexity theory. The point is that *kCNF* is more expressive than *k-term-DNF*, so it is easier to find a formula in *kCNF* which fits the training set.

In general, a bias is any prior choice affecting the capacity of a searcher or learner. The language is one such. Another is the heuristic guidance which the learner depends on, while it searches for new rules, concepts, or whatever. Of all possible biases, the choice of a language is the most significant.

### 8.3.1 EXPRESSIVE POWER

Each algorithm involves certain basic operations. Each operation can only be applied if its input is in suitable form. Therefore, the language used for expressing input must have features which match the algorithm's operations.

*Example: trigonometric integrals, again*

Recall the example of explanation-based learning which started from a solution for the indefinite integral

$$\int \cos^7 x \, dx$$

and generalized it to one for

$$\int \cos^k x \, dx$$

for any positive odd number $k$. In order to express it, we used the following techniques from school algebra:

(1)   arithmetic functions: addition, subtraction, multiplication, division;

(2)   exponentiation:

$$a^0 = 1 \qquad \text{and} \qquad a^{n+1} = a \times a^n,$$

(3)   integral notation;

(4)   substitution:

$$\sin x = u \qquad \text{and} \qquad \int f(\sin x) \cos x \, dx = \int f(u) \, du.$$

If we had followed the example to its end, we should also have used the notation for

(5)   summing finite series:

$$\sum_1^n \cdots$$

If any one of these notations were not available, then the example could not have worked. For instance, suppose that the language used did not include exponentiation. In that case, the given instance

$$\int \cos^7 x \, dx$$

can still be expressed in the available language as

$$\int \cos x \times \cos x \times \cos x \times \cos x \times \cos x \times \cos x \times \cos x \, dx$$

and so can every other stage of the solution for this instance. However, when we come to try to generalize, there is nothing there to be generalized. There is no number 7 which could be replaced by $k$. The key operation in this case is generalization of constants. It can only be applied when the learner's language includes exponentiation and the $\Sigma$ notation for summing polynomials, so that a suitable constant appears.

### 8.3.2   EXTENDING A LANGUAGE IS A LEARNING TASK

One of the cleverest aspects of human learning, as yet unmatched by our crude algorithms, is the skill to extend a language. Having said that, there are learning algorithms which do extend their languages. One such is Wolff's method for learning grammars, which is a form of clustering.

*Example: inference of grammars, again*

Recall, Wolff's learning algorithm is divided into stages:

WHILE the grammar is incomplete
*Parse* the input text;
*Revise* the grammar:
*Build* new AND symbols;
*Rebuild* over-general OR symbols;
*Fold* new OR symbols.

Its input is one grammar and a text and its output is another grammar. If the input grammar is empty, then the *fold* operation cannot work. It requires contexts which are found within broad parse trees and if the grammar has no AND symbols then the algorithm cannot construct parse trees. The *build* operation constructs precisely the AND symbols which the *fold* step depends on.

In this case, the grammar is a language for describing the algorithm's input text. The whole purpose of Wolff's algorithm is to extend this language until it suffices.

Very many kinds of situation can be described by context-free languages, but some cannot. Furthermore, there are situations which can be described by such languages, but correlation is not the most natural way to discover them. For instance, the exponential notation is justified because:

- it is succinct: it reduces the complexity of a description; and

- exponentiation satisfies some simple arithmetic laws.

A statistical method might perhaps invent exponentiation eventually, but not for the right reasons. An algorithm for learning notions such as exponentiation would follow a heuristic search for succinct notations which obey non-trivial laws. Wolff's algorithm does not do that. It lacks the right heuristic bias.

One can conceive of a system which is designed to extend its own description language for each task. It should choose extensions which match the task. To some extent, any system which invents concepts can be regarded as extending a language, but the extension only modifies bias if the new concepts can be employed in descriptions of subsequent inputs. Wolff's system does this. Of course, any such system is biased by its language for describing possible languages!

### 8.3.3  EXPLOITING SYMMETRY AND REDUNDANCY

One of the first and most profound studies in this area was made by Saul Amarel. He investigated the following problem.

*Example*: *missionaries and cannibals*

Three missionaries employ three cannibals as porters. They arrive at a river, where there is a boat which can carry up to two people. If cannibals outnumber missionaries at any stage, on either bank of the river or in the boat, then the cannibals will have a good time and the missionaries won't. How can they all cross the river safely?

This is not a learning situation, but it is a good exercise in choice of language. Amarel described this problem repeatedly, in a sequence of languages. Each description makes the problem appear simpler. Each new representation is arrived at by noting some symmetry in the problem or some redundancy in a description, and modifying the language so that the description is less redundant. Here in outline are some such descriptions (not quite the same as Amarel's).

*Description 1*
   Missionaries: $m_1, m_2, m_3$.
   Cannibals: $c_1, c_2, c_3$.
At any stage,
   Set of people on the left bank: *at-left*.
   Set of people on the right bank: *at-right*.
   Set of people in the boat: *in-boat*.
   Positions of the boat: *left-bank* or *right-bank*.
   Possible operations: *load-boat*, *cross*, *empty-boat*.
The language also includes cardinalities of sets and the predicate '$\leq$' between cardinalities.
   Start state: all missionaries, all cannibals, and the boat are at the left bank.
   Goal: all are at the right bank.
   Constraints:
   The set of missionaries on the left bank must be at least as large as the set of
      cannibals on the left bank or else this set of missionaries must be empty.
      (Similarly for the right bank and the boat.)
   No missionary ever appears or disappears.
   No cannibal ever appears or disappears.
   The boat can hold at most two people.
   The boat cannot cross empty. Someone has to row it.

Please excuse the length of this description. I have written it in some detail so that the later ones are manifestly better.

> *Symmetry*: any two missionaries are alike and any two cannibals are alike. Hence, the individuals are not significant. Only the numbers of each in each place matter.

*Description 2*
  Number of missionaries on the left bank: $M_L$.
  Number of missionaries on the right bank: $M_R$.
  Number of missionaries in the boat: $M_B$.
  Number of cannibals on the left bank: $C_L$.
  Number of cannibals on the right bank: $C_R$.
  Number of cannibals in the boat: $C_B$.
  Possible positions of the boat: *left-bank, right-bank*.
  Operations: …

One immediate advantage of this notation is that it makes the problem's constraints appear simple:

$$C_L \leq M_L \vee M_L = 0$$
$$C_R \leq M_R \vee M_R = 0$$
$$C_B \leq M_B \vee M_B = 0.$$

The numbers are also subject to the conditions

$$M_L + M_R + M_B = 3$$
$$C_L + C_R + C_B = 3$$
$$M_B + C_B \leq 2.$$

During the operation *cross*,

$$M_B + C_B \geq 1.$$

Amarel observed the following theorem.

*If $C_L \leq M_L \vee M_L = 0$ and $C_R \leq M_R \vee M_R = 0$ before and after a crossing and all the other conditions hold, then*

$$C_B \leq M_B \vee M_B = 0$$

*during the crossing.*

This non-trivial little result points to the following.

> *Redundancy*: the cannibalism constraint on numbers in the boat. The whole problem can be described by just listing the numbers of missionaries and cannibals at each stage on each bank before and after each transfer. Further, there is no need for explicit operations which load and unload the boat. Hence, we can formulate the next description.

*Description 3*

Number of missionaries on the left bank: $M_L$.
Number of missionaries on the right bank: $M_R$.
Number of cannibals on the left bank: $C_L$.
Number of cannibals on the right bank: $C_R$.
Possible positions of the boat: *left-bank, right-bank*
Operation: *cross*$(M_B, C_B)$.
This single operation:

- subtracts $M_B$ from $M_L$ if the the boat is at *left-bank*
  or from $M_R$ if the the boat is at *right-bank*;

- adds $M_B$ to $M_R$ if the the boat is at *left-bank*
  or to $M_L$ if the the boat is at *right-bank*;

- adjusts numbers of cannibals on the banks, similarly;

- changes *left-bank* to *right-bank* or vice versa.

In the third description of the problem, the new operation

$$cross(M_B, C_B)$$

subsumes the sequence of three operations *load-boat, cross, empty-boat* in Descriptions 1 and 2 and so the search space has fewer states.

*Theorem*

$$M_R = 3 - M_L \qquad \text{and} \qquad C_R = 3 - C_L.$$

*The cannibalism constraint on the right bank will be satisfied if and only if*

$$M_L \leq C_L \vee M_L = 3.$$

This was not true in the first two representations, because they included states in which the boat was not empty.

*Redundancy*: $M_R$ and $C_R$ can be omitted too

*Description 4*

Number of missionaries on the left bank: $M_L$.
Number of cannibals on the left bank: $C_L$.
Possible positions of the boat: $l, r$.
Operation: *cross*$(M_B, C_B)$.

This single operation:

- subtracts $M_B$ from $M_L$ if the boat is at $l$;

- adds $M_B$ to $M_L$ if the boat is at $r$;

- adjusts numbers of cannibals on the left bank, similarly;

- changes *l* to *r* or vice versa.

Constraints:

$$M_L \leq 3 \qquad C_L \leq 3$$
$$M_L = 0 \quad \vee \quad M_L = 3 \quad \vee \quad M_L = C_L$$
$$M_B + C_B \leq 2.$$

During the operation *cross*,

$$M_B + C_B \geq 1.$$

Thus, a state can be summarized as a triple

$$(M_L, C_L, b)$$

where *b* is either *l* or *r*. In this language, the start state is

$$(3, 3, l)$$

and the goal is

$$(0, 0, r).$$

This is as far as one can take the description language for individual states. Amarel succeeded in simplifying the problem further by analyzing its entire search space. His next observation was that it has symmetry.

*Symmetry*: the search task is symmetric between the two river banks. The symmetry

- interchanges *l* and *r*;

- changes $M_L$ to $3 - M_L$;

- changes $C_L$ to $3 - C_L$;

- interchanges the *start* and the *goal* states.

Furthermore, every operation in the space is reversible.

Hence, there is no need to explore the entire search space. Say this symmetry is called *T*. If there is a path in the search space from *start* to some state *s* for which either $Ts = s$ or $Ts = $ a state obtainable from *s* by one crossing, then the reverse of this path will lead from $Ts$ to the *goal*. In fact, $Ts$ is never the same as *s* because *T* moves the boat, so the original problem can be solved if (and only if) we can solve the final description.

*Description 5*

This is the same as description 4, with:

    Start state:   (3, 3, *l*);
    Goal:        any state *s* for which
                $Ts = $ a state obtainable from *s* by one crossing.

The search space for this task has half the depth of the original.

Amarel went on to explore generalizations of this problem, with any number $N$ of missionaries and an equal number of cannibals and a larger boat. He discovered properties of sequences of operators which simplify the search.

What lessons can be learned? Each of the simplifications above arises because there is a *theorem* about one representation, and this theorem lets us express some feature(s) in terms of other features. The theorem reveals *redundancy* in the representation, and the next representation eliminates it. *Symmetry* is a special kind of theorem. The redundant feature may be a term in the description language or an operator which can be incorporated into a composite sequence or an entire portion of the search space.

These lessons could be embodied as heuristic conditions in a system which invents new languages. They are widely applicable in problems which are naturally expressible as searching tasks.

## 8.4   THEORIES OF LANGUAGES AND LOGICS

There are so many ways of expressing a task. How can one set about choosing between them? How can one classify and relate them?

Most description languages consist of the following components:

- an *alphabet* of symbols which may occur in the language;

- a *grammar* which describes which sequences of symbols from the alphabet are meaningful formulas.

These two define the language. With it, you can write down a few formulas which describe the task. These formulas are *hypotheses*. For the eight tiles problem, the hypotheses will include descriptions of the start state and the goal state and of the four possible operations. Along with the hypotheses, which are specific to each task, there are also *axioms* which are always true for any task which you might ever describe in this language. When you have written a description of a task in the language, you will then want to study consequences of the description. For instance, in the eight tiles problem, you will want to decide what state a sequence of operations will lead to from the start state. These consequences are found by *inference rules*.

The *theory* of the task is the set containing all the hypotheses and axioms and also all the formulas which can be inferred from them by the inference rules.

A **formal system** consists of a language (alphabet plus grammar) and axioms and inference rules. When one 'adjusts the bias' of a task, one is translating its description from one formal system into another. Perhaps it is foolish to try to lay down hard and fast rules about the process of translation. If we did, then one day, someone somewhere would inevitably invent a valuable adjustment of bias which did not fit. However, there are certain kinds of translation which fall into a neat pattern. They cover many practical cases of bias adjustment and so they are outlined here.

The permitted translations should preserve the syntactic structures: each grammatical feature in the original language is translated to a unique corresponding feature of the new one. Say we call some such translation $\tau$. Then, for each symbol $a$ in the original alphabet, there should be some particular formula $\tau(a)$, for each production $p$ in the original, there should be a production $\tau(p)$, and so on. Similarly, each inference rule $i$ of the original system should correspond with one called $\tau(i)$ in the new system and all theorems in the old system should translate to theorems in the new.

Translations which conform to these criteria are the arrows of a category. There is a category of formal systems. Of itself, this remark is not very profound. The nice thing about categories is, the people who know about such things have invented a range of tools for manipulating categories. When the subject of bias becomes established on a sound footing, that footing will probably be an application of category theory.

### 8.4.1  META-LANGUAGES

Any theory of bias will start from a theory of languages and formal systems. This theory of systems will itself be expressed in some such system. There is a notion, dating back to David Hilbert, of meta-levels. Each system $L$ is described in an associated system called $L$'s meta-system, $\mathscr{M}L$. The meta-system could be used to describe many systems like $L$. If $L$ is inadequate (for example, if $L$ lacks exponentiation) then $\mathscr{M}L$ is the system which we can use to express a definition of exponentiation. Thus, starting from $L$, we could describe an extension of $L$ (perhaps called $XL$). Later, if $XL$ is also found lacking, we could define an extension of it too. All these 'object level' systems $L, XL, XXL, \ldots$ are described in $\mathscr{M}L$.

What if $\mathscr{M}L$ itself also proves inadequate? There are many ways to describe one object system, $L$. Each employs its own meta-system. Thus, there are many possible meta-systems for $L$. Perhaps our first choice for $\mathscr{M}L$ is not flexible enough to define all the extensions of $L$ which we shall need. In that case, we shall need a better $\mathscr{M}L$. In order to describe that, we shall need a meta-system for $\mathscr{M}L$, say $\mathscr{M}\mathscr{M}L$, or $\mathscr{M}^2L$. In turn, our first choice of $\mathscr{M}\mathscr{M}L$ may eventually prove inadequate, so we shall want to extend it too. Where does it all end?

There are two possible answers. The naïve one is, it never ends. Each system, at any meta-level, has its own meta-system at a higher level. Beyond individual levels, one can envisage an infinite sequence of systems at successive levels and a system called $\mathscr{M}^\omega L$ in which to describe the sequence. Then it will have its meta system, $\mathscr{M}^{\omega+1}L$. In general, there will be a meta-system $\mathscr{M}^nL$ for each ordinal number $n$, finite or infinite.

A clever learning program which can work at meta-levels might include a procedure for generating successive meta-systems. Whenever something within the program suggests that its present system is inadequate, the program could hop up a level, define an extension to its current object system, and then hop down again. In fact, there is much more that the program could do. A great deal of valuable generalization can go on at meta-levels. Also, some heuristic analysis depends on the complexity of descriptions and this cannot be expressed at the object level. Observe that such a program could never begin with knowledge of all these meta-level systems. It will have to invent each one as needed. Brian Cantwell Smith has invented such a system.

The second answer is simple but dangerous. It is this: the system $L$ is its *own* meta-system.

$$L = \mathcal{M}L.$$

(Smith's program uses such a system, although it distinguishes between different levels of processing.) The risk in such a procedure is that any theory expressed in $L$ may be inconsistent. People did not worry much about inconsistency until the end of the nineteenth century, when Frege attempted to put Cantor's set theory on a sound footing. Frege thought he had succeeded; but just as his book was going to press, he received a note from Russell which suggested consideration of the 'set'

$$R = \{x \mid \neg x \in x\}.$$

If $R \in R$ then $\neg R \in R$, but if $\neg R \in R$ then $R \in R$. Ever since, logicians have gone to great lengths to check that they are not talking nonsense.

Smith's system is a full programming language, but considered as a meta-system, it is not very expressive and, hence, it is probably safe. Others have been less cautious. Lenat wrote a notorious program called Eurisko which can invent concepts and heuristic rules about how to invent concepts and heuristic rules. He confesses readily that it crashes, but that does not seem to worry him very much. There is no very clear published account of Eurisko. Another ambitious program is described in outline by Ammon but, like Lenat, Ammon confesses that his program is unsound and his description is incomplete.

## 8.5   SOME OPEN ISSUES

To close, here is a short list of open research topics.

### 1.   *Training sequences which make learning easy*

As was remarked in the preface, teachers usually grade the examples and exercises which they set to pupils, with easy ones first. In any given learning situation, is there an optimum training sequence? Is there an optimum rate of increase in examples' complexities?

### 2.   *Description languages*

Each task has an associated set of languages in which it could be described. Does this set have a natural structure? What aspects of a language make solving the task easy? Are there any effective principles for finding a suitable language? How might one try to optimize the pay-off between the search for a good representation of the original task and the search to solve the task itself?

These questions apply to more than just tasks which concern learning. The answers will depend on the nature of the task.

### 3.   *The art of selective forgetting*

Real systems run in limited memory, so they cannot remember all the data which might help them learn, such as complete traces of all their attempts at search. Even in a system with plenty of memory, too much low-grade knowledge can hamper search. How might a system recognize that one datum is likely to be informative and another will not?

### 4.   *Conceptual clusterers which invent recursion*

Wolff's algorithm is one such. Recursion is not a notion which occurs in the classical theory of clustering, because our intuition suggests that distinct clusters are just subsets of a common set of all possible instances. This intuition is incomplete. There are probably many opportunities to learn recursive structure from noisy data.

### 5.   *Local optimization in conceptual clusterers*

It seems that clusterers which only optimize locally are effective and relatively fast. How local need they be? How might one measure locality and how does one compare the extents of the optimized regions in different clusterers? Can any clustering task be performed by a local clusterer?

### 6.   *Justifications for imposing structure on sets of clusters*

The hierarchy produced by a conceptual clusterer can be augmented:

- either to improve the statistical quality of the clusters;

- or to make it symmetric;

- or so that it includes features which allow observed trends to be explained by deductive methods.

Almost all clusterers to date only employ the first criterion. The Scott–Markovitch method uses the first and the second. It would be extremely interesting if someone could demonstrate a realistic implementation of the third.

### 7. Generalization subject to laws

As far as I know, this field is wide open. There should be plenty of scope here for applications of universal algebra and category theory.


### 8. Geometric applications of syntactic methods

The human eye is very good at generalizing visual patterns. Can this skill be simulated artificially? Neural nets can generalize scenes, but they cannot easily abstract the symmetry from regular scenes. (This problem can be treated as a special case of generalizing subject to laws. The laws in question are the laws of geometry.)


### 9. Program synthesis and the available functions

In any case of program synthesis by the method of Zhu and Jin, the available functions form a description language. What are the properties of this language and how does it affect the scope and efficiency of synthesis? The functions are subject to laws. Are there any useful general heuristic principles about how to apply these laws?


### 10. The laws relating synthesized functions

There must often be such laws. Is there a procedure (algorithmic or heuristic) for finding them?


### 11. Recursive application of program synthesis

The function synthesized in one case of program synthesis could be used in another case. Is it possible to design an elaborate system which synthesizes large programs by decomposing the large task into a sequence of smaller ones, synthesizing programs for the small ones, and then putting the small programs together?


### 12. The equivalence between sets of rules and programs

A logic program consists of clauses and an interpreter for them. On the one hand, logic programs are equivalent to knowledge bases with some form of conflict resolution. On the other, they have the same computational power as programs written in (say) FP. How can these equivalences be exploited? Can the algorithms which extend rule-based systems be applied to the synthesis of programs and vice versa?

### 13.   *Representing with functions or with predicates*

Syntactic generalizers can be used with representations in terms of predicates or with both predicates and functions, but they seem to work best with very specific choices of representation. What are the optimum choices? The discussion in Section 6.5.5 should be developed.

### 14.   *Properties of inductive inference*

As yet, not much is known about the relative merits of *FunG* and *PredG* in different circumstances, nor about which version of the algorithm in Section 6.5 is best for a particular task.

### 15.   *A test of systems which invent new representations*

How can one invent the notion of binary numerals from basic learning principles and a knowledge of unary arithmetic? Any system which invents new representations could be tested by giving it the goal of discovering a form of numbers which simplifies addition and multiplication. These operations are much quicker in binary and decimal notations. This is perhaps the easiest non-trivial self-contained exercise that one could pose to such a system.

I hope that this task will be solved, because it will probably throw light on a lot of heuristic aspects of representation. At the same time, I hope that students of representation will not study it to the exclusion of all else. Applications will be much more common in fields which do not require recursive representations and which do require a lot of background knowledge and constraints.

### 16.   *The pay-off between conditions and conflict resolution*

Knowledge about when to apply actions can be embodied either in conditions of rules or in a conflict resolution method. For instance, an instance of rule ordering could be replaced by extra conditions in the two ordered rules. If the knowledge is stored as extra conditions then conflict sets will be smaller, but the rule-ordering learning method works most effectively when there are already known instances of rule ordering, so if all such instances were replaced with extra conditions then learning might be slower. How is this knowledge most easily learned and where is it most effective?

### 17.   *Control of meta-levels*

A searcher which can work at different meta-levels must have some criterion for choosing which level to work at. Can this criterion be learned? A system which can learn at different levels might have opportunities to generalize its learned knowledge across levels. When can it do so safely?

### 18.   *The view of bias adjustment via category theory*

There is plenty of scope here. For instance, one might conjecture that the operation $\mathcal{M}$ on a formal system, which produces its meta-system, is a functor. When $\mathcal{M}$ is described carefully, one might even be able to show that it is a particular kind of functor called a monad. If it is, then all the problems about an unlimited sequence of meta-levels will disappear.


# FURTHER READING

A theory of learnability in neural nets is presented by Judd (1990). Lakshmivarahan (1981) studied certain Markov processes with pre-specified asymptotic behaviours. Learning in Markov processes was also studied by Watkins and Dayan (1992) who proved a convergence theorem. Gold's theorem appeared in Gold (1967). His idea has been developed to cover the inference of first-order theories from examples. It should have relevance to Plotkin generalization, as in Section 6.2, and invention of recursive rules by the method of Section 6.5. See Osherson, Stob, and Weinstein (1986).

Valiant (1984) originated the PAC theory of learnability. Our example comes from a paper by Blumer, Ehrenfeucht, Haussler, and Warmuth (1989), which is also the source of the connection between learnability and the Vapnik–Chernovenkis dimension. For more results in this vein, see that paper or Kearns (1990), or Schapire (1989), or Kearns, Li, Pitt, and Valiant (1987), or Pitt and Valiant (1986). The result about *k-term-DNF* and *kCNF*, quoted in Section 8.3, comes from these last two references.

The study by Langley and Iba (1993) shows that learning by extrapolation works better than standard PAC theory would predict, when one knows which attributes are relevant and which are irrelevant.

Several people realized that there should be a theory of learning based around the notions of computability and complexity. Li and Vitányi (1991, 1992, 1993) provide surveys and also show how these ideas can be incorporated into practical learning algorithms. Solomonoff (1964, 1978) was one of the originators. He acknowledges various other contributors, particularly Levin (Zvonkin and Levin, 1970) and Willis (1970). Dudley (1989), page 347 and Halmos (1974), page 157, give accounts of Kolmogorov's theorem about probability measures on product spaces such as $\Omega$.

There are many excellent references on Turing machines, such as Davis (1958) and Hopcroft and Ullman (1979). Unfortunately, they do not all agree precisely. For instance, some say that a single step of a machine involves:

- writing a character on to its tape; and then also

- moving the head one place either left or right along the tape.

Others treat these two operations as two distinct steps. Also, it is not always clear whether the machine's alphabet includes a blank character. These discrepancies can make it hard to understand just what is meant by the machine's 'input'. The account given here follows Solomonoff as far as possible.

Gács' theorem is quoted by Li and Vitányi (1992), who also offer a proof of the relationship between $P'$ and $K$. The MDL principle appears in Rissanen (1983). The class of simple measures is described by Li and Vitányi (1991). A good introduction and survey of more recent developments in complexity theory has been given by Flajolet (1992). He uses complex analytic power series.

Rendell, Seshu, and Tcheng (1987) and Schlimmer (1987) have written programs which adjust their own biases. Our account of the missionaries and cannibals example is derived from Amarel (1968). Other approaches to bias are presented by Kelly (1988) and by the various authors in Benjamin (1990). The current state of thinking about the category of formal systems, for this application, is presented in Giunchiglia and Walsh (1992). Category theory *per se* can be studied in various references, such as MacLane (1971), Lambek and Scott (1988), or Barr and Wells (1990).

Many researchers in artificial intelligence cite meta-languages and meta-level search. Some idea of the scope of the field can be gained from Maes and Nardi (1986). Emde and Morik (1989) have put meta-level data to practical use. Minton (1993) employs a meta-level to construct specific heuristic rules for each task. Baker and Smaill (1993) do likewise for theorem proving, and Stroulia and Goel (1993, 1994) use a meta-level to decide on credit assignment and to analyse algorithms. Smith (1984) describes a programming language with facilities for hopping up and down between meta-levels. One novel introduction to the area is given by Barwise and Etchemendy (1987). Giunchiglia and Serafini (1991) and Giunchiglia, Serafini, and Simpson (1992) study hierarchies of propositional meta-theories. Hutchinson (1991) shows how one can build a similar sequence of predicate meta-theories, with variables ranging over components of formulas.

The best available description of Eurisko is perhaps Lenat (1982–3). The paper by Ammon (1988) is somewhat shorter.


# EXERCISES

1.  Suppose that $\mathscr{C}$ and $\mathscr{D}$ are two families of sets and

$$A = \bigcup \mathscr{C} \qquad B = \bigcup \mathscr{D}.$$

Write $\mathscr{C} \times \mathscr{D}$ for the family of sets

$$\{C \times D \mid C \in \mathscr{C} \wedge D \in \mathscr{D}\}.$$

Suppose that $S \subseteq A$ and $\mathscr{C}$ shatters $S$, and $T \subseteq B$ and $\mathscr{D}$ shatters $T$.

(a)  If $\exists s \in \bigcup \mathscr{C}$ and $\exists t \in \bigcup \mathscr{D}$ then show that $\mathscr{C} \times \mathscr{D}$ shatters $S \times \{t\} \cup \{s\} \times T$ so in this case

$$\dim \mathscr{C} \times \mathscr{D} \geq \dim \mathscr{C} + \dim \mathscr{D}.$$

(b)  If $S = A$ and $T = B$ then $\mathscr{C} = \mathscr{P}A$ (the set of all subsets of $A$) and similarly $\mathscr{D} = \mathscr{P}B$. In this case, show that

$$|\mathscr{C} \times \mathscr{D}| = (|\mathscr{C}| - 1) \times (|\mathscr{D}| - 1)$$

so in this case

$$\dim \mathscr{C} \times \mathscr{D} < \dim \mathscr{C} + \dim \mathscr{D}.$$

(c)  If $s \in S$ and $t \in T$ then show that

$$\mathscr{C} \times \mathscr{D} \text{ shatters } (S \setminus \{s\}) \times \{t\} \cup \{s\} \times (T \setminus \{t\})$$

so in general

$$\dim \mathscr{C} \times \mathscr{D} \geq (\dim \mathscr{C} - 1) + (\dim \mathscr{D} - 1).$$

(d)   If $\mathscr{C} \times \mathscr{D}$ shatters a subset $U$ of $A \times B$, then let

$$S = \pi_1 U \qquad T = \pi_2 U.$$

Show that $\mathscr{C}$ shatters $S$ and $\mathscr{D}$ shatters $T$, so in general

$$\dim \mathscr{C} \times \mathscr{D} \leq \dim \mathscr{C} \times \dim \mathscr{D}.$$

(e)   If $A \cap B$ is empty then show that

$$\dim (\mathscr{C} \cup \mathscr{D}) = \max \{\dim \mathscr{C}, \dim \mathscr{D}\}.$$

(f)   Say $\mathscr{C} + \mathscr{D} = \{C \cup D \mid C \in \mathscr{C} \wedge D \in \mathscr{D}\}$.
      If $A \cap B$ is empty then show that

$$\dim (\mathscr{C} + \mathscr{D}) = \dim \mathscr{C} + \dim \mathscr{D}.$$

(g)   Suppose that there is a surjection $f{:}A \to B$ and that there is a number $k$ for which

$$\forall \, b \in B \ |f^{-1}(b)| \leq k.$$

Also, suppose that $\mathscr{D}$ is the family of sets of the form

$$\{f(c) \mid c \in C\}$$

where $C$ is in $\mathscr{C}$. Show that

$$k \dim \mathscr{D} \geq \dim \mathscr{C} \geq \dim \mathscr{D}$$

2.   Recall the three possible representations of the eight queens problem described in Section 2.3.1. The first represents a state as a set of pairs of coordinates, one pair for each square where there is a queen. The second depicts it as a list with eight entries and the third as a list of $k$ entries when there are queens in the first $k$ rows and none in later rows. In these representations, the forms of states are

R1:   $\{(i, j) \mid$ there is a queen in row $i$ and column $j\}$;

R2:   $[c_1 \quad c_2 \quad c_3 \quad \dots \quad c_8]$   where

$$c_i = \begin{cases} j & \text{if there is a queen in row } i \text{ and column } j \\ 0 & \text{otherwise;} \end{cases}$$

R3:   $[c_1 \quad c_2 \quad \dots \quad c_k]$ where in this state, just the first $k$ rows contain queens and $c_i = j$ if there is a queen in row $i$ and column $j$.

To these, we may add another representation whose states have the form

R$'$:   $[C_1 \quad C_2 \quad C_3 \quad \dots \quad C_8]$ where
        $C_i = \{ j \mid$ there is a queen in row $i$ and column $j \}$.

By analogy with Amarel's analysis, there are certain facts about sets and lists and this search task which suffice to show that representations R1 and R′ give rise to isomorphic search spaces. Write down these facts.

Representation R2 cannot depict all states representable in R′. Write down facts which suffice to show that any state on a path to a goal, which can be represented in R′, can be represented in R2.

Representation R3 cannot depict all states representable by R2, but any goal which lies on a path of states representable in R2 also lies on a path of states representable in R3. Write down facts which suffice to show this.

3.  Consider the following tasks:

   (i)    the eight queens problem;

   (ii)   the eight tiles problem;

   (iii)  the dwarf/wizard/pirate game (Section 5.6) in which the objective is to discover the effect of each action.

   In each case,

   (a)    describe a language for describing states of the task;

   (b)    describe the operators of the task;

   (c)    for each operator, describe any conditions on a state which must be satisfied if the operator can be applied to it;

   (d)    describe the task's goal and its start state.

   For the eight queens problem, carry out (a)–(d) for two representations, such as representations 1 and 3 in the example of Section 2.3.1.

   If you have not already done so, describe each of the languages by means of a Chomsky grammar and describe each operation as a partial function from states to states.

   Formulate meta-languages with which you can describe each of the other languages you have described so far. (*Hint*: start by writing down a description of what a Chomsky grammar is.)

# REFERENCES

References marked with a dagger $^\dagger$ are accepted references which I cannot remember ever having seen.

Aha, D. W., Lapointe, S., Ling, C. X., and Matwin, S. (1994). Inverting implication with small training sets. *Proceedings of ECML-94,* Catania, Italy, pp. 31–48. Springer, Berlin.

Aleksander, I. and Morton, H. (1990). *An introduction to neural computing.* Chapman and Hall, London.

Al-Mathami, S. (1990). *Evidential conceptual clustering.* Ph.D. thesis, King's College London.

Amarel, S. (1968). On representations of problems of reasoning about actions. In *Machine intelligence,* Vol. 3. (ed. D. Michie), pp. 131–71. Edinburgh University Press.

Ammon, K. (1988). Discovering a proof for the fixed point theorem: a case study. *Proceedings of ECAI,* Munich, pp. 613–18.

Andersen, N. and Jones, N, D. (1994). Generalizing Cook's transformation to imperative stack programs. In *Results and trends in theoretical computer science* (ed. J. Karhumäki, H. Maurer and G. Rozenberg), pp. 1–18. LNCS 812, Springer, Berlin.

Baader, F. (1990). Unification, weak unification, upper bound, lower bound, and generalization problems. SEKI report SR-90-02, Fachbereich Informatik Universität Kaiserlautern, Germany.

Backus, J. (1978). Can programming be liberated from the von Neumann style? *Communications of the ACM* **21** 613–41.

Baddeley, A. (1983). *Your memory—a user's guide.* Penguin, New York.

Baker, S. (1992).$^\dagger$ *Aspects of the constructive omega rule within automated deduction.* Ph.D. thesis, University of Edinburgh.

Baker, S. and Smaill, A. (1993). A proof environment for arithmetic with the omega rule. Research paper no. 645, Department of Artificial Intelligence, Edinburgh.

Barford, N. C. (1967). *Experimental measurements: precision, error and truth.* Addison-Wesley, London.

Barr, M. and Wells, C. (1990). *Category theory for computing science.* Prentice Hall, Englewood Cliffs, NJ.

Barwise, J. and Etchemendy J. (1987). *The liar.* Oxford University Press.

Benjamin, D. P. (ed.) (1990). *Change of representation and inductive bias.* Kluwer, Dordrecht.

Bergadano, F., Giordana, A., and Saitta, L. (1991). *Machine learning: an integrated framework and its applications.* Ellis Horwood, Chichester.

Bibel, W. and Jorrand, Ph. (1987). *Fundamentals of artificial intelligence.* Springer, Berlin.

Biermann, A. W. and Guiho, G. (eds) (1983). *Computer program synthesis methodologies.* Reidel.

Binmore, K. G. (1980). *Logic, sets and numbers.* Cambridge University Press.

Block, H. D. (1970). *Information and Control* **17** 501–22.

Blumer, A., Ehrenfeucht, A., Haussler, D., and Warmuth, M. K. (1989). Learnability and the Vapnik–Chervonenkis dimension. *Journal of the ACM* **36** (4) 929–65.

Bollobas, B. (1979). *Graph theory – an introductory course.* Springer, Berlin.

Booker, L. B., Goldberg, D. E., and Holland, J. H. (1989). Classifier systems and genetic algorithms. *Artificial Intelligence* **40** 235–82.

Boswell, R. (1989). Analytic goal regression: problems, solutions and enhancements. In *Machine and human learning* (ed. Y. Kodratoff and A. Hutchinson), pp. 9–23 Kogan Page/Chapman and Hall, London.

Brazdil, P. (1978). Experimental learning model. *Proceedings of the Joint Conference of the AISB/GI,* Hamburg (ed. Derek Sleeman), pp. 46–50.

Brazdil, P. (1981). A model for error detection and correction. PhD thesis, University of Edinburgh.

Brazdil, P. (1986). Transfer of knowledge between systems. *Proceedings of ECAI 86*, Brighton, Vol. 2, pp. 73–8.

Brazdil, P. (1989). Transfer of knowledge between systems: use of meta-knowledge in debugging. In *Machine and human learning* (ed. Y. Kodratoff and A. Hutchinson), pp. 229–48 Kogan Page/Chapman and Hall, London.

Brazdil, P., Gama, J., and Henery, B. (1994). Characterizing the applicability of classification algorithms using meta-level learning. *Proceedings of ECML-94*, Catania, Italy, pp. 83–102. Springer, Berlin.

Breuer, P. T. (1991). An analysis/synthesis language with learning strategies. Technical Report PRG–Tr–13–91, Programming Research Group, Oxford.

Bundy, A. (1983). *The computer modelling of mathematical reasoning*. Academic Press, London.

Bundy, A. (1985). Poof analysis—a technique for concept formation. *Proceedings of AISB-85*, Warwick University, UK, pp. 78–86.

Bundy, A., Burstall, R. M., Weir, S., and Young, R. M. (1980). *Artificial intelligence: an introductory course*. Edinburgh University Press, Edinburgh.

Bundy, A., Silver, B., and Plummer, D. (1985). An analytical comparison of some rule-learning programs. *Artificial Intelligence* **27** 137–81.

Buntine, W. (1991). Classifiers: a theoretical and empirical study. *Proceedings of the 12th IJCAI*, Sydney, Vol. 2 pp. 638–44.

Carbonell, J. (1989). Editor's introduction. *Artificial Intelligence* **40**,

Cerf, R. (1994). A new genetic algorithm. *C. R. Acad. Sci. Paris* **319**, Série 1, 999–1004.

Cestnik, B., Kononenko, I., and Bratko, I. (1987). ASSISTANT 86: a knowledge-elicitation tool for sophisticated users. In *Progress in machine learning*, Proceedings of EWSL 87 Bled; Yugoslavia (ed. I. Bratko and N. Lavrac), pp. 31–45. Sigma Press/John Wiley, Wilmslow, U. K.

Clocksin, W. F. and Moore, A. W. (1989). Some experiments in adaptive state-space robotics. *Proceedings of AISB 89*, Brighton, UK (ed. A. G. Cohn), pp. 115–25. Pitman, London; Morgan Kaufmann, San Mateo, CA,

Cooper, G. and Herskovits, E. (1991).[†] A Bayesian method for the induction of probabilistic networks from data. Technical Report KSL–91–02, Knowledge Systems Laboratory, Medical Computer Science, Stanford University.

Craw, S., Sleeman, D., Graner, N., Rissakis, M., and Sharma, S. (1992). CONSULTANT: providing advice for the Machine Learning Toolbox. In *Proceedings BCS Expert Systems 92* (Cambridge, UK), pp. 5–23. Cambridge University Press.

Davis, M. (1958). *Computability and unsolvability*. McGraw-Hill, New York.

DeGroot, M. H. (1970). *Optimal statistical decisions*. McGraw-Hill, New York.

Dewdney, A. K. (1989). Computer recreations. *Scientific American*, May, 138–41.

Doyle, R. (1986).[†] Constructing and refining causal explanations from an inconsistent domain theory. *Proceedings of AAAI–86*, Philadelphia, PA, pp. 538–44.

Dudley, R. M. (1989). *Real analysis and probability*. Wadsworth and Brooks/Cole, Monterey, CA

Emde, W. and Morik, K. (1989). Consultation–independent learning in BLIP. In *Machine and human learning* (ed. Y. Kodratoff and A. Hutchinson), pp. 93–104. Kogan Page/Chapman and Hall, London.

Ergezinger, S., and Thomsen, E. (1995). An accelerated learning algorithm for multilayer perceptrons: Optimization layer by layer. *IEEE Transactions on Neural Networks* **6(1),** 31–42.

Everitt, B. (1980). *Cluster analysis*. Heinemann, London.

Fikes, R. E., Hart, P. E., and Nilsson, N. J. (1972). Learning and executing generalized robot plans. *Artificial Intelligence* **3** 251–88.

Fisher, D. and Langley, P. (1985). Conceptual clustering and its relation to numerical taxonomy. In *Artificial intelligence and statistics* (ed. W. A. Gale), pp. 77–116, Addison-Wesley, Reading, MA.

Fitter, R. S. R. (1963). *Collins guide to bird watching*. Collins, London.

Flajolet, Ph. (1992). Analytic analysis of algorithms. *Proceedings of ICALP*, Vienna, Austria, pp. 186–210, Lecture Notes in Computer Science No. 623, Springer, Berlin.

Forsyth, R. (1989). *Machine learning: principles and techniques*. Chapman and Hall, London.

Fung, R. M. and Craword, S. L. (1990). Constructor: a system for the induction of probabilistic models. *Proceedings of AAAI-90*, Boston, MA, Vol. 2 pp. 762–9. MIT Press, Cambridge, MA.

Gazdar, G. and Mellish, C. (1989). *Natural language processing in Prolog*. Addison-Wesley, Reading, MA.

Geiger, D., Paz, A., and Pearl, J. (1990). Learning causal trees from dependence information. *Proceedings of AAAI-90*, Boston, MA, Vol. 2 pp. 770–6. MIT Press, Cambridge, MA.

Giunchiglia, F. and Serafini, L. (1991). Multilingual hierarchical logics. IRST-Technical report No. 9110–07, Instituto per la Ricerca Scientifica e Tecnologica, I 38100 Trento, Italy.

Giunchiglia, F. and Walsh, T. (1992). A theory of abstraction. *Artificial Intelligence* **57** 323–89.

Giunchiglia, F., Serafini, L., and Simpson, A. (1992). Hierarchical meta-logics: intuitions, proof theory and semantics. *Proceedings of Meta'92*, Uppsala, Sweden, pp. 171–85. Lecture Notes in Computer Science, Springer, Berlin.

Gold, M. E. (1967). Language identification in the limit. *Information and Control* **10** 447–74.

Gould, J. L. and Marler, P. (1987). Learning by instinct. *Scientific American* **256** (1) 62–73.

Gottlob, G. and Fermüller, C. (1991).[†] Removing redundancy from a clause. CD-TR 91/22, Austria.

Gribbin, J. (1985). *In search of the double helix*. Corgi,

Halliday, T. R. and Slater, P. J. B. (eds.) (1983). *Genes, development and learning*. Blackwell, Oxford.

Halmos, P. R. (1974). *Measure theory*. Springer, New York.

van Harmelen, F. and Bundy, A. (1988). Explanation-based generalisation = partial evaluation. *Artificial Intelligence* **36** 401–12.

Harris, M. and Coltheart, M (1986). *Language processing in children and adults*. Routledge and Kegan Paul.

Harris, M., Jones, D., Brookes, S., and Grant, J. (1986). Relations between the non-verbal context of maternal speech and rate of language development. *British Journal of Developmental Psychology* **4** 261–8.

Harris, M., Barrett, M., Jones, D., and Brookes, S. (1988). Linguistic input and early word meaning. *J. Child. Lang.* **15** 77–94.

Hartigan, J. A. (1975). *Clustering algorithms*. Wiley, New York.

Hartman, E. J., Keeler, J. D., and Kowalski, J. M. (1990). Layered neural networks with Gaussian hidden units as universal approximations. *Neural Computation* **2** 210–15.

Hinton, G. E. (1989). Connectionist learning procedures. *Artificial Intelligence* **40** 185–234.

Hinton, G. E., Sejnowski, T. J., and Ackley, D. H. (1984). Boltzmann machines: Constraint satisfaction networks that learn. Technical report CMU-CS-84-119, Carnegie-Mellon University, Pittsburgh, PA.

Hodges, W. (1977). *Logic*. Penguin, Harmondsworth.

Hopcroft, J. E. and Ullman, J. D. (1979). *Introduction to automata theory, languages, and computation*. Addison-Wesley, Reading, MA.

Huntbach, M. M. (1986). An improved version of Shapiro's model inference system. *Proceedings of the Third International Conference on Logic Programming* (ed. E. Y. Shapiro), pp. 180–7. Lecture Notes in Computer Science 225, Springer, Berlin.

Hutchinson, A. (1988). Building grammars from natural text. *Proceedings of EWSL 88*, Glasgow, UK (ed. Derek Sleeman) pp. 45–52. Pitman, London.

Hutchinson, A. (1991). *First order meta theories*. Technical report 91/4, Department of Computer Science, King's College London.

Johnson-Laird, P. N. (1983). *Mental models*. Cambridge University Press.

Johnstone, P. T. (1987). *Notes on logic and set theory*. Cambridge University Press.

Judd, J. S. (1990).[†] *Neural network design and the complexity of learning*. MIT Press, Cambridge, MA.

Kearns, M. J. (1990). *The computational complexity of machine learning*. MIT Press, Cambridge, MA.

Kearns, M., Li, M., Pitt, L., and Valiant, L. (1987). Recent results on Boolean concept learning. *Proceedings of the Fourth International Workshop on Machine Learning*, Irvine, California, pp. 337–52. Morgan Kaufmann, Los Altos, CA.

Keller, R. M. (1988).[†] Defining operationality for explanation-based learning. *Artificial Intelligence* **35** 227–41.

Kelly, K. T. (1988). Theory discovery and the hypothesis language. *Proceedings of the Fifth International Conference on Machine Learning*, University of Michigan, Ann Arbor, pp. 325–38. Morgan Kaufmann, San Mateo, CA.

Kodratoff, Y. (1988). *Introduction to machine learning*. Pitman, London.

Kodratoff, Y., Ganascia, J. G., Clavieras, R., Bollinger, T., and Tecuci, G. (1984). Careful generalization for concept learning. *Proceedings of ECAI-84*, Pisa, pp. 483–92. Elsevier, Amsterdam.

Kohonen, T. (1982). Self-organized formation of topologically correct feature maps. *Biological Cybernetics* **43** 59–69.

Kohonen, T. (1984). *Self-organization and associative memory*. Springer, Berlin.

Kolmogorov, A. N. (1965).[†] Three approaches to the definition of the concept 'quantity of information'. *Probl. Peredachi Informatsii* **1** (1) 3–11.

Kononenko, I. and Bratko, I. (1991). Information based evaluation criterion for classifier's performance. *Machine Learning* **6** (1) 67–80.

Korf, R. E. (1987). Planning as search: A quantitative approach. *Artificial Intelligence* **33** 65–88.

Kosko, B. (1992). *Neural networks and fuzzy systems*. Prentice-Hall, Englewood Cliffs, NJ.

Laird, J. (1984). *Universal Subgoaling*. Ph.D. thesis, Carnegie-Mellon University, Pittsburgh, PA.

Laird, J. E., Rosenbloom, P. S., and Newell, A. (1984). Towards chunking as a general learning mechanism. *Proceedings of AAAI 84*, pp. 188–92.

Laird. J. E., Rosenbloom, P. S., and Newell, A. (1986). *Universal subgoaling and chunking*. Kluwer, Boston, MA.

Lakshmivarahan, S. (1981) *Learning algorithms theory and applications*. Springer-Verlag, New York.

Lambek, J. and Scott, P. J. (1986). *Introduction to higher order categorical logic* (reprinted 1988). Cambridge University Press.

Langley, P. and Iba, W. (1993). Average-case analysis of a nearest neighbor algorithm. *Proceedings of IJCAI–93*, Chambéry, Vol. 2, pp. 889–94.

Langley, P. and Zytkow, J. M. (1989). Data-driven approaches to empirical discovery. *Artificial Intelligence* **40** 283–312.

Langley, P., Simon, H. A., Bradshaw, G. L., and Zytkow, J. M. (1987). *Scientific Discovery*. MIT Press, Cambridge, MA.

Lenat, D. B. (1982–3). The nature of heuristics. Part 1: *Artificial Intelligence* **19** 189–249; Parts 2 and 3: *Artificial Intelligence* **21** 31–98.

Li, M. and Vitányi, P. M. B. (1991). Learning simple concepts under simple distributions. *SIAM Journal of Computing* **20** (5) 911–35.

Li, M. and Vitányi, P. M. B. (1992). Inductive reasoning and Kolmogorov complexity. *Journal of Computer and System Sciences* **44** 343–84.

Li, M. and Vitányi, P. M. B. (1993). *An introduction to Kolmogorov complexity and its applications*. Springer, Berlin.

Lippmann, R. P. (1987). An introduction to computing with neutral nets. *IEEE Acoustics, Speech and Signal Processing (ASSP) Magazine*, April, 4–22.

Luckham, D. (1967). The resolution principle in theorem-proving. In *Machine intelligence*, Vol. 1 (ed. D. Michie), pp. 47–61. Edinburgh University Press.

Ludermir, T. B. (1990).[†] Automata theoretic aspects of temporal behaviour and computability in logical neural networks. Ph.D. thesis: Imperial College of Science and Technology, London.

Ludermir, T. B. (1991). Relating logical neural networks to conventional models of computation. Technical report no. 91/20 Department of Computer Science, King's College London.

MacLane, S. (1971). *Categories for the working mathematician*. Graduate Texts in Mathematics 5, Springer, Berlin.

Maes, P. and Nardi, D. (1986). Workshop on meta-level architectures and reflection, Alghero, Sardinia.

McCluskey, T. L. (1987). Combining weak learning heuristics in general problem solvers. *Proceedings of IJCAI 87*, Milan, Vol. 1 pp. 331–3. Morgan Kaufmann, Los Altos, CA.

Mendelson, E. (1987). *Introduction to mathematical logic*. Wadsworth and Brooks/Cole, Monterey, CA.

Michalski, R. and Stepp, R. (1983). Learning from observation: conceptual clustering. In *Machine Learning* (ed. R. S. Michalski, J. G. Carbonell, and T. M. Mitchell), pp. 331–63. Morgan Kaufmann, Los Altos, CA.

Michalski, R. and Stepp, R. (1986). Conceptual clustering of structured objects: a goal-oriented approach. *Artificial Intelligence* **28** 43–69.

Miller, G. A. (1956).[†] The magic number seven, plus or minus two: some limits on our capacity for processing information. *Psychological Review* **63** 81–97.

Miller, G. A. and Gildea, P. M. (1987). How children learn words. *Scientific American* **257** (3) 86–91.

Minsky, M. and Papert, S. (1969). *Perceptrons: an introduction to computational geometry*. MIT Press, Cambridge, MA.

Minton, S. (1988). *Learning search control knowledge*. Kluwer, Dordiecht.

Minton, S. (1993). An analytic learning system for specializing heuristics. *Proceedings of IJCAI–93*, Chambéry, Vol. 2, pp. 922–8.

Minton, S., Carbonell, Knoblock, Kuokka, Etzioni and Gil (1989). Explanation-based learning: a problem solving perspective. *Artificial Intelligence* **40** 63–118.

Mitchell, T. M. (1977).[†] Version spaces: a candidate elimination approach to rule learning. *Proceedings of IJCAI–77,* 305–10.

Mitchell, T. M. (1982). Generalization as search. *Artificial Intelligence* **18** 203–26.

Mitchell, T. M., Keller, R. M., Kedar-Cabelli, S. T. (1986). Explanation-based generalization: A unifying view. *Machine Learning* **1** (1) 47–80.

Moore, A. W. (1990). *Efficient memory-based learning for robot control*. Technical Report No. 209, Computer Laboratory, University of Cambridge.

Morik, K. (1991).[†] Balanced cooperative modelling. In *Proceedings of the first international workshop on multistrategy learning* (ed. R. S. Michalski and G. Tecuci). George Mason University, pp. 65–80.

Mostow, J. and Bhatnagar, N. (1987). Failsafe—a floor planner that uses EBG to learn from its failures. *Proceedings of IJCAI-87*, Milan, Vol. 1, pp. 249–55.

Muggleton, S. (ed.) (1992). *Inductive logic programming.* Academic Press.

Muggleton, S. and Buntine, W. (1988*a*). Machine invention of first order predicates by inverting resolution. *Proceedings of the Fifth International Conference on Machine Learning*, Michigan, pp. 339–52. Morgan Kaufmann, San Mateo, CA.

Muggleton, S. and Buntine, W. (1988*b*). Towards constructive induction in first-order predicate calculus. TIRM-88-031, Turing Institute, Glasgow.

Muggleton, S., and De Raedt, L. (1994). Inductive logic programming: theory and methods. *Journal of Logic Programming* **19, 20,** 629–679.

Newell, A. (1990). *Unified theories of cognition.* Harvard University Press, Cambridge, MA.

Nilsson, N. J. (1980). *Principles of artificial intelligence.* Tioga/Springer, Berlin.

Nilsson, N. J. (1990). *The mathematical foundations of learning machines.* Morgan Kaufmann, San Mateo, CA.

OGAI (1992). Reviews of technical reports. *Österreichische Gesellschaft für Art. Int.* **11** (1) 12.

Osherson, D., Stob, M., and Weinstein, S. (1986). *Systems that learn.* MIT Press, Cambridge, MA.

Owen, S. and Hull, R. (1988). The use of explicit interpretation to control reasoning about protein topology. *Proceedings of ECAI-88*, Munich, pp. 308–13.

Pankhurst, R. J. (1991). *Practical taxonomic computing.* Cambridge University Press.

Park, J. and Sandberg, I. W. (1991). Universal approximation using radial-basis-function networks. *Neural Computation* **3** pp. 246–57.

Parry, W. (1981). *Topics in ergodic theory.* Cambridge Tracts in Mathematics No. 75, Cambridge University Press.

Pearl, J. (1984). *Heuristics: intelligent search strategies for computer problem solving.* Addison-Wesley, Reading, MA.

Pitt, L. and Valiant, L. G. (1986).[†] *Computational limitations on learning from examples.* Technical report TR-05-86, Aiken Computation Laboratory, Harvard University.

Plotkin, G. D. (1969). A note on inductive generalization. In *Machine Intelligence* Vol. 5 (ed. B. Meltzer and D. Michie), pp. 153–63. Edinburgh University Press.

Poggio, T. and Girosi, F. (1990). *Extensions of a theory of networks for approximation and learning: dimensionality reduction and clustering.* AI Memo No. 1167, Massachusetts Institute of Technology.

Porat, S. and Feldman, J. A. (1991). Learning automata from ordered examples. *Machine Learning* **7** pp. 109–38.

Quinlan, J. R. (1979). Discovering rules by induction from large collections of examples. In *Expert systems in the micro-electronic age* (ed. D. Michie), pp. 168–201. Edinburgh University Press.

Quinlan, J. R. (1986).[†] Induction of decision trees. *Machine Learning* **1** (1) 81–106.

Quinlan, J. R. (1987). Generating production rules from decisions trees. *Proceedings of IJCAI 87*, Milan, Vol. 1, pp. 304–7.

Ragavan, H., Rendell, L., Shaw, M., and Tessmer, A. (1993). *Learning complex real-world concepts through feature construction.* Artificial intelligence technical report UIUC-BI-AI-93-03, University of Illinois, Urbana, IL.

Rajamoney, S. and DeJong, G. (1987).[†] The classification, detection and handling of imperfect theory problems. *Proceedings of IJCAI-87*, Milan, Vol. 1, pp. 205–7.

Razborov, A. A. (1990). Kolmogorov and the complexity of algorithms. From the obituary organised by D. Kendall, *Bulletin of the London Mathematical Society* **22** (1) 79–82 and 94.

Rendell, L. (1988). Learning hard concepts. *Proceedings of EWSL 88*, Turing Institute, Glasgow (ed. Derek Sleeman) pp. 177–200. Pitman, London.

Rendell, L. and Cho, H. (1990). Empirical learning as a function of concept character. *Machine Learning* **5** 267–98.

Rendell, L. and Ragavan, H. (1993). Improving the design of induction methods by analyzing algorithm functionality and data-based concept complexity. *Proceedings of IJCAI–93*, Chambéry, Vol. 2, pp. 952–8.

Rendell, L., Seshu, R., and Tcheng, D. (1987). More robust concept learning using dynamically-variable bias. *Proceedings of the Fourth International Workshop on Machine Learning*, University of California, Irvine, pp. 66–78. Morgan Kaufmann, Los Altos, CA.

Rich, E. and Knight, K. (1991). *Artificial intelligence*. McGraw-Hill, New York.

Rissanen, J. (1983). A universal prior for integers and estimation by minimum description length. *Annals of Statistics* **11** (2) pp. 416–31.

Robinson, J. A. (1965). A machine oriented logic based on the resolution principle. *J. Assoc. Comput. Mach.* **12**, 23–41.

Rolls, E. T. (1987). *Information representation, processing and storage in the brain: analysis at the single neuron level*. In *The neural and molecular bases of learning* (ed. J.-P. Changeux and M. Konishi) pp. 503–40. John Wiley, New York.

Rosenbloom P. and Newell A. (1983). The chunking of goal hierarchies. In *Machine learning*, Vol. 2 (ed. R. S. Michalski, J. G. Carbonell, and T. M. Mitchell) Chapter 10. Morgan Kaufmann, San Mateo, CA.

Ross, S. (1976). *A first course in probability*. Collier Macmillan, London.

Ruelle, D. (1989). *Chaotic evolution and strange attractors*. Cambridge University Press.

Schalkoff, R. (1992). *Pattern recognition: statistical, structural and neural approaches*. Wiley, New York.

Schapire, R. E. (1989). *The strength of weak learnability*. Technical memorandum MIT/LCS/TM-415, Massachusetts Institute of Technology.

Schlimmer, J. C. (1987). Incremental adjustment of representations for learning. *Proceedings of the Fourth International Workshop on Machine Learning*, University of California, Irvine, pp. 79–90. Morgan Kaufmann, Los Altos, CA.

Scott, P. D. and Markovitch S. (1989). Learning novel domains through curiosity and conjecture. *Proceedings of the Eleventh IJCAI*, Detroit, MI, pp. 669–74.

Scott, P. D. and Markovitch, S. (1991). Representation generation in an exploratory learning system. In *Concept formation. Knowledge and experience in unsupervised learning.* (ed. D. Fisher, M. Pazzani, and P. Langley) pp. 387–422. Morgan Kaufmann, Los Altos, CA.

Scott, P. D. and Markovitch, S. (1992). Experience selection and problem choice in an exploratory learning system. To appear.

Shalin, V. L., Wisniewski, E. J., Levi, K. R., and Scott, P. D. (1990). A formal analysis of machine learning systems for knowledge acquisition. In *Machine learning and uncertain reasoning.* (ed. B. R. Gaines and J. H. Boose) pp. 75–92. Academic Press, London.

Shannon, C. E. and Weaver, W. (1949). *The mathematical theory of communication.* University of Illinois Press, Champaign, IL.

Shapiro, E. Y. (1983). *Algorithmic program debugging*. MIT Press, Cambridge, MA.

Shorrocks, B. (1972). *Drosophila*. Pergamon Press/Ginn, London.

Siddall, J. N. (1982). *Optimal engineering design: principles and applications*. Marcel Dekker, New York.

Siekmann, J. H. (1986, 1989). Unification theory. *Proceedings of ECAI 86*, Brighton, UK, Vol. 2, pp. vi–xxxv. Reprinted in an extended form in the *Journal of Symbolic Computation* **7**, 207–74.

Sleeman, D. (1994). Towards a technology and a science of machine learning. *AI Communications* **7(1)**, 29–38.

Smith, B. C. (1984). *Reflection and semantics in LISP*. Report No. CSLI-84-8, Center for the Study of Language and Information, Stanford, CA.

Solomonoff, R. J. (1964).† A formal theory of inductive inference. *Inform. and Control* **7** 1–22.

Solomonoff, R. J. (1978). Complexity-based induction systems: comparisons and convergence theorems. *IEEE Transactions on Information Theory* **IT-24** (4) 422–32.

van Someren, M. W. (1984). A heuristic for rule learning. *Sixth European Conference on Artificial Intelligence Proceedings ECAI-84*, Pisa, (ed. Tim O'Shea), pp. 493–6.

Steels, L. and Van de Velde, W. (1989). Learning in second generation expert systems. In *Machine and human learning* (ed. Y. Kodratoff and A. Hutchinson), pp. 53–75. Kogan Page/Chapman and Hall, London.

Sternberg, M. J. E., King, R. D., Lewis, R. A., and Muggleton, S. (1994). Application of machine learning to structural molecular biology. *Phil. Trans. R. Soc. Lond. B* **344**, 365–371.

Stroulia, E., and Goel, A. K. (1993). *Functional representation and reasoning for reflective systems*. College of Computing, Georgia Institute of Technology, Atlanta, GA.

Stroulia, E., and Goel, A.K. (1994). Learning problem-solving concepts by reflecting on problem solving. *Proceedings of ECML-94*, Catania, Italy, pp. 287–306. Springer, Berlin.

Thornton, C. J. (1992). *Techniques in computational learning*. Chapman and Hall, London.

Utgoff, P. E. (1989). Incremental induction of decision trees. *Machine Learning* **4**(2) 161–86.

Utgoff, P. E. and Mitchell, T. M. (1982). Acquisition of appropriate bias for inductive concept learning. *Proceedings of the National Conference on Artificial Intelligence*, Pittsburgh, PA, pp. 414–17. Morgan Kaufmann, Los Altos, CA.

Valiant, L. G. (1984). A theory of the learnable. *Communications of the A.C.M.*, **27** (11) 1134–42.

Vere, S. A. (1977). Induction of relational productions in the presence of background information. *Proceedings of IJCAI*, Cambridge, MA. Vol. 1, pp. 349–55.

Waldinger, R. (1977). Achieving several goals simultaneously. *Machine intelligence* Vol. 8 (ed. E. W. Elcock and D. Michie), pp. 94–136. Ellis Horwood/Wiley, New York.

Watkins, C. J. C. H., and Dayan, P. (1992). Q-Learning. *Machine Learning* **8**, 279–292.

Whitley, D. (1989). Applying genetic algorithms to neural network learning. *Proceedings of AISB 89*, Brighton, UK (ed. A G Cohn), pp. 137–44. Pitman, London/Morgan Kaufmann, San Mateo, CA.

Williams, J. G. (1991). *Instantiation theory*. Lecture Notes in Artificial Intelligence No. 518, Springer, Berlin.

Willis, D. G. (1970).† Computational complexity and probability constructions. *J. Ass. Comput. Mach.* April, 241–59.

Wilson, R. (1985). *Introduction to graph theory*. Longman,

Wilson, S. W. (1987). Hierarchical credit allocation in a classifier system. *Proceedings of IJCAI-87*, Milan, Vol. 1 pp. 217–20.

Winston, P. H. (1992). *Artificial intelligence* (3rd edn). Addison-Wesley, Reading, MA.

Wolff, J. G. (1982). Language acquisition, data compression and generalization. *Language and Communication* **2** (1) 57–89.

Wolff, J. G. (1990). Simplicity and power—some unifying ideas in computing. *Computer Journal* **33** (6) 518–34.

Wolstenholme, D. E. (1989). DLGMS: a dependency-based lemma generation and maintenance system. *Proceedings of the BCS Expert Systems Conference*, London.

Young, R. M., Plotkin, G. D., and Linz, R. F. (1977). Analysis of an extended concept-learning task. *Proceedings of IJCAI*, Cambridge, MA (ed. R. Reddy), p. 285.

Zhu, Hong and Jin, Lingzi (1991). A knowledge-based approach to program synthesis from examples. *J. of Comput. Sci. and Technol.* **6** (1) 47–58.

Zurada, J. M. (1992). *Introduction to artificial neural systems*. West Publishing Company, St Paul, MN.

Zvonkin, A. K. and Levin, L. A. (1970). The complexity of finite objects and the development of the concepts of information and randomness by means of the theory of algorithms. *Russian Mathematical Surveys* **25** (6) 83–124.

# Index